Construction Planning and Scheduling

Third Edition

Jimmie W. Hinze
Rinker School of Building Construction, University of Florida

Upper Saddle River, New Jersey
Columbus, Ohio

Library of Congress Cataloging in Publication Data

Hinze, Jimmie.
　Construction planning and scheduling / Jimmie W. Hinze.—3rd ed.
　　p. cm.
　ISBN-13: 978-0-13-238562-6
　ISBN-10: 0-13-238562-7
　1. Building—Superintendence. 2. Production scheduling. I. Title.
　TH438.4.H55 2008
　690.068—dc22

2007013101

Editor in Chief: Vernon Anthony
Acquisitions Editor: Eric Krassow
Associate Managing Editor: Christine Buckendahl
Editorial Assistant: Nancy Kesterson
Production Coordination: Suganya Karuppasamy, GGS Book Services
Production Editor: Holly Shufeldt
Design Coordinator: Diane Ernsberger
Cover Designer: Ali Mohrman
Cover photo: Superstock
Production Manager: Deidra Schwartz
Director of Marketing: David Gesell
Executive Marketing Manager: Derril Trakalo
Marketing Assistant: Les Roberts

This book was set in Minion by GGS Book Services. It was printed and bound by Courier Westford, Inc. The cover was printed by Phoenix Color Corp.

Copyright © 2008, 2004, 1998 by Pearson Education, Inc., Upper Saddle River, New Jersey 07458. Pearson Prentice Hall. All rights reserved. Printed in the United States of America. This publication is protected by Copyright and permission should be obtained from the publisher prior to any prohibited reproduction, storage in a retrieval system, or transmission in any form or by any means, electronic, mechanical, photocopying, recording, or likewise. For information regarding permission(s), write to: Rights and Permissions Department.

Pearson Prentice Hall™ is a trademark of Pearson Education, Inc.
Pearson® is a registered trademark of Pearson plc
Prentice Hall® is a registered trademark of Pearson Education, Inc.

Pearson Education Ltd.
Pearson Education Singapore Pte. Ltd.
Pearson Education Canada, Ltd.
Pearson Education—Japan

Pearson Education Australia Pty. Limited
Pearson Education North Asia Ltd.
Pearson Educación de Mexico, S.A. de C.V.
Pearson Education Malaysia Pte. Ltd.

10 9 8 7 6 5 4 3 2

ISBN-13: 978-0-13-238562-6
ISBN-10:　　0-13-238562-7

Preface

Many textbooks have been published on construction scheduling. Many provide excellent information on a variety of scheduling subjects. Unfortunately, they are often limited in their scope, often omitting scheduling subjects that might be of particular interest to the reader. This text is written to provide broad coverage on all major scheduling subjects.

My first employment with a construction contractor was a summer job in the 1960s. Although I was given many different assignments, I have vivid memories of the arrow diagram network that I was asked to draw by hand. Fortunately, I had recently taken a class on the fundamentals of arrow diagramming, so the scheduling assignment was reasonably easy for me. A few years later while working for a different firm, my primary responsibility was scheduling. The scheduling effort was largely focused on the coordination of subcontractors on several different projects for which I used the precedence diagramming method. It was during this period that I developed a strong appreciation for the value of effective scheduling and the use of precedence diagrams.

There are many approaches to providing scheduling information. Some of these are described briefly in Chapter 1. This textbook is written with a major emphasis on precedence diagramming, with only the last chapter addressing arrow diagramming. Although most scheduling is done with precedence diagrams, I also believe that an introduction to arrow diagramming is appropriate. In academic settings, I have found that students can grasp arrow diagramming more easily if they have not already been exposed to precedence diagrams. For this reason, instructors who plan to lecture on arrow diagrams should consider jumping to Chapter 16 after the first two chapters are covered. Chapter 3 explains the fundamentals of precedence diagrams. Regardless of the scheduling technique used, successfully using scheduling information begins with accurate time estimates for activity durations as discussed in Chapter 4.

One topic seldom addressed in scheduling texts, especially in detail, is that of contract provisions related to scheduling. This text devotes an entire chapter to this subject (Chapter 5). Resource leveling and resource allocation are described in Chapter 6. The impact of scheduling provisions on cash flow is also addressed (Chapter 7). Manual solutions are described for solving problems related to resource utilization and cash flow. Although such problems are often solved by computer, it is helpful for schedulers to understand the process of arriving at a solution in order to fully comprehend computer solutions.

Schedules are management tools. It is through the proper use of schedules that management is able to make informed decisions about scheduling activities. This use includes updating the schedules when the schedule information ceases to be useful for making informed decisions. This process is described in Chapter 8.

Chapter 9 addresses computer applications. This chapter is not a user's manual, nor is it a proponent for any particular scheduling software. The more widely used software programs are described, but no endorsement of any particular product is made. The purpose of the chapter is to familiarize the reader with some of the basic scheduling concepts that are addressed by computer software.

Chapter 10 describes earned value concepts. Project schedules are generally adversely impacted by changes in the project. Chapter 11 provides information for quantifying such impacts. Such information is often required when a claim is prepared. Chapter 12 presents a brief discussion of the value of schedules in litigation.

The treatment of short-interval schedules in Chapter 13 is far more extensive than any known writings on the subject. The use of short-interval schedules is vital to the successful completion of many construction projects. Although concepts of their use and application are simple, the subject warrants a discussion in any serious text on scheduling.

Linear scheduling (discussed in Chapter 14) is a relatively new scheduling technique used in the construction industry. Linear scheduling is a viable method for projects that would otherwise be difficult to schedule. Schedulers should consider using linear scheduling on projects that lend themselves to this technique. The use of probabilistic duration estimates is described in Chapter 15. Although the use of PERT is perhaps minimal in the construction industry, the basic concepts should be understood.

SUPPLEMENTS

To access the online Instructor's Manual, instructors need to request an access code. Go to www.prenhall.com, click the Instructor Resource Center link, and then click Register Today for an instructor access code. Within 48 hours after registering you will receive a confirming e-mail including an instructor access code. Once you have received your code, go to the site and log on for full instructions on downloading the materials.

ACKNOWLEDGMENTS

It is perhaps rare for a text to be written entirely by one person. I certainly can make no such claim. Others have provided valuable assistance in helping me compile all of the information for this text. Dr. Ian Flood also offered valuable comments as the initial text was being finalized. The efforts of Bruce Jamieson were instrumental in compiling the information on short-interval scheduling, and Rory McCarty contributed to the chapter on litigation. A considerable amount of the material on linear scheduling was developed by Greg Hanby, Phil Nelson, Brendan Kennedy, and H. C. Phillips. Dr. Robert Shawcroft contributed significantly by providing me with some scheduling class notes that eventually became part of this text. Most of all, I must thank my good friend John Gambatese, who offered suggestions for the third edition. He made many suggestions for changes and improvements in the third edition. This version of the text is a true credit to him. Of course, as in the first edition, Chapter 9 is wholly his contribution. Helpful suggestions were also provided by Dr. Phillip Dunston, Dr. Douglas Lucas, and Dr. Richard Smailes.

Contents

CHAPTER 1   INTRODUCTION — 1

Bar Charts — 2
    Shortcomings of Bar Charts — 3
    The Sports Facility Project — 7
    Value of Bar Charts — 7
Other Scheduling Approaches — 10
Work Breakdown Structure — 13
Reasons for Planning and Scheduling in Construction — 20
Review Questions — 21

CHAPTER 2   DEVELOPING A NETWORK MODEL — 22

Steps in Building a Network Model — 22
    Defining Activities — 23
    Ordering Activities — 25
    Drawing the Network Diagram — 28
    Assigning Durations to Activities — 31
    Assigning Resources and Costs — 32
    Calculating Early and Late Start/Finish Times — 32
    Identify the Critical Path — 32
    Scheduling Activity Start/Finish Times — 33
Final Comments — 33
Review Questions — 33

CHAPTER 3   PRECEDENCE DIAGRAMS — 35

Precedence (Activity-on-Node) Networks — 35
Activity Relationships — 36
    Basics about Precedence Diagrams — 45

Calculations on a Precedence Network 49
Independent Float and Interfering Float 55
Computations for Different Activity Relationships 58
Final Comments 61
Review Problems 61

CHAPTER 4   DETERMINING ACTIVITY DURATIONS   72

Estimating 72
Types of Estimates 73
    Conceptual Estimates 73
    Detailed Estimates 73
Conducting a Detailed Estimate 73
Estimating Durations 79
Scheduling Issues 82
Factors Influencing Choice of Activity Schedules 83
    Weather and the Schedule 84
    Uncertainty in Duration Estimates 85
Final Comments 88
Review Problems 88

CHAPTER 5   TIME IN CONTRACT PROVISIONS   90

Time Is of the Essence 91
Requirements for Project Coordination 91
Cooperation 92
Progress Schedule 92
Ownership of Float 98
Notice to Proceed 100
Time of Completion 101
Units of Time: Working Days or Calendar Days 102
Liquidated Damages—Damages for Late Completion 105
Weather 106
Use of Completed Portions of the Work 106
Substantial Completion 107
Notice of Delays 108
Avoidable Delays 108
Unavoidable Delays 109
Extension of Time (Avoidable Delays) 109
Extension of Time (Unavoidable Delays) 109

Submittals	111
Progress Payments	112
Payment for Materials	114
Final Payment	114
Suspension	114
Termination by Contractor	115
Final Comments	116
Review Questions	116
CHAPTER 6 RESOURCE ALLOCATION AND RESOURCE LEVELING	**117**
The Management of Resources	117
When Resources Are Limited (Resource Allocation)	118
The Manual Solution for Resource Allocation	120
The Brooks Method of Resource Allocation	129
When Project Duration Is Fixed (Resource Leveling)	136
The Manual Solution for Resource Leveling	138
The Sports Facility Project	146
Final Comments	147
Review Problems	149
CHAPTER 7 MONEY AND NETWORK SCHEDULES	**164**
Cash Flow	165
The Time Value of Money	165
Interest Rates	165
Contractor Cash Disbursements	166
Contract Provisions That Impact Cash Flow	168
Owner Policies and Practices That Impact Cash Flow	170
The Cash-Flow Analysis	170
The Sports Facility Project	172
The Present Worth of Cash Flow	172
The Value of Cash-Flow Analysis	173
Time–Cost Trade-Offs	176
Direct Costs	177
Indirect Job Costs (Job Overhead)	177
Overhead (Company Overhead)	177
Profit	177
Four Different Solutions for Each Network	181
Logically Reducing Project Duration	182
Final Comments	193
Review Problems	194

CHAPTER 8   PROJECT MONITORING AND CONTROL — 199

Construction Time — 199
Effective Scheduling — 201
Monitoring Project Status — 202
Difficulties in Assessing Progress — 206
Updating the Schedule — 207
Controlling the Project — 208
    The Sports Facility Project — 210
As-Built Schedules — 211
Final Comments — 214
Review Questions — 214

CHAPTER 9   COMPUTER SCHEDULING — 216

Computer Scheduling Terms — 218
Scheduling Software — 220
    Primavera (P3®) — 221
    SureTrak Project Manager — 221
    Microsoft Project — 222
    Web-Based Programs — 222
Creating a Schedule — 222
Updating a Schedule — 226
Presenting a Schedule — 226
Useful Software Features — 228
    Sorting and Filtering — 228
    Global Editing — 229
    Cash-Flow Analysis — 229
    Resource Leveling — 229
Linking to Other Project Management Software — 229
Final Comments — 231
Review Questions — 231

CHAPTER 10   EARNED VALUE: A MEANS FOR INTEGRATING COSTS AND SCHEDULE — 232

The Earned Value Concept — 233
Difficulties in Integrating Cost and Schedule Systems — 238
Final Comments — 241
Review Questions and Problems — 242

CHAPTER 11 THE IMPACT OF SCHEDULING DECISIONS ON PRODUCTIVITY 246

Working Overtime 246
Increasing the Workforce (Crowding) 249
Increasing the Number of Starting Points 251
Identifying the Causes of Delays 253
Interruption of Work on Multiple Units (Impact of Lost Learning) 254
Learning Applied to Individual Units 257
Learning Applied to Cumulative Average Units 260
What Happens When Work Is Interrupted? 262
Other Sources of Lost Productivity 265
Final Comments 266
Review Problems 266

CHAPTER 12 CPM IN DISPUTE RESOLUTION AND LITIGATION 270

Going to Court 270
Types of Schedules 277
Impact of Changes 278
Impact of Delays 279
Final Comments 280
Review Questions 280

CHAPTER 13 SHORT-INTERVAL SCHEDULES 281

Short-Interval Schedules in the Literature 283
How Contractors Use Short-Interval Schedules 283
Other Short-Interval Schedules 288
Final Comments 296
Review Questions 296

CHAPTER 14 LINEAR SCHEDULING 297

What Is Linear Scheduling? 300
Example 1: Project to Replace a State Park Walkway 304
    Production Rate Diagrams 304
    Buffers 306
    Generating the Linear Schedule 306
Example 2: Project to Construct 500 Tract Housing Units 308
Final Comments 312
Review Questions 312

CHAPTER 15 PERT: PROGRAM EVALUATION AND REVIEW TECHNIQUE — 313

Uncertainty in Activity Duration Estimates — 313
Uncertainty in the Duration Estimates of an Activity Chain — 318
Uncertainty in the Duration Estimates of Projects — 320
Monte Carlo Simulation — 321
Final Comments — 323
Review Problems — 323

CHAPTER 16 ARROW DIAGRAMS — 326

Activity Relationships — 327
The i–j Notation of Activities — 330
    Dummies — 330
Performing Time Calculations with Arrow Diagrams — 336
Float Values — 343
Understanding Free Float and Total Float — 349
Computations for Different Activity Relationships — 352
Final Comments — 354
Review Problems — 355

REFERENCES — 365
ADDITIONAL REFERENCES — 367
INDEX — 371

# 1

Introduction

It's about time.

Planning can be thought of as determining "what" is going to be done, "how," "where," and by "whom." In scheduling, this information is needed in order to determine "when." In construction projects the "plans" (blueprints) and specifications for the project generally define both the end product and, often, the general time frame in which to complete the project. However, they normally do not specifically identify the individual steps, their order, and the timing followed to achieve the end product. Thus, when we discuss planning and scheduling in the construction process, we must address the "how" and, therefore, the "what," "who," "where," and "when."

When we discuss scheduling, we are usually interested in some aspect of the time element of the plan. In essence, a schedule is a timetable of activities, such as of "what" will be done or "who" will be working. Such a timetable can be looked at in two ways: the first is focusing on an activity, such as determining "when" a certain task will be performed relative to other activities. The second is concentrating on a specified time frame and then ascertaining "who" will be working (or needed) or "what" should be occurring at a particular time. All of us are involved in planning and scheduling on an ongoing basis. The degree to which we carry it out and the techniques we use vary depending upon the complexity of our situations and our needs and objectives. In summary, planning relates to developing the logic of how a project will be constructed, while scheduling consists of integrating that plan with a calendar or a specific time frame.

We all do planning and scheduling on a regular, albeit informal basis. For whatever undertaking, we mentally determine a plan and schedule, such as what we will do in the next half-hour or how and when we will accomplish that task, such as a homework assignment. Often it is necessary for us to go a step beyond this level by creating a "to-do list." None of us can retain the organization of all the tasks we have to do on a daily basis, so we document what needs doing by writing down the information. This is also helpful if we are coordinating with other parties. By writing down the list of items, and perhaps copying and distributing it, we have documented a basis of agreement. We may also prioritize this list by writing the items in the order in which they will be done.

As the number of items increases and/or the time frame expands, we find we have to put our to-do list in the context of time. Normally, we do this using an appointment book or calendar. The driving forces typically are to avoid scheduling multiple things at the same time, to ensure that we allow sufficient time to prepare for an event, and/or to provide a record of what activities we undertook and when and how long we spent on them.

BAR CHARTS

In 1917 Henry Gantt developed a method of relating a list of activities to a time scale in a very effective manner, by drawing a bar (or Gantt) chart such as those shown in Figures 1.1 and 1.2. Activities are represented as bars on the chart, while across the top or bottom of the chart is a time line. For each activity, a bar is drawn from the activity's starting time until its ending time. The Gantt chart has been widely used in depicting schedules for construction projects and has some very useful characteristics. Its primary advantage is that its simple graphic representation allows one to grasp schedule information quickly and easily.

Bar charts are simple presentations that show how major work activities are scheduled. A major advantage is that they are easily prepared as time-scaled presentations. Bar charts are the most commonly employed and readily recognized scheduling models in use today. In recognition of their creator, the terms *bar chart* and *Gantt chart* are used interchangeably by many schedulers.

The widespread use of bar charts can best be attributed to the ease with which they can be understood with only a cursory examination. The bar chart in Figure 1.1 is a good example: the activity sequencing is apparent, and one can surmise easily when

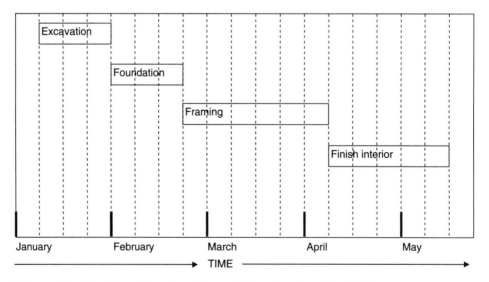

Figure 1.1 Bar Chart Showing General Construction Work Tasks

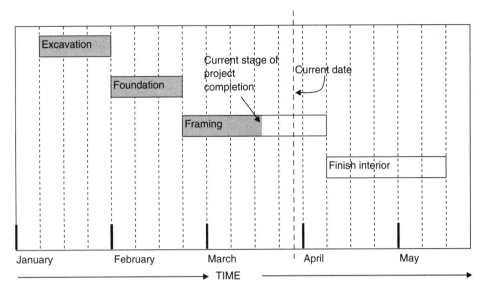

Figure 1.2 Bar Chart Showing Scheduled Versus Actual Performance

each activity is to begin and when it is to be completed. This simple example shows at a glance how the different activities relate to each other. Note that the activities are time-scaled and that they have been superimposed over a calendar.

With the time-scale presentation, a bar chart shows operations and the time consumed by each operation. In addition, it can show the scheduled versus actual progress. This is demonstrated in Figure 1.2. The heavy dashed vertical line represents the current date, and the shaded portions of the activities indicate the amount of work that has been completed by the current date. It is obvious that the project is slightly behind schedule. The progress on the framing activity has not met expectations. Adjustments to the schedule may be warranted if the delay in project completion (about one week) is not acceptable. In this simple depiction, it is evident that the project can be completed on time by accelerating the work effort on framing, finishing interiors, or both. This information is easy to grasp from this bar chart, and there is little chance of misinterpretation.

Shortcomings of Bar Charts

Despite the wide usage and appeal of bar charts, they do possess some features that make them difficult to use in certain settings. It is particularly when projects become more complex that bar charts begin to fail to provide the type of information that is often so valuable for planning and scheduling. An example with a bar chart will illustrate some of these shortcomings. First, consider the simple schedule shown in Figure 1.1. It can easily be shown in network form; see Figure 1.3.

It is simple to see the relationship between the bar chart shown earlier and the same information presented in network form, in this case as a precedence diagram. This diagram is not time-scaled in that the lengths of the activities do not correspond to the durations.

4   Chapter 1

Figure 1.3   Simple Precedence Diagram of a Project Showing Activity Sequences

The bar chart in Figure 1.4 shows a project overlaid on a calendar, similar to the bar chart already discussed. Its additional feature is that several of the activities are shown as occurring simultaneously. The first general criticism of bar charts is that they do not show clear dependencies between activities. In the previous bar chart, most observers may presume that the activities follow in sequence, just as shown in Figure 1.3. That is, as soon as the excavation work is completed, foundation work can begin, and so forth. A similar deduction might be made for some of the activities in Figure 1.4. For example, Activities B, C, and D appear to begin as soon as Activity A is completed. It also appears as if Activity E is scheduled to begin as soon as Activities B and D are completed. But what about Activity F? Is the start of Activity F linked to the completion of Activity B, Activity D, or both? The start of Activity F could also be linked to Activity C. This cannot be ascertained from the bar chart. Similar dilemmas exist for Activities G, H, and I. The failure to show the interrelationships between activities is a major shortcoming of bar charts, mandating that more sophisticated scheduling techniques be utilized on complex projects.

The bar chart shown in Figure 1.4 shows the relative status of completion. Note that Activity G is ahead of schedule and Activity F is on schedule, but Activity E is behind schedule by about a week. While the relative schedule status of each activity

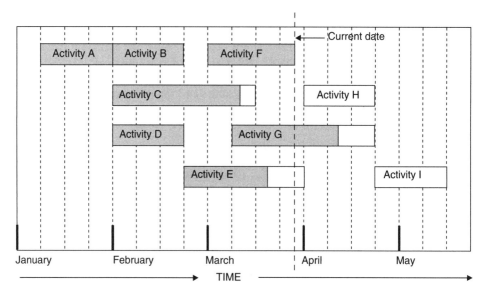

Figure 1.4   Bar Chart Showing General Construction Work Tasks

can be determined from the bar chart, the actual status of the project cannot be readily determined. The big question for the scheduler would be, "Does the behind-schedule status of Activity E or C compromise project completion?" This cannot be determined from the bar chart. Therein lies another shortcoming of bar charts: although the status of individual activities can be readily ascertained, the overall status of a project cannot be determined when some activities are not on schedule. This makes it difficult to assess the need for making scheduling adjustments, and it also makes it difficult to determine the appropriate activities to target for acceleration.

Even a change in the logical sequencing of the activities in a bar chart cannot be readily made, especially when many activities are involved. The information shown in the bar chart in Figure 1.4 is shown in the precedence diagram in Figure 1.5. Note that the relationships of the activities are consistent with the information shown in the bar chart, but the precedence diagram could not have been, with certainty, developed solely from the bar chart. The precedence diagram is not time scaled. While one can see the relationships between the information shown in the bar chart and the precedence diagram, it is not generally necessary to develop both types of schedules. Many of the computer scheduling programs available today readily enable the conversion (automatic) of network information into bar charts.

While some of the disadvantages of bar charts have been noted, the shortcomings are not overwhelming. In fact, bar charts are the most popular means by which scheduling information is communicated in the construction industry. This is even true of projects that are complex and involve many activities. Figure 1.6 shows a bar chart schedule for the construction of a metal barn. The level of detail is such that the general nature of the activities can be readily understood. Many of the interdependencies between the activities, although not specifically noted, can be surmised with relative confidence based on the logical relationships that generally exist between activities. Of course, one would need to know more about the project itself to be certain about some interpretations.

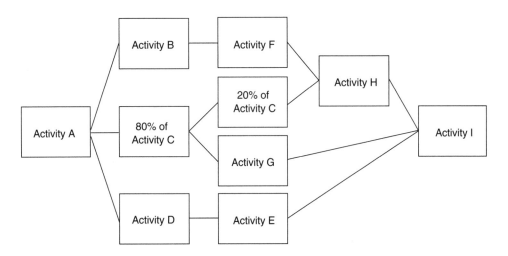

Figure 1.5 Precedence Diagram of the Project Shown Earlier in Bar Chart Form

Activity Description	Dur	1	2	3	4	5	6	7	8	9	10	11	12	13	14	15	16	17	18	19	20	21	22	23	24	25	26	27	28	29	30	31	32	33	34	35	36	37
Mobilization	2	■	■																																			
Clearing, grubbing and grading	4			■	■	■	■																															
Site electric	2					■	■	>	>																													
Dig footings, rebar and concrete	6							■	■	■	■	■	■																									
Form slab, rebar and concrete	3													■	■	■																						
Erect steel framing	3																■	■	■																			
Roof framing	5																			■	■	■	■	■														
Rough electrical	3																								■	■	■	>										
Metal roofing	4																								■	■	■	■										
Metal siding	4																								▨	▨	▨	▨	>									
Exterior doors and windows	2																											■	■	>	>							
Finish electric	2																													▨	▨	>	>	>	>			
Exterior painting	3																													▨	▨	▨	>	>	>			
Sidewalk and asphalt paving	4																													■	■	■	■					
Landscape	5																																	■	■	■	■	■
Final inspection and cleanup	3																																					■

Figure 1.6 Bar Chart for the Construction of a Metal Storage Barn

Note: Dark shaded activities are critical as they must occur as scheduled or the project duration will be adversely impacted. Light shaded activities are not critical as they can be delayed at least one day without having an adverse impact on the project duration. The ">" symbols designate days that an activity duration could be extended, commonly known as free float.

The forgoing shortcomings of bar charts are intended to apply specifically to traditional or handdrawn bar charts. These shortcomings virtually disappear when scheduling software is used. As already noted, most popular computer-generated schedules permit easy conversion between precedence diagrams and bar charts. Consequently, with scheduling software the changing or updating of schedules is done on the precedence diagram, with any changes automatically being included in the bar chart. There is another good feature of computer-generated bar charts that should be noted. Bar charts can often be displayed with the dependencies also shown; i.e., link lines are drawn between the different bars in the bar chart. Thus, when scheduling software is used, the bar charts have essentially no shortcomings that are of any concern.

The Sports Facility Project

Figure 1.7 depicts a bar chart of a sports facility project. It is evident that the schedule depicted can be interpreted with relative ease, even though the specific configuration of the project is not known. Another bar chart of this project is shown in Figure 1.8, which shows the bar chart schedule for the sports facility project as presented by computer scheduling software. The similarities between the bar charts in Figures 1.7 and 1.8 are quite apparent. Note that the computer-generated bar chart (Figure 1.8) shows "necked-down" portions of the activities where weekends fall. The project that is depicted consists of constructing a tennis court, bleachers, roadway, and landscape. The dependencies of the activities can be inferred with reasonable confidence if the nature of the activities is examined. This project will be used in some of the subsequent chapters to make certain points and especially to demonstrate different scheduling-related techniques.

Value of Bar Charts

The broad use of bar charts in the construction industry is clear evidence that they are a strong communication tool regarding scheduling information. The value of bar charts should not be underestimated. Their usefulness is not eliminated by their deficiencies. While bar charts make it difficult to maintain accurate schedules and make significant schedule changes, one of their major strengths is the ability to clearly and quickly present the status of a project. The key to the use of bar charts is that details and complexity are not readily compatible with the use of most bar charts. Instead, bar charts can be used to convey—often to upper management—the overall status of a project. The details of what is to take place on a given day are generally of little concern to upper management. Rather, they want to establish a quick sense of how a project is doing. The same type of information might also be conveyed to a subcontractor. The subcontractor may not be concerned with those activities that are unrelated to the work in question, but he or she may focus only on the work related to a particular specialty trade. This might be shown quite well on a bar chart.

The truly wonderful aspect of bar charts is that no extensive training is required to learn how to extract information from them. They are in such common use and present information in such a simple manner that a cursory inspection of a bar chart is

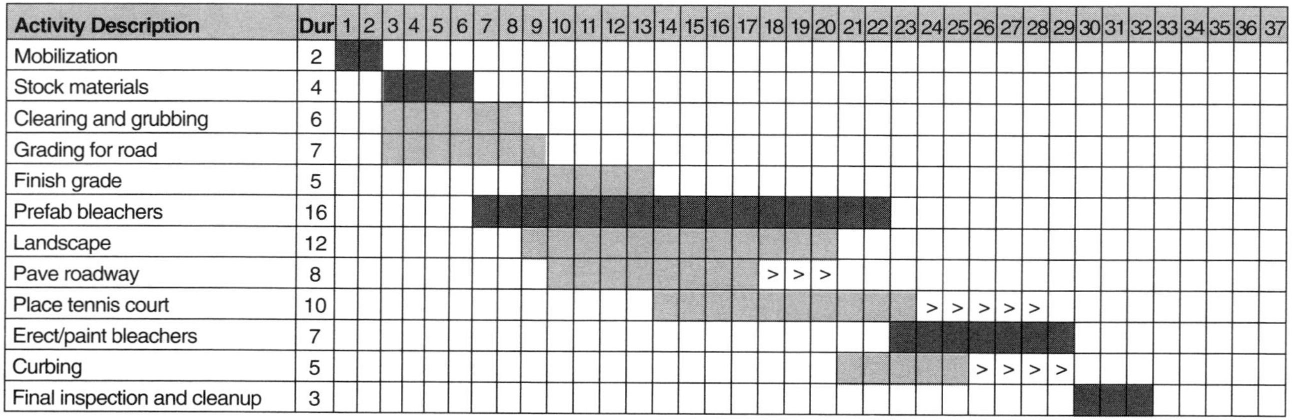

Activity Description	Dur	1	2	3	4	5	6	7	8	9	10	11	12	13	14	15	16	17	18	19	20	21	22	23	24	25	26	27	28	29	30	31	32	33	34	35	36	37
Mobilization	2	■	■																																			
Stock materials	4			■	■	■	■																															
Clearing and grubbing	6			■	■	■	■	■	■																													
Grading for road	7									■	■	■	■	■	■	■																						
Finish grade	5																■	■	■	■	■																	
Prefab bleachers	16							■	■	■	■	■	■	■	■	■	■	■	■	■	■	■	■															
Landscape	12																					■	■	■	■	■	■	■	■	■	■	■	■					
Pave roadway	8																	■	■	■	■	>	>	>														
Place tennis court	10																								>	>	>	>	>									
Erect/paint bleachers	7																								■	■	■	■	■	■	■							
Curbing	5																									>	>	>	>	>								
Final inspection and cleanup	3																															■	■	■				

Figure 1.7 Bar Chart for the Construction of a Sports Facility

Introduction 9

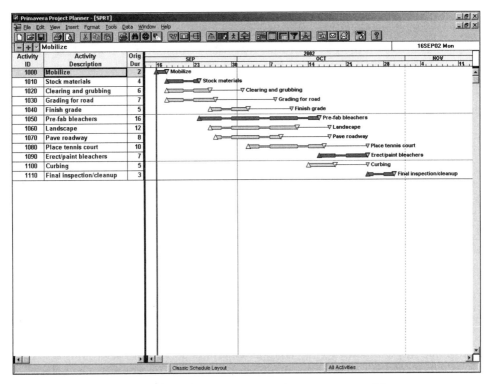

Figure 1.8   Bar Chart of the Sports Facility Project as Presented by a Computer Software Program *(Primavera Project Planner®)* (Reprinted with permission from Primavera Systems, Inc.)

often sufficient for conveying the desired information to the viewer. The shortcomings of the bar chart as a scheduling tool are overcome through the explicit definition of interrelationships among activities and through the representation of those relationships by drawing a network of how activities relate to one another.

This approach to developing network schedules began in the late 1950s and has evolved over the years. Bar charts have been surpassed by the use of networks as scheduling tools, especially on larger projects. A network of the activities that comprise a project represents a model, or plan, of the project as it is proposed to be undertaken. Estimating the length of time required to complete each activity in the network then makes it possible to determine how long the project will take and when each activity can be expected to start and finish. There are actually two similar network modeling techniques that were developed independently, but simultaneously: the Critical Path Method (CPM) and Program Evaluation and Review Technique (PERT).

Within the construction industry, CPM is the far more important of the two. In this technique, each activity is assigned a specific duration, and calculations through the network provide a single, specific duration for the project as a whole. Each activity also has times calculated for it that specify the earliest it can be expected to start and finish (relative to the start of the project) and the latest it can start if the project is to be completed in some specified amount of time. When reviewing CPM data, it is important to recognize the distinction between *duration* and *event*. The duration of an activity is the period of time that will be consumed in completing a task. An event is the point in time or an instant at which the status of completion of a project or activity can be defined. The starting time for an activity defines that point in time at which an activity can begin. The duration of the activity includes the time at which the activity begins and continues to the point in time at which the activity is completed.

CPM identifies those chains of activities (the critical paths) in the project that control how long the project will take. There are two variations of CPM. The traditional technique is called *Activity-on-Arrow* (A-on-A), or an arrow diagram, because activities are represented in the network as arrows or lines. The alternative approach is *Activity-on-Node* (A-on-N), more commonly referred to as the *Precedence technique*. This method defines the activities as boxes (nodes in the network), which are connected together (their relationships identified) by lines (links). The precedence approach has risen in popularity in recent years because of the availability of microcomputer software, which makes it very easy to use and which represents complicated dependencies between activities more easily.

PERT can really be thought of as a "generalized" CPM. The major difference between the approaches is that PERT assumes that an activity's duration cannot be precisely predetermined. Instead, it requires the planner to specify three separate durations for each activity: the most likely, the optimistic, and the pessimistic duration estimates. These three values are then statistically manipulated to come up with an "expected" duration for each activity and, consequently, an expected project duration, by calculating times through the network. This expected project duration is that which will be met or bested 50% of the time and exceeded 50% of the time. This process is termed *probabilistic*, as it deals with project duration and activity start and end times in terms of the probability that certain dates or times will be reached. In contrast, CPM is said to be *deterministic* in that specific durations, dates, and times are determined through its use.

PERT was developed and is used primarily in research and development or other undertakings where insufficient experience or historical data is available for estimating the durations of individual activities in a project. It is recognized that even on construction projects, estimates of activity durations may not be very reliable. However, even the greater analytical opportunities afforded by PERT do not outweigh the much greater data requirements and technical understanding and interpretation required by the technique.

OTHER SCHEDULING APPROACHES

Most of the previous paragraphs have discussed bar charts and precedence diagrams as the primary types of schedules that are employed in the construction industry. While most construction personnel will be able to understand a bar chart and while precedence diagrams are common on almost all large construction projects, there are other methods that are employed to schedule and control the construction process. One such method is the matrix schedule. A matrix schedule is one in which a spreadsheet is used to show all of the activities on a particular project. This type of matrix approach is used most often on housing projects. This approach has some benefits as it assumes that each project (housing unit) will consist of the same activities. This allows the contractor or developer to show all of the houses being built on a single matrix. To each of the major activities, a scheduled date is assigned. Notes can be made directly on the matrix, or the cell associated with each activity for a project can be highlighted to indicate that it is completed. Figure 1.9 shows a portion of a matrix schedule. There is nothing graphical in the matrix. Since the activities are very similar from project to project, it is not essential to include minute details.

Turkey Creek

Activities	Unit Number	#1	#2	#3	#4	#5	#6	#7
	Plan	2b	3b	3bx	4b	1b	3b	4b
Layout and trench		01-Jul	01-Jul	01-Jul	01-Jul	01-Jul	01-Jul	02-Jul
Pour footings		02-Jul	02-Jul	02-Jul	02-Jul	02-Jul	02-Jul	03-Jul
Set forms		03-Jul	03-Jul	03-Jul	03-Jul	03-Jul	03-Jul	05-Jul
Rough plumbing		10-Jul	10-Jul	10-Jul	10-Jul	11-Jul	11-Jul	11-Jul
Inspection		11-Jul	11-Jul	11-Jul	11-Jul	12-Jul	12-Jul	12-Jul
Cable		17-Jul	17-Jul	17-Jul	17-Jul	18-Jul	18-Jul	18-Jul
Inspection		19-Jul	19-Jul	19-Jul	19-Jul	22-Jul	22-Jul	22-Jul
Pour slab		22-Jul	22-Jul	22-Jul	22-Jul	23-Jul	23-Jul	23-Jul
Pour garage		23-Jul	23-Jul	23-Jul	23-Jul	24-Jul	24-Jul	24-Jul
Utility trenching		30-Jul	30-Jul	30-Jul	30-Jul	31-Jul	31-Jul	31-Jul
Rough framing		20-Aug	20-Aug	20-Aug	22-Aug	22-Aug	22-Aug	20-Aug

Figure 1.9   Partial Matrix Schedule for a Housing Project

Turkey Creek

Activities	Unit Number	#1	#2	#3	#4	#5	#6	#7
	Plan	2b	3b	3bx	4b	1b	3b	4b
Plumbing top out		22-Aug	22-Aug	22-Aug	26-Aug	26-Aug	26-Aug	22-Aug
Hvac rough		03-Sep	03-Sep	03-Sep	05-Sep	05-Sep	05-Sep	03-Sep
Electrical rough		12-Sep	12-Sep	12-Sep	16-Sep	16-Sep	16-Sep	12-Sep
Roof dry-in		16-Sep	16-Sep	16-Sep	18-Sep	18-Sep	18-Sep	16-Sep
Siding		18-Sep	18-Sep	18-Sep	20-Sep	20-Sep	20-Sep	18-Sep
Inspection - frame		19-Sep	19-Sep	19-Sep	23-Sep	23-Sep	23-Sep	19-Sep
Frame walk-through		20-Sep	20-Sep	20-Sep	24-Sep	24-Sep	24-Sep	20-Sep
Insulation		24-Sep	24-Sep	24-Sep	26-Sep	26-Sep	26-Sep	24-Sep
Inspection		25-Sep	25-Sep	25-Sep	27-Sep	27-Sep	27-Sep	25-Sep
Hang drywall		01-Oct	01-Oct	01-Oct	03-Oct	03-Oct	03-Oct	01-Oct

Figure 1.9 *(continued)*

The form of the schedule will be dictated to a large degree by the type of information it is to convey. The matrix schedule has the potential of providing considerable detail about the scheduling of a project or a series of projects. A form of scheduling that shows less detail is the "horse blanket" schedule. An example of a "horse blanket" schedule is shown in Figure 1.10. From this schedule, it is apparent that this is a very graphical schedule as it shows the relative levels of effort or time that will be committed to the various stages of the project as it evolves. It is easy to see at a glance how much effort is to

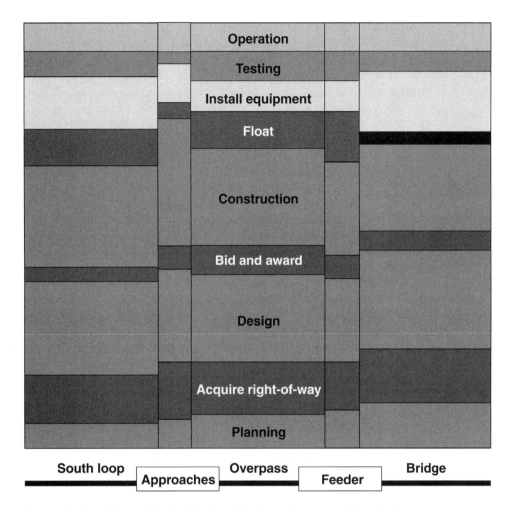

Figure 1.10 "Horse Blanket" Schedule for a Transportation Project

be expended on a project through its evolution from planning to final occupation and use. It shows more than the construction phase, as it can incorporate the planning phase and be extended to the maintenance and use phase of a project. The particular needs of the end user will dictate the types of stages that will be portrayed in the schedule.

WORK BREAKDOWN STRUCTURE

When projects are simple, consisting of few defined activities, it might be possible to grasp the total construction effort with little difficulty. Unfortunately, most projects for which formal schedules are prepared tend to be defined with dozens or even hundreds or thousands of activities.

When a schedule is being developed, it is important that the logic be carefully laid out and that all important tasks are included in the schedule. A person may be inclined to approach the development of the schedule by starting with the first task to be performed. After this, the next task or tasks would be scheduled. This progression would be followed until the last task in the project is included in the schedule. While this may appear to be a straightforward approach, it is not generally a good one to take, especially when a project consists of several hundred activities. Under this approach, some tasks or activities will surely be excluded from the schedule. This is because the schedule ostensibly evolves in a logical fashion, but the schedule eventually becomes so involved that it is virtually impossible to think of every task that is to be incorporated in the schedule.

When the tasks become numerous, the schedule development becomes more haphazard and omissions are sure to occur. The saying that "you eat an elephant by taking one bite at a time" does not really apply here as the smaller tasks (like bites) must be scheduled in a logical sequence. If one or a series of activities are left out, the adverse impact on the schedule could be quite significant. A more logical tactic is the divide-and-conquer approach.

The schedule consists of tasks that must be performed to complete the project. Thus, the schedule is dominated by verbs, things to do, or tasks. The schedule development must take place without the omission of any major tasks. This can be virtually assured if the preparation of the schedule is done in a systematic manner. One such approach is to use a work breakdown structure (WBS) to describe the entire scope of work. This will ensure a thorough and complete schedule. The use of a WBS facilitates the management of schedule and it makes for better control of the construction project.

WBS is a systematic way of describing the components pieces of a project schedule. For example, in a typical building the WBS would contain such systems as the sitework, structure, MEP (mechanical electrical and plumbing), interior finishes, roofing system, etc. Developing the WBS begins with the definition of the major systems or components of a project. Then each system is defined in greater and greater detail until there exists a discrete or measurable piece of work and a single responsibility. These tasks are work packages. Each work package is a single discrete scope of work that is to be performed by a single responsible party. Each work package has a definite starting and ending point. These work packages can be viewed as mini-projects that are contained within the entire project. If greater detail is desired in the schedule, the specific tasks required to perform the general tasks can be identified. The level of detail should be carefully selected and will depend primarily on the size, type, and needs of the project. Ultimately, the level of detail that is needed should be determined on the basis of the level of information needed to properly control the project. The WBS provides the framework for organizing the tasks to complete an entire project.

The WBS is not to be developed in a vacuum. Information should be provided by various individuals, such as the project manager, job superintendent, and others. The development of the WBS should be from the top down and not from the bottom up. Consider the project in terms of phases, components, discipline, functional areas, and so on. Continue to break down each successive level into more detailed elements until work packages are clearly identified. As the information is gathered, also collect cost/budget information and schedule data.

In essence the work breakdown structure divides and subdivides a project into different components, whether by area, phase, function, or other descriptive means. The highest level in the work breakdown structure may consist of a single element, the project. At the next level there may by only five or ten elements or items. Naturally, the further one goes down the work breakdown structure, the greater the amount of detail. Regardless of the means used to define the elements, the schedule activities are to be defined for the lowest level in the hierarchy or at the greatest level of detail that is required to adequately manage and control the construction process. The level of detail will be dictated by the project complexity and the scheduling needs. Of course, the level of detail desired may also change with the recipient of the information. For example, the owner of a house being constructed will be most interested in the anticipated completion date of the project, but a drywall subcontractor will be primarily interested in information related specifically to the drywall in the project. The needs will vary for different parties involved in the construction process.

By using the WBS approach, the project will become easier to comprehend. Otherwise, the schedule is developed as a large number of tasks and this effort can become overwhelming. It should be noted that the WBS will be utilized throughout the project duration, as it is more than an approach to organizing a project schedule. The WBS will be used to provide further definition to the tasks to be performed. For example, it will serve as the basis for determining the task durations and to estimate the costs. As such, the WBS becomes the backbone of the project control or project tracking system. The WBS ultimately serves as a benchmark for measuring performance and it also serves as the historical database for the project. It will become the means by which the various parties on the project will communicate, including workers, accountants, project managers, superintendents, schedulers, etc.

At this point, the importance of the WBS should be obvious. It should be pointed out that if ten different individuals were to develop a WBS for a project, there might very well be ten different variations of the WBS. While the general logic of each approach might be evident and they may all bear some comparison to each other, some nuances will invariably exist. There is no single correct solution. Individuals with different experiences will simply use different approaches and this does not present problems as long as the project is properly scoped with the WBS.

The WBS is an orderly presentation of the tasks that must be performed to complete a particular project. Remember that the WBS documents the scope of work involved in the project but not the plan of how the project will be actually be constructed. That is, the elements of the WBS are not presented in any chronological sequence. Assume that a WBS is prepared for a simple project consisting of constructing a white board fence. While there are various ways to present the information, the WBS might be represented with the following major categories:

- Build fence
- Survey work
- Clean-up

Note that the WBS need not be in any particular order. In fact, the survey work could be listed first, last, or anywhere in the WBS, regardless of when this work is actually done. The WBS should not be confused with the schedule. The WBS will be utilized to develop the schedule, which will show the logical sequence of when the various tasks are to be performed. A breakdown of the major categories in the WBS shows the tasks to be performed in each category:

- Survey work
  - Establish legal boundary
  - Lay out fence line
  - Mark post locations
- Build fence
  - Dig post holes
  - Plumb and set posts
  - Attach boards to posts
  - Paint fence
- Clean-up
  - Touch up paint
  - Smooth out area along fence line

The above example was deliberately formulated as a simple presentation of the WBS. On a typical construction project of a building, it is common for the WBS to consist of numerous categories, depending on the complexity of the building. The WBS is used to document the scope of the various bid packages. While there will be exceptions, some of the major categories will include the scope of work to be performed by individual specialty contractors. For example, a major category for a building project might include electrical work, heating ventilation and air conditioning (HVAC), plumbing systems, sprinkler system, site work, and so on. The WBS of a relatively small and simple construction project is shown in Figure 1.11. This is the WBS for a storage facility which shows the major categories for the work packages.

The above WBS will be examined in slightly greater detail as it pertains to the construction of this storage facility (see Figures 1.12 and 1.13). For this project, the systems identified in the WBS might be as follows:

- Sitework
  - Drainage
  - Grading
  - Paving
  - Landscaping
- Site Security
  - Fencing
  - Security Lights
- Structure
  - Exterior
    - Slab-on-Grade
    - CMU Block Walls

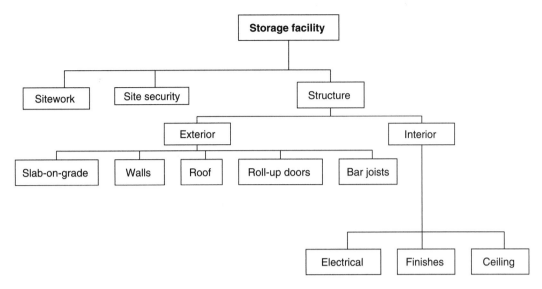

Figure 1.11 Graphical Presentation of the Work Breakdown Structure for the Storage Facility Project

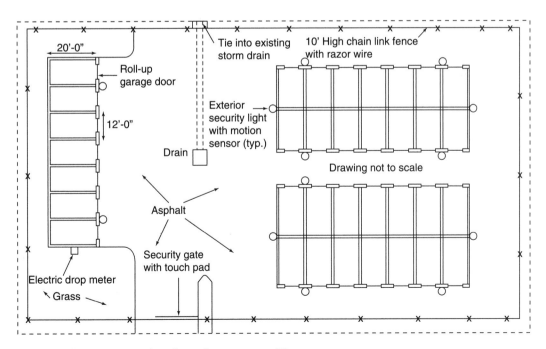

Figure 1.12 Site Plan of Storage Facility

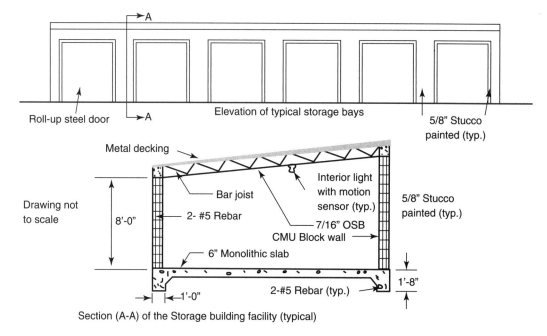

Figure 1.13 Elevation and Section of Storage Buildings

- Bar Joists
- Roofing
- Roll-Up Doors
- Interior
  - Interior Finish
  - Electrical
  - Ceiling

The WBS may appear to be somewhat related to the sequence with which different components will be constructed, but closer examination will reveal that this is not true. For example, the sitework will involve grading work that will take place very early in the project, while the landscaping will most likely be done nearer to project completion. In general terms, the WBS can be viewed as a representation of the physical breakdown of the work to be performed and it scopes out the construction portion of the schedule. Information such as procurement activities, preconstruction efforts, and general conditions is included in the schedule, but they are generally not represented in the WBS.

As already stated, the WBS is used for various purposes. The estimate can be prepared from the WBS. Even the task durations can be computed through the use of the WBS. For example, Figure 1.14 shows a work item breakdown of the grading work that will be scheduled for the storage facility. Note that this single sheet, entitled the Activity

Activity Data Sheet

Project: *Storage Facility*

Activity: *Grading*   Activity Code: *S103*

Responsibility: *GT&B Grading Specialists – Subcontractor*

Scope of Work:

Reference: *Site Plan Sheets 1, 2, and 3*
Description:
- *Properly obtain a finish grade for the site as shown on the plans and detailed in the Specifications*
- *Properly dispose of all waste created during the performance of the work*
- *Maintain full compliance with the OSHA regulations while on the site*
- *Responsible for all equipment needed to properly execute the work*

Work Plan and Assumptions:
  There will be no site encumbrances that will impede work during the days that the grading work is scheduled to be performed.

Duration Calculations:

Two Operators needed — two eight-hour days
Two Laborers needed — two eight-hour days

Total: *64 labor hours needed*

Reference:

Cost Calculations:

Labor:
Two operators at $47.00 per hour for 32 hours = $1,504
Two laborers at $34.00 per hour for 32 hours = $1,088

Equipment:
One front-end loader at $85.00 per hour for 16 hours = $1,360
One motor grader at $70.00 per hour for 16 hours = $1,120

Materials:
None

Total Cost: *$5,072*

Figure 1.14   Example of Background Information for WBS Item—Grading

Data Sheet, contains information on the company designated to perform the work, the activity code (unique number the company will use for this task in the schedule), the scope-of-work information, the computation of the task duration, and the cost computations. Similar information must be prepared for each item of work that must be performed to construct the storage facility. It is evident that considerable information will be contained in these completed documents.

When the WBS is prepared in a systematic manner, considerable information will be compiled. The information of interest here is related to the schedule. Since assumptions and other types of explanations about the work can be added to the work sheets, the information will serve as a historical record that will be referenced whenever questions arise.

REASONS FOR PLANNING AND SCHEDULING IN CONSTRUCTION

Owners and contractors may use more than one system to plan and schedule a project. However, most construction projects are complex enough and the tools (computers and software) readily available so that a vast majority of the construction projects would benefit from being planned and scheduled with the use of network models, namely, CPM. The reasons for using such a formal method of planning and scheduling vary from one firm to another and vary between projects. Moreover, many project contract documents expressly require contractors to submit a CPM analysis prior to commencing work. In some cases, a CPM analysis is required along with the bid.

Historically, CPM has been most widely used in the planning of projects prior to their construction. Its preparation requires the owner or contractor to develop a clear idea of the operations needed and their sequencing and timing. As the model is developed, it may alert the user to potential problem areas, such as excessive resource demands, and allow them to manipulate the schedule to ameliorate these problems. When completed, the model provides a well-reasoned estimate of the project's duration and schedule of when individual activities will occur. Thus, coordination of the various activities and resources can be planned ahead.

Money is always of special importance to those involved in construction projects. CPM allows the user to examine the trade-off between the time and cost required to carry out a project. Knowing the critical path through a network enables the planner to systematically determine the costs (or savings!) involved in getting the project done more quickly than currently scheduled. The prospect of incurring damages or penalties for late completion of a project or the receipt of bonuses for completing work ahead of schedule provides significant incentive to a project manager to carefully monitor project progress. The project manager must evaluate, from a cost perspective, whether it makes sense to attempt to recover from any delays or otherwise speed up or slow down production.

The monitoring and control of projects is now performed more effectively than in the past. Monitoring implies the recording of actual start and finish dates for activities while the project is underway, and control relates to the analysis of the impacts of any schedule deviations and evaluation of what remedial actions, if any, should be taken.

Monitoring information on a project may show that some activities are either ahead of or behind schedule. By accurately monitoring project status, decisions can be made regularly regarding the viability of the schedule. Once a project is "off" schedule, the need to update the schedule will be apparent. This updating must take place frequently if the project status deviates significantly from the project schedule. Only with an accurate portrayal of the project status can the schedule be used as a management tool. The implementation of monitoring and control using a network model also leads to the ability to employ CPM as an objective tool in resolving claims and disputes, whether they go into litigation or not.

Recently, the use of CPM in construction has increased rapidly, not from changes in the technique but rather from the proliferation of lower-cost computers and project management software for carrying out CPM calculations and procedures. It is now practical for owners and contractors, large and small, to apply the techniques presented here because they can do so much with greater ease and in far less time than five or ten years ago. This does not mean that every construction project will benefit from the rigorous application of the full range of CPM techniques and applications. However, most parties in the construction process can benefit from making the use of CPM a routine element of the management of all of their projects, provided they apply it in a manner commensurate with the needs of the project.

REVIEW QUESTIONS

1. Describe the disadvantages of scheduling a project with the use of bar charts.
2. Explain why the use of bar charts is popular in the construction industry.
3. Why is it often common to use bar charts and precedence diagrams to depict schedules on the same construction project?
4. What is the difference between planning and scheduling?
5. Compare bar charts and precedence diagrams in terms of the ease with which activities can be added to the schedule.

2

Developing a Network Model

Well begun is half done.

The development of the network model is perhaps the most important step in using a network as a scheduling tool. An ill-prepared model serves no useful purpose and may compromise construction progress. The model must bear a reasonably accurate portrayal of the actual steps to be followed in constructing a facility. The fundamental concept of a network-based planning and scheduling tool is that the network represents a time-oriented model of a system, not unlike a physical scale model of a project. The associations of the various tasks required to construct a project must be fully understood and these relationships must then be clearly conveyed to the user in the final network. If these requirements are not satisfied in the network model, it may never be adopted and will soon be abandoned as a project's management tool or be subjected to considerable modification to improve its utility.

A network model can actually be developed in many ways. Much depends on the experience of the individual preparing the model. If that scheduler develops the network model without input from others, he or she can dictate the steps. If others are involved in the process, the step sequence will often depend on the input of the various individuals. Nonetheless, regardless of the order, a number of steps are required to fully build and employ this model:

STEPS IN BUILDING A NETWORK MODEL

1. Define activities
2. Order activities
3. Establish activity relationships and draw a network diagram

4. Determine quantities and assign durations to activities
5. Assign resources and costs
6. Calculate early and late start/finish times
7. Compute float values and identify the critical path
8. Schedule activity start/finish times

While these are listed in sequence, the process of network development is more likely to be iterative. For example, steps 1 through 3 may be performed simultaneously by drawing the network as activities are defined. This process generally continues throughout the life of a project; the schedule must be changed to reflect changes made by the owner and to redefine the sequencing of activities. These steps can be accomplished on many scheduling software programs (which are commercially available). It is not unusual (and, in fact, is often advisable) to develop the network through all eight steps with the expectation of returning to earlier steps, taking into account all of the constraints that must be reflected in the network. Many of these scheduling constraints may not be recognized until the network has begun to take shape. Chapter 3 discusses the final calculations for activity scheduling (steps 6, 7, and 8). Resources and costs are assigned (step 5) only when resource utilization and/or cash flow issues are to be addressed, a step not used by many schedulers.

If one follows these eight basic steps, our final network should be a reasonable representation of the actual project. However, the resulting model will be an approximation or best guess of how the various tasks will be performed on a project. Many unanticipated incidents may occur that will cause the project to proceed in a manner not reflected in the model. For example, the owner may issue changes that must be incorporated in the network, or, for a wide number of reasons, the project might run behind schedule. Whether there are labor shortages, delays in material deliveries, performance problems with a subcontractor, or differing site conditions, it is likely that the schedule will be impacted. Once the model no longer reflects the actual construction activities, it is prudent to update the network. This is an important aspect of the scheduling process and should follow essentially the same steps used to develop the initial schedule network.

Defining Activities

Anything that must be accomplished (whether by the owner, contractor, subcontractor, supplier, inspector, or other party) in order for the project to be completed may warrant inclusion in the network. One must recognize the different types of activities that can come into play in a construction project.

Production/Construction: These are activities that relate directly to the physical effort of creating the project. These are the most readily understood activities as they include much of the labor effort that results in the completion of a facility. These activities use traditional resources of labor and materials. In most instances the resource utilization will involve labor, but this is not always the case. The resource of particular interest is time. Thus, the curing of concrete may be a viable activity to be included in the network, even though it consumes essentially no resource other than time. Other activities may very well be dependent on the curing of concrete before they can begin, justifying the inclusion of curing concrete as a separate activity.

Procurement: These activities include arranging for the acquisition of materials, money, equipment, and manpower. These may include a number of activities that influence the timing of production activities. Readily available items generally do not require separately defined procurement activities. Special-order, long, or uncertain lead-time items should always have procurement activities incorporated into the network. This permits appropriate analyses to be performed if changes or delays occur. The inclusion of long lead-time items will also function as a reminder to take appropriate procurement actions on selected items before they begin to delay the project.

Management: Support or administrative tasks often directly impact the project schedule. The scheduler should always be aware of those required "extra" or management activities that are not included in the above two categories. Activities such as preparing inspection reports, processing shop drawing approvals, tracking submittal approvals, developing as-built drawings, providing certifications on factory tests performed, and a variety of similar tasks may need to be included.

When the model is being developed, a variety of issues must be considered.

Objective of the Model: If a model is being prepared simply to satisfy a contract requirement, it probably will not be used as a management tool. In that case, one would not need or want to spend the time to model the project in a lot of detail unless required by the contract. On the other hand, if one is building a model that will permit him or her to identify when certain resources or subcontractors will be needed, the detail must be greater.

1. Users: If the model will be used to show upper management a general overview of the project, then the activities will likely be defined in much broader terms than if the intended users are field personnel performing direct project management.
2. Frequency of Use: The activities must have durations compatible with the frequency of the project monitoring being performed. For example, if the model is to be used to monitor weekly progress on the project, it would not be prudent to include activities with durations defined in months. Conversely, monthly monitoring may not warrant going to the detail of scheduling activities with durations defined in portions of a day.
3. Products: Some potential products, such as cost reports or requests for progress payments, dictate that activities be defined in a way that allows necessary data to be compiled and reported. For cost reporting, it is necessary to relate activities to the estimating and/or accounting system units used for compiling cost estimates. Progress payments may require that activities relate appropriately to pay items and permit progress to be measured in a manner mutually agreeable to owner and contractor. For example, suppose that prior to the start of construction, the owner and contractor agreed on the

value of a concrete slab as $20,000. Once the slab is completed, the contractor would want to quickly submit a request for payment for that work.

But what constitutes an appropriate activity? If a project involves building a small shed, is it appropriate to define an activity as "Frame the walls," or should there be four activities, such as "Frame North wall" or how about "Hammer nail number 574"? Obviously, the level of detail in the last activity is not at a usable scale. The activity is far too short in duration; one could never realistically plan and schedule at this level of detail. Nor would one want to, because the complexity of the network would very quickly overwhelm him or her and the time spent on preparing and managing the network would not be appropriate. But what about the definition of "Frame" versus "Frame North wall"? Is one inherently correct and the other incorrect? The scheduler must decide.

To determine the appropriate detail, the scheduler must think about how the model will be used. One must be careful to develop the model with a clear goal of generating a schedule that will serve as a management tool rather than simply a very detailed depiction of how a project will be assembled.

It is preferable to err on the side of providing too much detail in the activities rather than too little. With today's project management software, it is easy to consolidate activities into "sub-networks" or "super-activities" by using simple coding schemes. Thus one can develop an "abstracted" network for those uses that do not require greater detail. If the original network activities are too gross, it will require additional manual effort to break these activities down into smaller ones, determine interrelationships, and assign durations, costs, and resources before obtaining a usable network. Remember, activities can be aggregated easily, but they cannot be disaggregated.

Ordering Activities

The order of activities is based on the timing of some activities relative to the occurrence of other activities. For each identified activity, the following must be determined:

- Which activities must precede it?
- Which activities must follow it?
- Which activities can be concurrent with it?

Actually, all of these questions can be addressed by simply identifying, for each activity, all of the immediately preceding activities (IPAs). If the IPAs of each activity are considered, the network will automatically address the other questions, provided no resource constraints will be imposed on the project. When the IPAs for all of the activities are described, each activity's succeeding activities will have been generated, along with the activities that can occur at the same time. Thus, the success of the model depends on defining the set of activities that "control" when each activity can start, because an activity can only begin if all of its predecessors are completed. Once the IPA list is completed, the network can be created.

Ordering activities is not necessarily as straightforward as it may seem on the surface. The reason why two activities must be done in a particular order can be termed a constraint. Without constraints on a project, all activities can theoretically begin on the first day of construction. But constraints do exist in the real world and they must be considered in order for a network to be useful. Constraints are of a variety of types.

Physical Constraints: Physical constraints exist due to the physical process of construction, such as the need to erect forms before concrete can be placed. These are logic constraints that include those defined by "how" (construction methods) the project is to be carried out.

Resource Constraints: These constraints are conditions of limited availability that dictate that certain activities cannot be performed simultaneously because insufficient resources are available. For example, having only one crane available that must be used on two otherwise independent activities might require that the activities be scheduled so that they do not occur at the same time. Similarly, the amount of concrete that can be placed in a single day may be dictated by the production capacity of the concrete batch plant.

Safety Constraints: Safety requirements may dictate that activities not occur simultaneously (e.g., overhead and ground level work in the same area, drilling and blasting taking place concurrently) or that a specified sequence occur (e.g., erection of safety barriers before allowing work in an area). Safety considerations may also dictate defining nonworking days for extremely hot or cold days. Project lighting requirements may also be dictated by safety concerns.

Financial Constraints: Monetary constraints can include the staggering of high-cost activities to minimize cash requirements during construction or the necessity of securing loans prior to undertaking certain portions of a project. Large cash flow items might also be scheduled so that they are incurred in a particular "tax year."

Environmental Constraints: Environmental constraints can include the need to carry out mitigation procedures prior to other activities and may also address restrictions such as not working in certain areas during such times as spawning season, fish runs, or eagle nesting.

Management Constraints: Sometimes referred to as "arbitrary," these can be defined simply as additional constraints not otherwise categorized here. They may relate to requirements of supervisory time, consequences of tax strategy decisions, cash flow needs, or the demands of other projects not reflected in the network. Management may elect to consider the days between Christmas and New Year's Day as holidays. At any rate, these reflect decisions of management that are deemed to result in a reasonable benefit to the company.

Contractual Constraints: The owner may impose constraints on the construction process. The owner of a condominium project may require that a particular phase of the project be fully completed and occupied prior to beginning construction of the next phase. On a remodeling project, the owner may require that construction noise or dust be kept to a minimum if a portion of the facility will remain occupied and in operation.

Regulatory Constraints: Governmental agencies are also known to impose constraints on the construction process. This includes regulations that are enforced by such federal agencies as the Environmental Protection Agency, land use restrictions that might be imposed by a municipal government, and other regulations enacted at the municipal, county, state, or federal levels. Some of these constraints would also be included among the environmental constraints.

In the initial definition of a network, it is desirable to minimize the number of constraints introduced into the network during the ordering of activities. This is because the introduction of excessive constraints in network logic can have the following impacts on a project:

- Reduce scheduling flexibility
- Lengthen project duration
- Generally increase project cost
- Confuse basic scheduling logic

If the logic of a network is not questioned, the imposition of constraints tends to result in a linear ordering of activities. Such a network will tend to have fewer activities running parallel to each other, leading to longer chains of activities, and making the project longer and more costly.

Only physical constraints should be entered during the early development stage of the project model. Other constraints can be deferred until the actual scheduling of activities, at which time the network should be inspected to determine, first, if the constraint is not met by the calculated schedule and, second, if it can be addressed by the shifting of activities within their available "float" times. Any constraints not handled in this way can be addressed in another iteration of the network development process.

Once the basic network is completed, the constraints should be imposed in some sequence. Since this will generally be done with scheduling software, the impact of particular constraints can be readily evaluated. When the impact of a particular constraint is especially severe, the scheduler will probably reassess the need for the particular constraint. If the constraint is unavoidable, the logic of the network itself may come under close scrutiny. In many instances, a change in scheduling logic may accommodate a constraint without seriously compromising construction progress.

Illustrated in Figure 2.1 is a sample activity list and associated list of immediate preceding activities (IPAs). Notice that procurement activities ("Purchase steel reinforcement" and "Order concrete") tend not to have predecessors because normally they can be done without anything preceding them. Note that this does not mean that all of the materials will be procured at the very beginning of the project. It simply means that procurement of materials can be scheduled for the very beginning of a project. At this point only the logic of the process is being defined.

	Sample Activity List with IPAs Project to Construct Concrete Footing	
Activity Label	Activity Description	IPAs
A	Lay Out Foundation	—
B	Dig Foundation	A
C	Place Formwork	B
D	Place Concrete	G, H
E	Purchase Steel Reinforcement	—
F	Cut and Bend Steel Reinforcement	E
G	Place Steel Reinforcement	C, F
H	Order Concrete	—

Figure 2.1   Sample Activity List with IPAs

Drawing the Network Diagram

There is one commonly used type of network diagram, the precedence diagram. In the 1960s and 1970s, there was also a considerable use of arrow diagrams, but such diagrams are used very little today. Most popular software that is currently used for computer scheduling supports only precedence diagrams. For this reason and for some inherent disadvantages of arrow diagrams, arrow diagrams are no longer used extensively in the construction industry. Precedence diagrams and arrow diagrams present the same basic information about the sequence of activities that must take place to complete a project. While this text addresses primarily precedence diagrams, parties interested in arrow diagrams are directed to Chapter 16, which provides a description of arrow diagrams and their use.

The precedence diagram depicts activities as nodes with logic link lines that depict the dependencies that exist between the activities (see Figure 2.2). The relationships of the activities in a precedence diagram are conveyed with relative simplicity by the use of the logic link lines.

The precedence diagram is "read" from left to right. With the logic link lines shown in the precedence diagram, it is a simple procedure to show the proper sequence of activities to construct a project or a component within a project. In Figure 2.3, it is readily observed that the first activity is "Lay out foundation." Once this has been completed, the "Dig foundation" activity can begin. It is important to recognize that every activity consumes time and generally other resources as well, while the link lines consume no time or resources. Events also do not consume time, but merely describe the project status at a point in time. For example, the point at which a building is "dried in" would be considered an event, and the point in time at which the project is completed is also an event.

Developing a Network Model 29

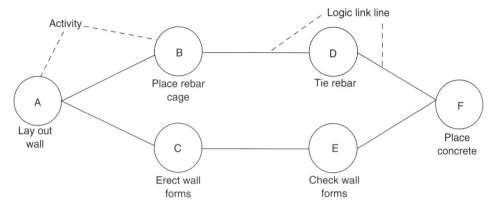

Figure 2.2 Simple Example of a Precedence Diagram for Erecting a Concrete Wall

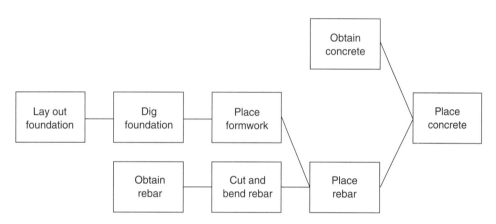

Figure 2.3 Example of a Precedence Diagram for Constructing a Concrete Footing

The metal storage barn project that was depicted as a bar chart in Figure 1.6 is shown in Figure 2.4 in a precedence diagram format. This diagram does not depict the durations of the activities as was the case in the bar chart, but the precedence network clearly shows the dependencies of the different activities and the durations could be easily shown on each activity in the precedence diagram.

One must recognize that activities consume time and events simply occur at a point in time. The point in time at which an activity can start may be considered an event. The point in time or instant at which an activity is completed can also be considered an event. For most activities, the simple computation of the start and finish times of activities is sufficient for a scheduler to effectively manage a project. In some instances it is important to identify events of interest or milestone occurrences. A milestone event might occur when a building has been "dried in" or when a building has "topped out." It could also consist of the receipt of a permit such as a construction

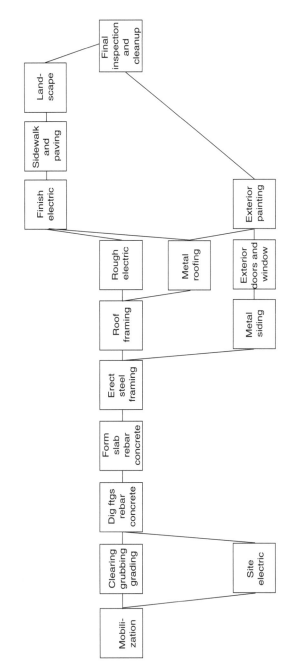

Figure 2.4 Precedence Diagram for the Construction of a Metal Storage Barn

permit, blasting permit, or dumping permit. While events are not included in many networks, their value should not be underestimated.

Assigning Durations to Activities

The duration of an activity is the estimated time that will be required to complete it. Typically, in construction, the unit of time is days and it is assumed that work is performed on a continuous and uniform basis within the standard workday and work week. Time might also be measured in months, weeks, shifts, or even hours. Defining durations, as in defining the activities themselves, is less of a science than it is an art, largely due to the range of variables that can impact a construction project and progress on it. The time unit appropriate for a project depends on the duration of the project and the nature of the work. Hours would typically be used for shorter projects lasting only several days. For example, on shutdown or turnaround projects that might be completed within a few hours, days, or weeks, it is common to schedule the work in hours. This is largely necessitated because these types of projects frequently have a large number of workers on the project and the work is performed in shifts, with work being performed continuously (7 days a week) until the project is completed. For long-term projects that last for several years, days, weeks, or even months may be more appropriate.

Activity durations frequently are tied directly to the resources applied to them (e.g., crew size and equipment) and the productivity of these resources. Consequently, durations can frequently be directly related to labor cost estimates for the tasks. Cost estimates may be developed as follows:

$$\frac{\text{quantity of work}}{\text{qty/crew-hour}} \times \frac{\$}{\text{crew-hour}} = \$ \text{ labor cost}$$

Given this formula the duration of the activity is clearly assumed to be:

$$\frac{\text{quantity of work}}{\text{qty/crew-hour}} = \text{crew-hours} \quad \text{or} \quad \frac{\text{crew-hour}}{\text{hour/day}} = \text{days}$$

The primary issue here is the reliability of the productivity factor (quantity/crew-hour). These factors may be based on standard published rates (e.g., from R. S. Means publications), perhaps adjusted by regional or project-specific factors. Many people discount the accuracy of such "standard" rates, however, because they are simply "averages" of widely varying numbers. Furthermore, some activities do not involve performing large quantities of work and, therefore, do not lend themselves to productivity rate analysis.

Whether or not durations are developed from productivity rates, an alternative is to use historical data, meaning quantitative data from actual projects that the firm has undertaken previously. Even if such data are available, the activity must be carefully examined to determine whether the data can be applied to it, as it may be felt that the project being planned is atypical and thus the historical data are not comparable.

In the absence of sound quantitative data, the scheduler often "guesstimates" activity durations based on his or her experience. Someone with extensive experience in comparable projects may be able to establish very reliable durations, while younger, less experienced persons are ill-suited to the challenge. Such persons may need to seek the professional or experienced counsel of others through interviews.

Larger owners, such as public agencies, and contractors involved in a large number of similar projects may also find it possible to develop automated algorithms (procedures) for estimating activity durations. Dealing with many similar types of projects over a period of time allows these schedulers to establish first, a "template" network (standard model) for their projects, and second, a relationship between project characteristics and the durations of individual activities within the network.

Assigning Resources and Costs

Frequently, CPM analyses include the evaluation of the temporal (time) distribution of resources and costs. This associates each activity with the amount of resources (labor, equipment, and materials) it requires and its associated costs. Such information should be derivable directly from the estimating process, as costs for the project depend upon the quantity of materials and amount and type of labor and equipment resources applied. The major requirement for the effective assignment of resources and costs to individual activities is a clear description of the relationship between the CPM activities and the units or phases of work by which the estimate is developed.

In simplest terms, using CPM with resources and costs permits the amount of each resource needed during specified time periods to be examined. Beyond this, there are a number of analytical techniques associated with evaluating, altering, and attempting to "optimize" the distributions. (See Chapters 6, "Resource Allocation and Resource Leveling," and 7, "Money and Network Schedules," for discussions of these techniques.)

Calculating Early and Late Start/Finish Times

Steps 1 through 4 must be completed (limited computer solutions are possible without step 4) in order for calculations to be made on the network. The initial computations that are of particular interest include the early and late start times of an activity and the early and late finish times. The early start time of an activity is the earliest time that the activity can begin based on the relationships that exist in the schedule. The late start time is the latest that an activity can start without adversely impacting the date of project completion. Similarly, the early finish time is the earliest that an activity is expected to be completed, and the late finish time is the latest that an activity can be completed. If the late finish date is exceeded (runs over), the project duration can be expected to be increased by the same number of days that the activity is completed beyond the late finish time. If the early and late start times differ, the activity is said to have flexibility, or "float."

Identify the Critical Path

If the early and late start dates of an activity are the same, it is evident that the activity has no flexibility, or "float." In other words, if the activity starts later than the assigned date or if the activity takes longer to complete than the assigned duration, the project completion date will be extended by the same amount of time. When an activity start date is fixed in this way, the activity is said to have no float. Such activities are said to be "critical," meaning that any delay in the start date or completion date will extend the project duration. The critical path is the path from the beginning of the network to the end of the network on which all activities are critical. (These concepts will be discussed in considerable detail in subsequent chapters.)

Scheduling Activity Start/Finish Times

Once the calculations have been made to further define the network activities, the management process can begin. The network and the information generated for each of the activities will be useful for management to execute the project requirements. Management decisions essentially revolve around the use of any flexibility or float that the activities possess. The key is that the scheduling information must be used in order for the network to be a management tool.

FINAL COMMENTS

The development of a network model consists of the completion of several fundamental steps. While some steps occur in sequence, others may take place concurrently. The nature of the activities and their relationships to other activities dictate the sequencing of the steps to a considerable extent. Although some steps can take place in a dependent fashion, others are performed independently. The step sequence depends on the preference of the scheduler and the extent to which computerized scheduling is being utilized. The accuracy or quality of the network model determines whether it can serve as an effective scheduling tool.

REVIEW QUESTIONS

1. What is the difference between an activity and an event?
2. What are examples of two different types of constraints that might have an adverse impact on a scheduling network?
3. What is the fundamental difference between an arrow diagram and a precedence diagram?
4. Why should many of the constraints be imposed on a network after the basic logic network has been completed, rather than incorporating them as the network is being developed?
5. Describe two different network schedules that might be generated for the same project. Explain why.

6. What is an example of a procurement activity that might be a valuable inclusion in a network? Conversely, what is an example of a procurement activity that has little merit for inclusion in a network?
7. Why is the initial schedule to be viewed as a best guess of how a project will be constructed rather than a definitive statement about how it will be done and when it will be completed?
8. Describe potential users of a construction schedule.
9. Give an example of a safety constraint that might be imposed in a schedule.
10. Describe factors that might impact the level of detail used in a schedule.

3

Precedence Diagrams

*If you can't afford to do it right the first time,
you won't be able to afford to do it twice.*

PRECEDENCE (ACTIVITY-ON-NODE) NETWORKS

The most common type of network schedule in use today is the precedence diagram. Because of the popularity of precedence diagrams as opposed to arrow diagrams, this text is focused primarily on the use of precedence diagrams (arrow diagrams are discussed in Chapter 16). Also, most scheduling software in use today requires the user to input the information in the form of a precedence diagram. In precedence diagrams, activities are represented by nodes. The nodes are linked by lines that show the relationships that exist between the activities. The basic elements can be described as follows:

Nodes: These represent "Activities" and may be drawn in any shape desired. Often they are drawn as circles, boxes, or other common geometric shapes (see Figure 3.1). While the diagram presents only basic information about the construction of a pole barn, the logic should be apparent. The diagram is read from left to right and the link lines show clear dependencies between the various activities.

Lines: Lines represent "Activity links," which are used to represent the dependencies between activities. Links are interpreted from left to right to indicate the dependency relationship. There will be one link in the network to denote each dependency between an activity and one of its immediately preceding activities.

The precedence diagram consists of a series of nodes with lines (links) connecting them to illustrate the activities to be executed and the sequence in which they must be done. The normal convention in the use of the precedence technique is that an activity can be started after *all* links drawn to its left (starting) side have been "traversed." This occurs when each of an activity's predecessors has been completed. After

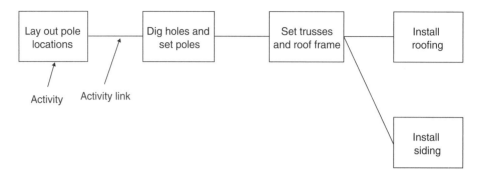

Figure 3.1   Precedence Diagram of a Simple Pole Barn

the activity is completed, all links leaving from its right-hand (finish) side can be traversed. Most of the discussion on the use of precedence diagrams will assume that all relationships between activities are *finish-to-start* in nature. In a finish-to-start relationship, once an activity is finished, its successor can start. For manual solutions, this dramatically simplifies the computations. Naturally, computer programs using the precedence diagram approach can easily make the various computations for whatever types of relationships are indicated. While most of the discussion on activity relationships will be focused on the finish-to-start relationship, the other types of relationships should be well understood.

Normally, the traversing of a link does not require any time, but shows the dependency between activities. However, it is possible to assign a time to a link to indicate that the following activity cannot be started until this specified amount of time after the preceding activity has passed. This time will be referred to as *lead time*. A link line can consume no resource other than time.

ACTIVITY RELATIONSHIPS

In the past, relationships between activities were defined as being finish-to-start (FS), whereby the preceding activity must be finished before the succeeding activity can start. This assumption makes it much easier and simpler to develop schedules by manual means. This type of relationship was assumed to exist in the schedule shown in Figure 3.1. Since most scheduling is now done with the aid of computers, this limitation no longer poses a constraint or concern. Other types of relationships are also encountered, namely *start-to-start* (SS), *finish-to-finish* (FF), and *start-to-finish* (SF). The use of these various types of relationships between activities makes it possible for schedules to be prepared that accurately portray the true relationships between activities. The limitations of the scheduling software will dictate the restrictions to be placed on defining the relationships between activities. These relationships are used to organize activities that overlap to some degree or that have a delay time of some sort between them.

In a finish-to-start relationship, the start of each activity depends on the completion of its preceding activity. That is, if an activity depends on another, it will not be able

Figure 3.2   A Typcial Sequence of Finish-to-Start Relationships

to start until the preceding activity has been completed. For example, a door jamb must be installed before the door can be hung. Similarly, footing excavation must be completed prior to placing concrete for the footing. For many activity relationships, this assumption is accurate or adequate. For manual computations on networks, this assumption is essentially necessary. While this finish-to-start relationship between activities is generally viable, these other types of relationships are also encountered. In various instances, the finish-to-start relationship is convenient for illustration purposes, but it may not be an accurate portrayal of the actual relationship that exists between two variables. Examples of these other types of relationships will be described briefly.

A series of examples have been developed to show how different relationships might be encountered in a project schedule. The actual site conditions must be evaluated in virtually every instance to determine the appropriate relationship between any two activities. In some instances there will be more than one way to protray the activity relationships. Consider the simple series of tasks that involve placing concrete in forms, curing the concrete, and stripping the forms, shown in Figure 3.2. These activities and their durations are shown with the traditional finish-to-start relationship.

At first glance, this sequence seems quite logical. The activity "Cure concrete" is typically one that does not entail the use of resources, other than time itself. With only time being consumed, the use of the activity "Cure concrete" is simply a means of putting a delay activity between placing the concrete and stripping the forms. These activities (placing concrete and stripping forms) might also be shown as being related by *finish-to-start with a delay*. Note that the start of stripping forms depends on the completion of the placement of the concrete followed by a designated delay. This is shown in Figure 3.3. Rather than creating the delay as a separate activity, it is made a component of the finish-to-start relationship; namely, the concrete forms can be stripped 28 days after the concrete has been placed. The logic remains the same, but the presentation of the relationship is slightly modified.

While manual computations would be cumbersome with this type of delay, many scheduling software programs now have the capabilities of handling such situations

Figure 3.3   A Finish-to-Start Relationship with a 28-Day Delay

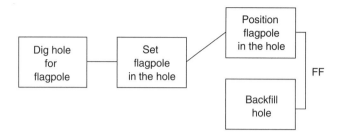

Figure 3.4 Simple Task Showing the Finish-to-Finish Relationship

with ease. Other types of relationships occasionally encountered in construction may also be readily handled by such programs.

One type of relationship that might be encountered on occasion is the finish-to-finish relationship, in which two activities are related by the fact that they must both be completed at the same time. Assume a simple task of setting a flagpole. The task will consist of digging a hole for the pole, setting the pole in the hole, positioning the pole in the hole, and backfilling the hole with the excavated materials. Note that the pole must be held in position until the backfilling is completed. Thus, the backfilling task and the positioning the pole task will be finished at the same time, as shown in Figure 3.4.

Consider a portion of a building construction project in which masonry work is performed on the exterior of the building. A few windows are also to be included in the exterior walls of the building. When windows are installed in the masonry walls, additional masonry work must be performed around the window frames to complete the installation. In a conventional finish-to-start relationship, this would be depicted as shown in Figure 3.5. Essentially, the bricklaying activity is broken into two distinct activities in order to demonstrate the need to delay some bricklaying work until after the windows have been installed.

These activities may be reasonably close to describing the actual work sequence. The level of detail of the activity "Masonry work around windows" may not be consistent with the more general descriptions used for the other activities. While this may be suitable for most schedulers, consideration might be given to describing the activities

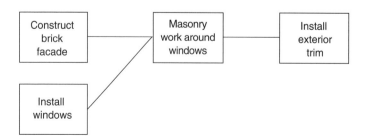

Figure 3.5 Sequence of Finish-to-Start Relationship for Window Installation

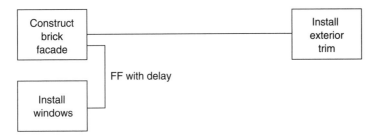

Figure 3.6 Sequence of Finish-to-Finish with a Delay Relationship

as a *finish-to-finish with a delay* relationship. This might be accomplished with the representation shown in Figure 3.6.

Note that the "Install windows" activity is now associated with the finish of the "Construct brick facade" activity. The exterior trim work cannot begin until both of these activities are completed, but there is an added stipulation that the window installation must be completed a certain amount of time before the brickwork is completed. It is during this delay period that the necessary masonry work around the windows would take place.

In Figure 3.7, the network shows where a series of activities have finish-to-finish with a delay relationships. The removal of the old paint must be completed one day prior to the completion of the activity involving sanding and prepping the wood. In addition, the wood sanding and prepping must be completed two days prior to the completion of the painting activity.

The start-to-start relationship is another type of activity relationship that is encountered during construction operations. Let's consider the task of putting in a new

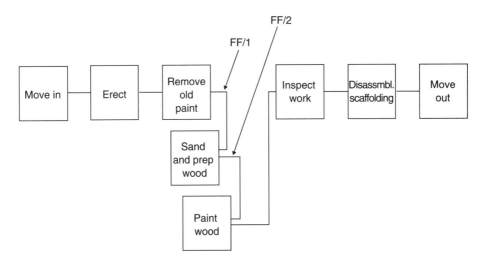

Figure 3.7 Activities with Finish-to-Finish with a Delay Relationships

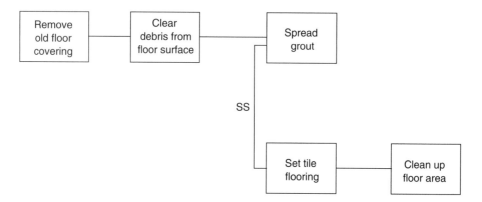

Figure 3.8 Activities with a Start-to-Start Relationship

tile floor in the foyer of a commercial building. Some of the activities to accomplish this are shown in Figure 3.8. In this example, one key activity is to spread grout on the floor to receive the tiles and the other activity of interest is to set the floor tiles. Because of the short set time of the grout, there is essentially no delay between applying the grout and setting the tiles; i.e., tiles must be set in the fresh grout. While the tile-setting task must wait a few moments for the grout to be initially applied, for all practical purposes, they must begin at the same time.

Another relationship that occurs in construction is the *start-to-start with a delay*. First, consider the construction of a simple stud wall that will receive insulation and drywall (for simplicity, it is assumed there is no plumbing or electrical wiring in the walls). Suppose several rooms are being constructed for an office building. The conventional finish-to-start relationship might show the activities as in Figure 3.9.

These activities show that the insulation must be started before the drywall installation can begin. However, once the insulation work has begun, the drywall installation can

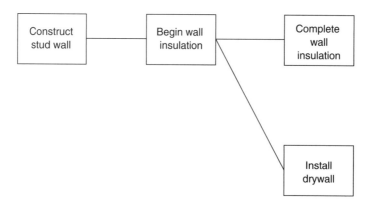

Figure 3.9 Sequence of Finish-to-Start Relationship for Wall Construction

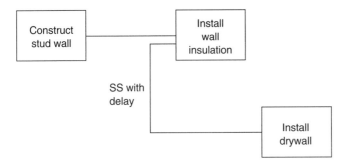

Figure 3.10 Sequence of Start-to-Start with a Delay Relationship

also begin. Note that it is assumed that the insulation installation requires considerably less time than the drywall installation. That is, once the insulation work begins, it is assumed that this work will easily outpace or stay ahead of the drywall installation. The above logic might also be presented as a start-to-start with a delay as depicted in Figure 3.10.

With this depiction, it can be seen that the drywall installation can begin a stipulated amount of time after the wall insulation has begun. Since several offices are being constructed, it is assumed that, after the insulation is installed in the walls of one office, the drywall can be installed. The specifics of the particular offices involved are not shown in the schedule, as this might be regarded as needless detail.

Figure 3.11 shows the activities of a simple painting project of erecting scaffolding, removing old paint, preparing wood for repainting, painting wood, inspecting the work, and disassembling the scaffolding. The three activities of primary interest are removing the old paint, preparing the wood for repainting, and painting the wood. There is a start-to-start relationship with a one-day delay between removing the old paint and preparing the wood for repainting. There is also a start-to-start relationship with a two-day delay between preparing the wood for repainting and painting the wood.

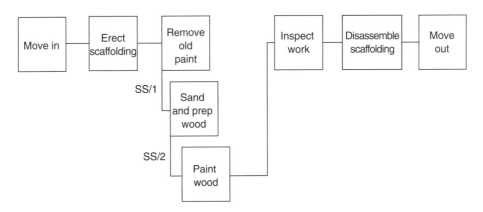

Figure 3.11 Activities with Start-to-Start with a Delay Relationships

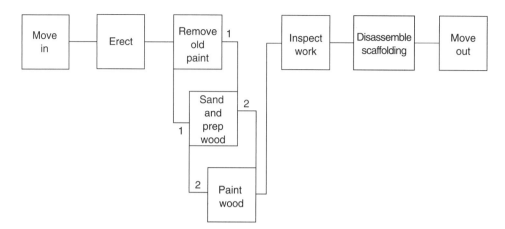

Figure 3.12 Activities with Start-to-Start and Finish-to-Finish Relationships

The information in Figure 3.11 might be shown in yet another manner. Figure 3.12 shows these same activities, but with their relationships shown as combinations of both start-to-start (SS) and finish-to-finish (FF) relationships, perhaps the most realistic portrayal of the actual relationships between these activities. While quite realistic, this is not a very common way of actually presenting these activity relationships. Activities that have SS and FF relationships must be carefully paced in order that all criteria can be satisfied. Some computer software programs can recognize all of these relationships, although showing a combination of SS relationships and FF relationships for the same two activities is not as common in the available software. In this example, the start-to-start relationship might be preferred if the three activities in question have progressively longer activity durations. On the other hand, the finish-to-finish relationship might be preferred if the successive activities normally have shorter durations.

Another type of relationship that might exist is the start-to-finish sequence. At first, this type of relationship might appear to be illogical or even irrational. However, there are numerous situations for which the start-to-finish (typically with a delay) relationship is appropriate to illustrate the relationships between different activities. In reality, this is actually a fairly common relationship that exists between many activities. A few of these types of relationships will be shown to demonstrate how common they really are. Consider the traditional finish-to-start relationships depicted in Figure 3.13. From this information, it is evident that the concrete forms are set prior to tying the reinforcing steel and that the concrete is placed after the rebar has been tied and after the concrete has been ordered from the supplier. This information may appear to be logical, but it is not entirely accurate. For example, the ordering of the concrete is not dependent on setting the concrete forms.

A more realistic portrayal of the information would be created if the start-to-finish relationship with a delay was used, as shown in Figure 3.14. It is assumed that the local concrete supplier has stipulated that orders for ready-mix concrete should be made at least five days prior to the day the concrete is to be delivered. That is, the concrete

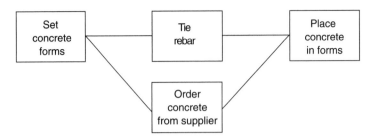

Figure 3.13 Sequence of Finish-to-Start Relationship for Concrete Placement

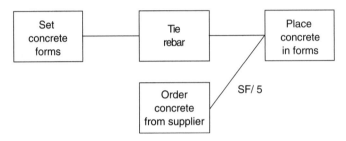

Figure 3.14 Sequence of Start-to-Finish with a Delay Relationship for Concrete Placement

placement will start five days after the concrete has been ordered. The relationship between the ordering of the concrete and the concrete placement activities is a finish-to-start with a delay. It should be apparent that the concrete order should be placed prior to the concrete placement, but one might be tempted to order the concrete after the rebar has been tied. Note that this might delay the actual concrete placement due to the five-day lag between ordering the concrete and having the concrete delivered.

Of the different types of relationships, perhaps the least used in the past was the start-to-finish (SF) relationship. Consider the start-to-finish relationships shown in Figure 3.15. These SF relationships are shown with delays, but similar examples could just as well exist without delays. For example, the purchase of materials that require no lead time would be typical SF relationships without delays. Training could be shown as an SF relationship without a delay, but this would give the training personnel no flexibility in scheduling the training session.

By examining the examples in Figure 3.15, it should be apparent that a distinct difference exists in the start-to-finish relationships than in the other types of relationships. One apparent difference is that the activity that must be finished first is generally not associated with any other activity in the network; i.e., the essence of the relationship is that the activity is linked specifically to one activity. In Figure 3.15 (a), the example shows that setting the prefabricated trusses can start after the trusses have been ordered 45 days earlier. This lag will ensure that the truss manufacturer has the necessary time to fabricate the trusses, assuming that 45 days is the stipulated lead time. In

44   Chapter 3

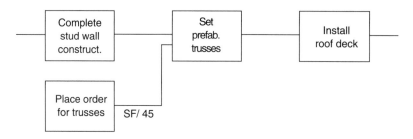

(a) Start-to-finish with a delay relationship for long lead-time purchases

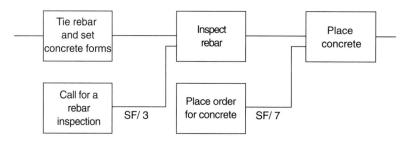

(b) Start-to-finish relationship for requested inspections and for specific delivery times

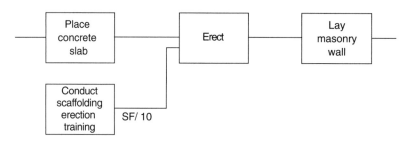

(c) Start-to-finish relationship involving training

Figure 3.15   Examples Demonstrating the Use of Start-to-Finish with a Delay Relationships

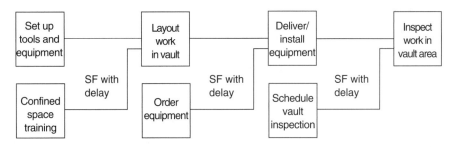

Figure 3.16 Sequence of Start-to-Finish with a Delay Relationship for Vault Work

Figure 3.15 (b), concrete rebar is inspected, but only after the inspection has been requested three days earlier. In this case, inspectors are not on-site every day, and they will only conduct an inspection when they are given three days' advance notice. Naturally, the request for the inspection is not directly linked to any other activity, so the start-to-finish relationship is appropriate. This figure also shows that the concrete placement can start seven days after the concrete delivery has been requested. It is common for concrete deliveries to be scheduled in advance. In this instance, a seven-day advance notice of the delivery is required. It should be understandable why the placing of the order for the concrete is not directly linked to the satisfactory inspection but rather to the placement of the concrete. Figure 3.15 (c) shows that scaffold erection can begin ten days after the necessary scaffold erection training has been completed. In this instance, the ten days may not be a necessary requirement, but this might allow the company sufficient time to schedule the training with its own personnel.

There may be numerous activities on a single project that involve the purchase of materials. If they are included in a precedence diagram, they might be best presented in the form of a start-to-finish with a delay. Figure 3.16 demonstrates how several different start-to-finish relationships might be encountered in a single network. Note that, before the layout work will be done in the vault, a training session will be held on confined space entry work. Prior to the delivery and installation of the equipment, the equipment must be ordered, possibly with a long lead time. Finally before the vault area is inspected, the inspection must be scheduled, typically with a few days' advance notice.

The finish-to-start with a delay, finish-to-finish with a delay, start-to-start with a delay, and start-to-finish with a delay are commonplace in computer programs. While no manual approaches will be presented for computing information for such relationships, popular computer programs have little difficulty performing this task.

Basics about Precedence Diagrams

Precedence diagrams are easy to draw, and it is a simple procedure to add an activity to a network. When viewing a precedence network diagram, it is important to avoid confusing the link lines with activities. The link lines represent dependencies only. Delays between two successive activities can be represented on the link lines. In the examples in this text, unless clearly noted otherwise, the assumed link line delay values are 0.

Act. I.D. #	Activity Description		Activity Cost
SS	Activity Duration		SF
FS	Activity Resources		FF
ES	LS	EF	LF
FF	TF	Int. F	Ind. F

Figure 3.17 Example Node Designation that Contains Considerable Activity Detail

Note that the nodes or precedence activities can be denoted simply by a single character. The nodes can be identified in considerably more detail, as shown in Figure 3.17. The format for node designation is not standardized, so one need only adopt a format that is comfortable for a particular application. The incorporation of all of the details shown in Figure 3.17 on a precedence diagram would make the diagram quite unwieldy.

While the node designations should not be too complex, it is still common for them to show various items of information. The node depicted in Figure 3.18 is perhaps fairly typical of that used in most computer schedules. These can generally be customized to the user's convenience, so the format can be adjusted to suit the specific needs of the scheduler. In this node, the activity identification is most prominent. Immediately below this (in the center of the node) is the duration of the activity. Along the left side of the node is information on the early start date (ES) and the early finish date (EF) of the activity. The early start of an activity is the earliest that the activity can begin, and the early finish is the earliest that the activity can be completed. The right side of the node shows information on the late start date (LS) and the late finish date (LF) of the activity. The late start of an activity is the latest that the activity can begin, and the late finish is the latest that the activity can be completed. Immediately below the duration is information on the free float (FF) and total float (TF) of the activity. As will be described later, the FF and TF values relate to the flexibility that an activity has

	Activity		
ES	Duration		LS
EF	FF	TF	LF

Figure 3.18 A Simplified Yet Informative Format for a Precedence Activity

relative to the scheduling of that activity. Other than the duration, all of this information must be computed on the precedence diagram. These concepts and how their values are computed will be explained in the remaining portion of this chapter. This will be vital information to obtain in order for field personnel to understand the scheduled occurrence times for all of the activities in the precedence diagram.

The procedures for creating a precedence network have already been discussed, namely, identify activities, order them, draw the network, and assign durations. A preliminary or rough network can be generated by positioning each activity (denoted as a box, circle, or other shape) relative to other activities and then drawing lines between each set of related activities. Link lines may cross in order to accurately depict activity relationships. The diagram can then be redrawn to better place the activities to minimize crossing lines and other clutter. It is even possible to draw a time-scaled version of the precedence diagram, in which the geometric shapes are elongated to coincide with the time-scaled durations of the activities.

Alternatively, a more systematic approach using sequence steps can be employed. Each activity is assigned to a particular sequence step. Activities in a chain are assigned to different sequence steps. All activities without predecessors are said to be on step 0. Activities immediately following step 0 activities are on sequence step 1 and so on. The use of sequence steps in laying out a precedence diagram assures an orderly and structured presentation of the schedule logic.

The following list of activities and their immediately preceding activities (IPAs) is presented as an example for determining sequence steps and drawing a precedence diagram (see Figure 3.19). The precedence diagram depicting these activities and their relationships is shown in Figure 3.20.

The precedence diagram in Figure 3.20 shows a network with three starting activities. This is certainly one way to present the project logic. Many schedulers would prefer to have a network beginning with a single activity. If a single starting activity is added at the beginning of the network, the project start date need only be defined for

Activity	IPAs	Sequence Step
A	-	0
B	A	1
C	B	2
D	G, H	4
E	-	0
F	E	1
G	C, F	3
H	-	0

Figure 3.19   Activities and Their Immediate Preceding Activities (IPAs)

48   Chapter 3

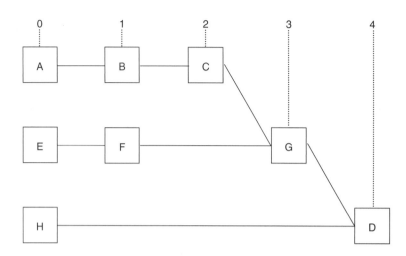

Figure 3.20   **Activities Organized in a Network with Sequence Steps**

that starting activity. This will ensure that all activities are somehow associated, either directly or through a sequence of activities, with the project start date.

Computations using the precedence technique involve straightforward calculations through a network (which will be described). Information of particular interest for each activity will be the early start date (ES), the late start date (LS), the early finish date (EF), and the late finish date (LF). Information on free float (FF) and total float (TF) is often computed as well.

Lag is time associated with a link line. It indicates the difference between the early finish of the activity preceding the link and the early start of the activity following the link. Thus it indicates how much delay, if any, can occur in the preceding activity before the start of the following activity is affected. Link lags are used to calculate free float and total float, and these values clearly indicate which relationships control the start time of an activity. Note that lag is *not* the same as lead time or delay assigned to a link. Thus, links will often be coded to indicate whether a link line has a positive lag or whether it is 0. If a link line is on the critical path, a different designation may be employed (see Figure 3.21). The examples that appear later in this chapter use this form of notation for the nodes, but the link lines will not use the designations as shown in Figure 3.21, primarily because simple networks can be readily understood without showing this representation. As described later in this chapter, independent and interfering float values can be determined for each activity in the precedence diagram, but these values will not be shown or computed.

As will become apparent, link lag values are useful in determining the free float and total float of an activity, but the link lag values should not be confused with float. This is often a common misconception about link lags. Note that free float and total

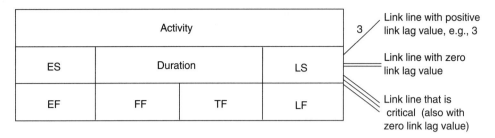

Figure 3.21 Link Line Notation to Indicate Relevant Scheduling Information

float are assigned to activities. The link lines themselves are not activities, but merely provide a means of showing the association of one activity to another; i.e., lag values are associated with link lines, not the activities they link.

CALCULATIONS ON A PRECEDENCE NETWORK

The most tedious, yet most important, aspect of network models is that they permit the planner/scheduler to calculate the total time that a project is projected to take and the times at which each activity can and must start in order for the project to be completed in the estimated amount of time. While in practice most calculations are now performed by scheduling software programs, it is still important for the scheduler to have a thorough understanding of the specific nature of the computations that are performed in determining the schedule information. Developing a manual solution of the network information will help considerably in understanding how these values are derived. Several terms are used to describe the scheduling information.

Early Activity Start (ES): The earliest time that an activity can start as determined by the latest of the early finish times of all immediately preceding activities.

Early Activity Finish (EF): The earliest time that an activity can finish. It is determined by adding the duration of the activity to the early start time of that activity (EF = ES + duration).

Late Activity Start (LS): The latest time that an activity can start without delaying the project completion. It is determined by subtracting the duration of the activity from its late finish (LS = LF − duration) and it can also be computed by adding the total float of an activity to the early start of the activity or (LS = ES + TF).

Late Activity Finish (LF): The latest time that an activity can be finished without delaying the entire project completion. It is equal to the earliest of the late starts of the immediately succeeding activities. It is computed by adding the total float to the early finish date (LF = EF + TF).

Early Event Occurrence Time: The earliest that an event can occur, determined by the latest of the early finish times of those activities that immediately precede the event. Remember that events do not consume time, but they exist at a point in time.

Late Event Occurrence Time: The latest that an event can occur, determined by the earliest of the late start times of those activities that immediately follow the event.

Free Float: The amount of time that an activity can be delayed before it impacts the start of any succeeding activity. It is computed as the smallest link lag value of all link lines that occur immediately after the activity.

Total Float: The amount of time that an activity can be delayed before it impacts the completion date of the project. Computations are simplified by setting the total float of the last activity in the network to 0. The total float of any activity is determined as being the sum of the total float of the following activity and the associated link lag value. If more than one activity follows the smallest sum value is determined to be the total float.

LAG: The amount of time that exists between the early finish of an activity and the early start of a specified succeeding activity ($LAG_{AB} = ES_B - EF_A$).

To illustrate the computations on a precedence diagram, a simple schedule will be utilized. This is shown in Figure 3.22. Since this is being solved manually, all relationships are assumed to be finish-to-start relationships.

EARLY TIMES (EARLY START [ES] AND EARLY FINISH [EF])

1. Assign 1 as the early start date of the first activity (using the beginning-of-day convention). Note that the early start could be assigned as 0, meaning that the end-of-day convention would be used. For simplicity, this text will use only the beginning-of-day convention.

2. Calculate the early finish time for the activity (EF = ES + duration). Note that if the first activity is assigned an early start time of day 1 and has a duration of one day, the early finish will be day 2. Since the beginning-of-day convention is being used, this means that the activity will be completed on the beginning of day 2, which is the same as being completed at the end of day 1. Note that most computer programs will show the start times of activities with the beginning-of-day convention, and the finish times are reported with the end-of-day convention. It is only when performing manual solutions that the convention being used must be kept in mind.

3. The early start of activities (other than the first activity or activities) will be determined by the early finish times of their preceding activities. The ES and the EF of the activities in the network are determined by beginning at the first activity and proceding to the immediately following activities. This procedure is followed until the ES and EF of the last activity are determined. The ES of an activity is computed as the EF of the immediately preceding activity. If an activity is preceded by more than one activity, its ES will be computed as the latest of the EF times of the immediately preceding activities, as all preceding activities must be completed before the activity can

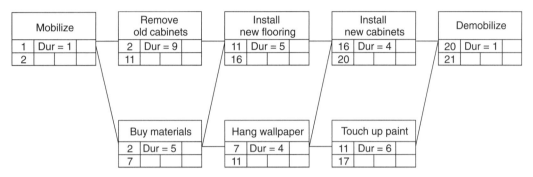

Figure 3.22 Network for a Simple Remodeling Project Showing ES and EF for Each Activity

begin. If lead time is specified for a link, then the ES of the succeeding activity for that link would be the EF of the preceding activity plus the LT (lead time) of the link. In Figure 3.22, note how the ES of "Install New Flooring" is determined. Since the new flooring cannot be installed until both its preceding activities are completed, the ES of "Install New Flooring" is 11, the latest EF of the preceding activities of "Remove Old Cabinets" and "Buy Materials." A similar computation is made for "Install New Cabinets," which is preceded by two activities.

   4. Repeat steps 2 and 3 for each activity in the network until the ES and EF have been determined for the last activity.

FREE FLOAT (FF)

   5. Calculate lags for each link by determining the difference between the early start of each activity that follows a link line and the early finish of the activity that precedes it ($LAG_{AB} = ES_B - EF_A$). The link lag values will be either 0 or some positive number, because link lag values cannot be negative. Note that the sum of the durations and link lag values of all paths between the starting and ending activities will be the same sum total. Note that the nonzero link lag values are shown in Figure 3.23. Note that the ES of "Install New Flooring" is 11 and the EF of "Buy Materials" is 7, so the link lag value for the link line between these activities is 4. While the link lag values of 0 can be shown directly on the link lag lines, this is not essential for simple networks. It should be noted that every activity has at least one preceding link line that has a value of 0, so if an activity has only one link line connecting it to a preceding activity, the link lag value will be 0.

52   Chapter 3

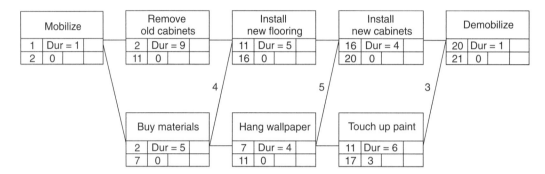

Figure 3.23   Network for a Simple Remodeling Project Showing ES, EF, and FF

6. Determine the free float of an activity as being equal to the smallest lag value of all links *leaving* from that activity node. Each activity will be followed by at least one link line (the last activity in the network is the only exception, as it has no link line following it). The free float of each activity is computed as the smallest link lag value among those link lines that follow it (those that emanate from it). The FF of the last activity in the network has no real meaning as it has no succeeding activity or link line, so its FF is set at 0. The determination of the free float (FF) values for all the activities in the example network is shown in Figure 3.23. For this project, only one activity ("Touch up Paint") has an FF value that is not 0.

TOTAL FLOAT (TF)

7. Assign a value of 0 total float (TF) to the last activity in the network.

8. The total float values for the activities in the network can now be determined. Beginning with the last activity, a backward pass is made through the network, in which the total float of the last activity (assigned as 0) is added to the link lag value of the link line between the last activity and an immediately preceding activity to derive the TF of the immediately preceding activity. This sum of the TF of the following activity and the link lag value of the link line that connects them is computed for all activities in the network.

9. Proceed in a backward-pass fashion until you have computed the TF for all activities. When an activity is connected to two or more activities that follow it, you will compute several sums of the TF and link lag values. Always assign the smallest sum as the TF for the activity in question. The determination of the TF of all the activities

Precedence Diagrams 53

TF = TF of immediately following activity + the link lag value that connects them (if an activity is followed by more than one activity, the TF is the smallest sum that is computed)

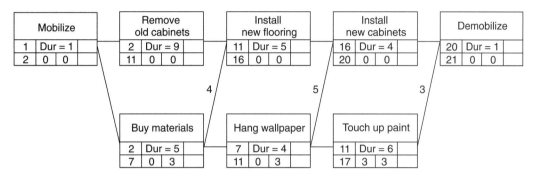

Figure 3.24 Network for a Simple Remodeling Project Showing ES, EF, FF, and TF

in the example network is shown in Figure 3.24. Note that "Hang Wallpaper" is followed by two activities, namely "Install New Cabinets" and "Touch-up Paint." The TF for "Hang Wallpaper" is determined to be 3, the lesser value of sums computed as 5 (link lag value plus TF associated with "Install New Cabinets") and 3 (link lag value plus TF associated with "Touch up Paint"). Note that there is a string of activities between the first and the last activity for which the TF values are all 0. There must be at least one such string of 0 TF activities in every network, assuming that the TF for the last activity is set at 0.

LATE TIMES (LATE START [LS] AND LATE FINISH [LF])

10. This step consists of determining the late start times for all of the activities. This is a simple matter since the total float values have already been determined. The total float value is added to the early start of each activity to determine the late start of each activity. Make these computations for all activities in the network (LS = ES + TF).

11. Similarly, the total float is added to the early finish of each activity to determine the late finish time of each activity. Add the TF to the EF of each activity to determine the LF of each activity (LF = EF + TF). The LF of an activity can also be computed by adding the activity duration to the LS of the activity. The determination of the LS and the LF of all the activities in the example network is shown in Figure 3.25.

Once the early and late start times, early and late finish times, free float, and total float of all the activities have been determined, the calculations are completed.

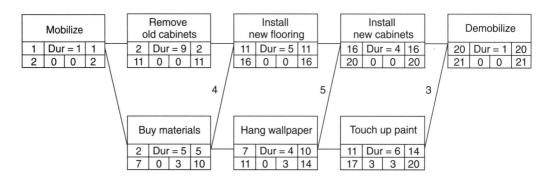

Figure 3.25 Network for a Simple Remodeling Project Showing ES, EF, FF, TF, LS, and LF

The information is shown in Figure 3.25. Of particular interest will be the identification of the critical path, that sequence (or sequences) of activities for which the total float of each activity in the path is 0. In the sample network, it can be quickly established that there is a single critical path consisting of five activities, namely "Mobilize," "Remove Old Cabinets," Install New Flooring," "Install New Cabinets," and "Demobilize." The "Buy Materials" activity can be delayed for up to three days before the delay would impact the project duration, but any delay would impact the start time of "Hang Wallpaper." The "Hang Wallpaper" activity can be delayed up to three days before it will impact the project duration, but any delay will impact the start date of "Touch up Paint." The "Touch up Paint" activity can be delayed up to three days before the project duration is impacted and before any other activity is impacted.

Since this is a relatively simple network, it is easy to understand the general relationships that exist between the different activities. A quick computation can be made of the durations of the activities that fall on the critical path to determine the project duration. This sum is 20. Note that the computed early finish and late finish of the project was determined to be 21. This is because the beginning-of-day convention is being used. That means that the project will be completed at the beginning of the workday on day 21. This is the same as completing the project by the end of day 20. When performing manual computations, it is important to fully understand the starting day convention that is being used. Most computer programs that generate scheduling information use the beginning-of-day convention for starting activities and the end-of-day convention for completing them.

The steps that have been outlined in this example illustrate the basic computations associated with precedence diagrams. These steps are followed in a similar manner for

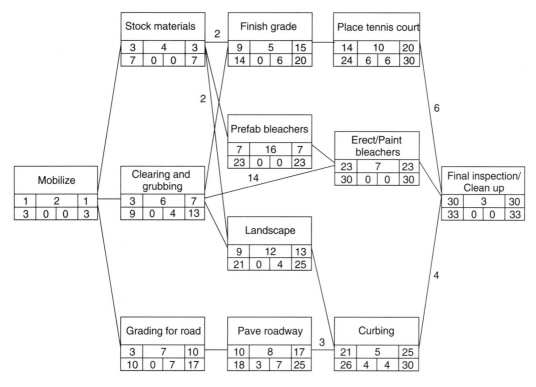

Figure 3.26 Example of Computed Values on a Tennis Court Project

actual projects, regardless of the network complexity or size. Note that the calculations for relationships other than finish-to-start require adjustments to these procedures.

A slightly more involved project is shown in Figure 3.26. It shows the precedence diagram for constructing a tennis court facility.

INDEPENDENT FLOAT AND INTERFERING FLOAT

While total float and free float are the most commonly used types of float, there is merit in at least having some knowledge of two other forms of float, namely *independent float* and *interfering float*. While these are seldom used, some scheduling software programs do make the computations for their values.

Independent float is float that is "owned" exclusively by one activity. As such, the independent float of one activity is not available for use by any other activity. It can be said that this float of an activity is "independent" of the late finish times of preceding activities and of the early start times of succeeding activities. Thus, independent float, also called *"safe" float*, is flexibility that is ascribed to the specific use by one activity. The independent float of an activity will not be jeopardized by the float of any other

56   Chapter 3

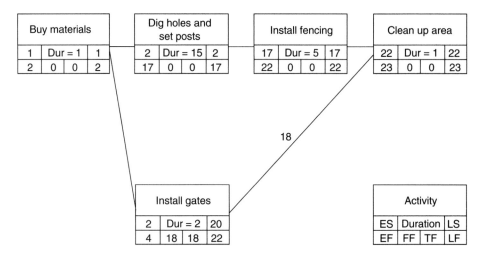

Figure 3.27 Simple Network to Determine Independent Float

activity being used. An activity can have independent float only if it has free float (Harris, 1978; Pilcher, 1992; Stevens, 1990). Figure 3.27 shows an activity, "Install Gates," with independent float. The independent float of an activity can be computed with the following general equation:

$$\text{Independent Float}_{\text{Activity A}} = \text{ES}_{\text{Successor}} - \text{LF}_{\text{Predecessor}} - \text{Duration}_{\text{Activity A}}$$

An examination of the simple network in Figure 3.27 will show that all activities are critical with the exception of one, namely "Install Gates." Furthermore, the independent float value of "Install Gates" is 18, the same value as its free float and total float.

Independent Float of Activity "Install Gates"
  = ES of "Clean Up Area" − LF of "Buy Materials" − Duration of "Install Gates"
  = 22 − 2 − 2 = 18 days

Interfering float (also called *shared float*) is the float an activity might have that is subject to use by other activities. Thus, interfering float is held in common by at least two or more activities. Essentially, the interfering float of an activity is that portion of the total float of that activity that is shared with other activities. The interfering float of an activity is the amount of time that its early start can be delayed without delaying the project completion but which will delay at least one other succeeding activity. When a chain of noncritical activities is encountered in a network, they will all have the same total float value and, except for the last activity in the chain, will all have the same value of interfering float. The interfering float of an activity is the difference between its total float and its free float (Harris, 1978; Horowitz, 1967; Stevens, 1990).

$$\text{Interfering Float}_A = \text{TF}_A - \text{FF}_A$$

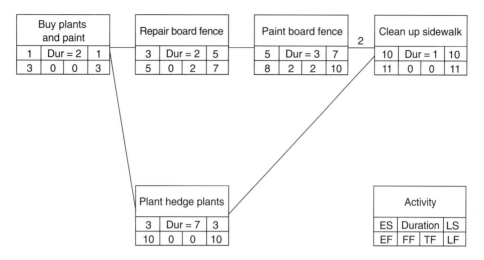

Figure 3.28 Simple Network to Determine Interfering Float

A network will be used to illustrate information about interfering and total float. The project is a simple one in which a hedge is to be planted on one side of a sidewalk leading up to a house and an existing white board fence is to be repaired and repainted on the other side of the sidewalk. The durations are noted in hours. The network is as shown in Figure 3.28.

Note that two activities, "Repair Board Fence" and "Paint Board Fence," have positive total float values of two hours. In addition, "Paint Board Fence" is the only activity with any free float, also two hours. Since the interfering float of an activity is the difference between the total float and the free float of that activity, the interfering float of "Repair Board Fence" is 2 and of "Paint Board Fence" is 0.

Interfering Float of Activity "Repair Board Fence"
$$= TF_{\text{Repair Board Fence}} - FF_{\text{Repair Board Fence}}$$
$$= 2 - 0 = 2$$

Interfering Float of Activity "Paint Board Fence"
$$= TF_{\text{Paint Board Fence}} - FF_{\text{Paint Board Fence}}$$
$$= 2 - 2 = 0$$

Since "Repair Board Fence" has two hours of interfering float, it means that the total float of "Repair Board Fence" is shared with "Paint Board Fence." If "Repair Board Fence" is delayed by one hour, this will "interfere" with the scheduled start time of "Paint Board Fence" by one hour, but the project will still finish on time.

From the various start and finish times provided in Figure 3.25, the independent and interfering float values can be determined. Note that activities "Hang Wallpaper" and "Touch up Paint" are the only activities for which these computations will be made. From the values shown, the independent float and interfering float values are determined as follows:

FOR ACTIVITY "HANG WALLPAPER"

Independent Float of Activity "Hang Wallpaper"
$$= ES_{\text{Touch-up Paint}} - LF_{\text{Buy Materials}} - Duration_{\text{Hang Wallpaper}}$$
$$= 11 - 10 - 4 = -3 \text{ days (reported as 0)}$$

Interfering Float of Activity "Hang Wallpaper"
$$= TF_{\text{Hang Wallpaper}} - FF_{\text{Hang Wallpaper}}$$
$$= 3 - 0 = 3 \text{ days}$$

FOR ACTIVITY "TOUCH UP PAINT"

Independent Float of Activity "Touch up Paint"
$$= ES_{\text{Demobilize}} - LF_{\text{Hang Wallpaper}} - Duration_{\text{Touch up Paint}}$$
$$= 20 - 14 - 6 = 0 \text{ days}$$

Interfering Float of Activity "Touch up Paint"
$$= TF_{\text{Touch up Paint}} - FF_{\text{Touch up Paint}}$$
$$= 3 - 3 = 0 \text{ days}$$

Note that the interfering float for activity "Hang Wallpaper" is determined to be three days and its independent float is –3 days. While this is the accurate computation, the concept of negative days of independent float is perhaps illogical. It is generally suggested that negative independent float days be interpreted as zero independent float days. For activity "Touch up Paint," the interfering float and independent float values are computed as being 0.

COMPUTATIONS FOR DIFFERENT ACTIVITY RELATIONSHIPS

The different types of activity relationships were described earlier in the chapter. While the computations for the determination of early/late start times and early/late finish times were provided for only the finish-to-start relationships, calculations can also determine the scheduling information for other activity relationships. This will be demonstrated through the use of a very simple network involving three activities. A simple finish-to-start relationship of this network, consisting of Activities I, J, and K, is shown in Figure 3.29. As a quick review, the determination of the early/late start times and the early/late finish times will be given. The values of interest are the early start/finish times, which are computed by conducting a forward pass through the network; late start/finish times are computed by conducting a backward pass through the network. The computations are made as follows:

Basic forward pass calculations:
$$ES_J = EF_I \qquad\qquad EF_J = ES_J + \text{duration}_J$$

Figure 3.29 Simplified Precedence Relationships with No Lag Values

Basic backward pass calculations:
$$LS_J = LF_J - duration_J \qquad LF_J = LS_K$$

The simple network is made slightly more complicated when the subject of lag is introduced to the activity relationships. The addition of lag to the relationships is shown in Figure 3.30. While the computations will be impacted on whether the lag values are positive or negative, the formulas remain the same. The computations for a finish-to-start relationship with the incorporation of lag values are as follows:

$$ES_J = EF_I + lag_{IJ}$$
$$EF_J = ES_J + duration_J$$
$$LS_J = LF_J - duration_J$$
$$LF_J = LS_K - lag_{JK}$$

If activities I, J, and K were described by start-to-start relationships, the activities could be depicted as shown in Figure 3.31. Lag values might also exist, and these could be either positive or negative in value. The lag notations on the figures (direction of the link lines with arrows) indicate the conditions for which the lag values would be positive. The computations would be as follows:

$$ES_J = ES_I + lag_{IJ}$$
$$EF_J = ES_J + duration_J$$

Figure 3.30 Finish-to-Start Activity Precedence Relationships with Lag Values

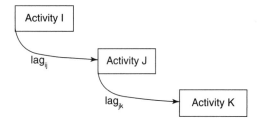

Figure 3.31 Start-to-Start Precedence Relationships with Lag Values

60   Chapter 3

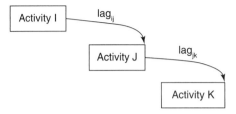

Figure 3.32   Finish-to-Finish Activity Precedence Relationships with Lag Values

$$LS_J = LS_K - lag_{JK}$$
$$LF_J = LS_J + duration_J$$

If finish-to-finish relationships existed between activities I, J, and K, the network could be as shown in Figure 3.32. As in all activity relationships, the lag values could be positive, negative or zero. The primary computations for the scheduling variables would be as follows:

$$ES_J = EF_I + lag_{IJ} - duration_J = EF_J - duration_J$$
$$EF_J = EF_I + lag_{IJ}$$
$$LS_J = LF_K - lag_{JK} - duration_J = LF_J - duration_J$$
$$LF_J = LF_K - lag_{JK}$$

For start-to-finish activity relationships between activities I, J, and K, the activities are depicted in Figure 3.33. The basic computations for the scheduling information would be as follows:

$$ES_J = ES_I + lag_{IJ} - duration_J = EF_J - duration_J$$
$$EF_J = ES_I + lag_{IJ}$$
$$LS_J = LF_K - lag_{JK}$$
$$LF_J = LF_K - lag_{JK} + duration_J = LS_J + duration_J$$

With the common usage of computer programs, manual computations are rarely performed to determine the early/late start times or the early/late finish times, regardless of

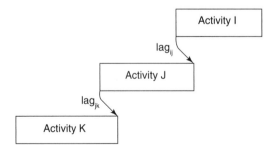

Figure 3.33   Start-to-Finish Activity Precedence Relationships with Lag Values

the activity relationships. Nonetheless, it is important to understand the basic concepts of how the scheduling information is actually derived.

FINAL COMMENTS

The network type of choice of many construction professionals is the precedence diagram. Once the logic is worked out, the values of early start, early finish, late start, latse finish, total float, and free float for each activity are easily computed. These values serve as the basis for making various types of management decisions. With this information, projects can be monitored and controlled. With the detailed scheduling information about each activity, informed decisions can be made when the need arises to accelerate the schedule, make changes in the network, or redefine the scheduling logic.

REVIEW PROBLEMS

1. The following network shows a precedence diagram for a reroofing project. Using the steps outlined in this chapter, solve the network for the basic scheduling information. For each activity, show the computed value of early start, early finish, late start, late finish, total float, and free float.

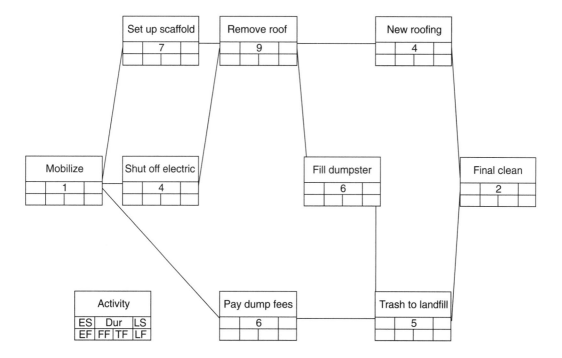

2. Draw the precedence diagram for the following project.

Activity List for an Apartment Project

	Activity Description	Duration Days	Predecessors*
1	Building layout	2	
2	Set forms	3	1
3	Plumbing R-I	3	1
4	Electrical R-I	2	1
5	Beam trench	3	3, 3 FF/1, 4 FF/1
6	Fine grade and screed	1	5
7	Visqueen	1	6
8	Rebar	2	7
9	Pour/finish concrete	4	8
10	Strip forms	1	9/2
11	Frame 1st floor	5	10
12	Plumb R-I 1st floor	3	11
13	Joist 2nd floor	2	11
14	Deck 2nd floor	1	12, 13
15	Frame 2nd floor	6	14
16	Plumb R-I 2nd floor	3	15
17	Pour Lt Wt 2nd floor	2	22, 16
18	Electrical R-I 1st floor	3	19
19	HVAC duct 1st floor	3	14
20	Roof trusses	3	15
21	Fascia	2	20
22	Roof deck	3	21
23	Fireplaces	2	20
24	Gyprock 1st floor	2	11
25	Gyprock 2nd floor	3	15
26	Electrical R-I 2nd floor	3	27
27	HVAC duct 2nd floor	3	17
28	Telephone & TV R-1	2	20
29	Build stairs	3	14
30	Set patio doors	4	24, 25FF/2
31	Set windows	5	24, 25FF/3
32	Roofing and sheet metal	4	22, 23
33	Furr downs 1st floor	2	14, 19SS/1
34	Siding and exterior trim	4	30, 31
35	Exterior doors	1	34
36	Soffitt	2	34
37	Balcony rails	4	34
38	Steel stair rails	3	29
39	Insulation	3	18, 26, 28, 32
40	Drywall	8	29, 35, 33, 39SS/1, 70
41	Tape and float	4	40 FF/1
42	Texture and acoustic	1	41
43	Ceramic tile	2	51 FF/1

#	Activity	Duration	Predecessors
44	Heater door and trim	2	42
45	Paint interior	6	44 SS/1
46	Paint exterior	3	35, 36, 37, 38
47	Masonry	7	46 SS/1
48	Prefinish shelves	1	45
49	Threshold and W.S.	1	45
50	Wallpaper	3	43
51	Prefinish cabinets	3	45
52	Prefinish doors	3	57 SS/2
53	Prefinish base	1	52
54	Mirrors	1	50
55	Paint touch-up	2	48, 53, 49
56	Finish hardware	4	52 SS/1
57	Vinyl tile	4	50
58	Finish HVAC	5	57 SS/2
59	Finish TV	2	57
60	Appliances	3	57 FF/1
61	Finish electrical	4	60 SS/1
62	Finish plumbing	5	60 SS/1
63	Carpet	4	54, 55, 58, 59, 61, 62
64	Cleanup	2	47, 56, 63FF/1
65	Drapes	2	64 SS/1
66	Turn on utilities	2	63
67	Punchout	2	66, 64 SS/1
68	Corrections	2	67, 65 FF/1
69	Final inspection	1	68
70	Furr downs 2nd floor	2	17, 27

*Under Predecessors, the following notation is used:

x/#: Activity x must finish # days before this activity can begin.

xSS/#: The # of days after activity x starts that this activity can begin.

xFF/#: This activity cannot finish until # days after x is completed.

3. Draw the precedence diagram for the following project.

Activity List for Reroofing Project

	Activity Description	Duration Days	Predecessors*
A	Move in	2	
B	Set up scaffolds	3	A
C	Tear off old shingles	3	B/1
D	Inspect roof truss and deck	2	C SS/1
E	Remove old shingles to recycling yard	3	C FF/2
F	Purchase new shingles	1	G SF/2
G	Install new shingles	5	C, D
H	Clean up	3	G
I	Move out	2	H

*Under Predecessors, the following notation is used:

x/#: Activity x must finish # days before this activity can begin.

xSS/#: The # of days after activity x starts that this activity can begin.

xFF/#: This activity cannot finish until # days after x is completed.

XSF/#: This activity must finish # days before x can start.

4. Solve for the scheduling data for the following site development project. Compute the relevant information and include the information on the figure.

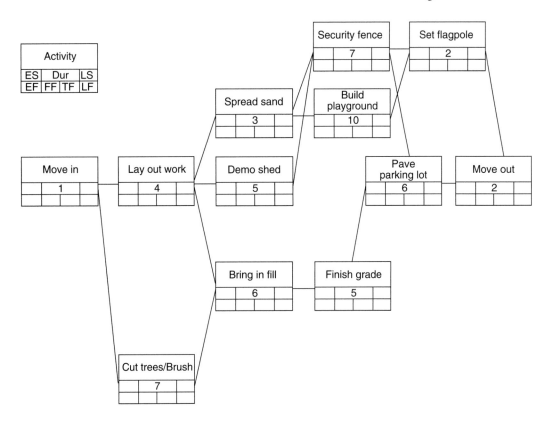

5. Solve for the scheduling data for the following site improvement project. Compute the relevant information and include the information on the figure.

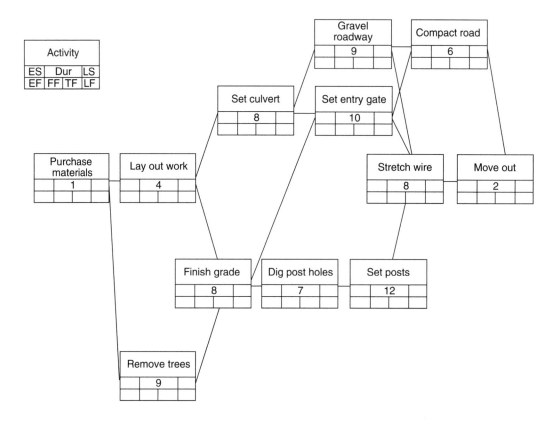

6. Solve for the scheduling data for the following barn addition project. Compute the relevant information and include the information on the figure.

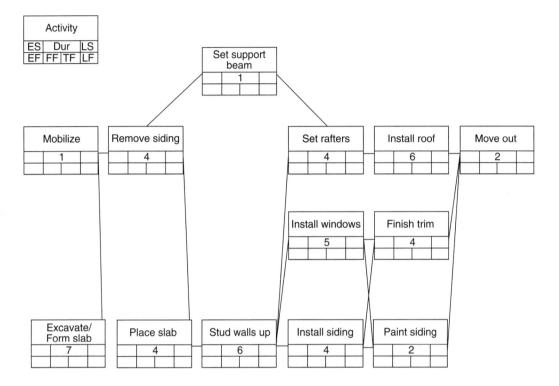

68   Chapter 3

7. Solve for the scheduling data for the following housing development project. Compute the relevant information and include the information on the figure.

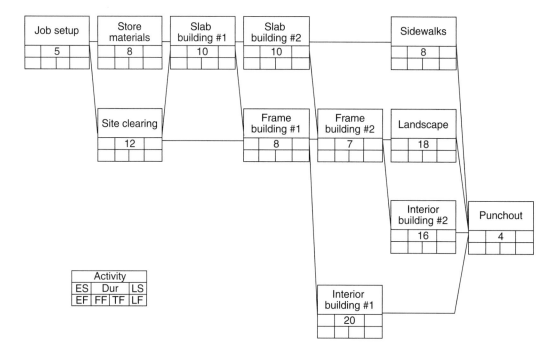

8. Solve for the scheduling data for the following sewer repair project. Compute the relevant information and include the information on the figure. Note that time units are in hours.

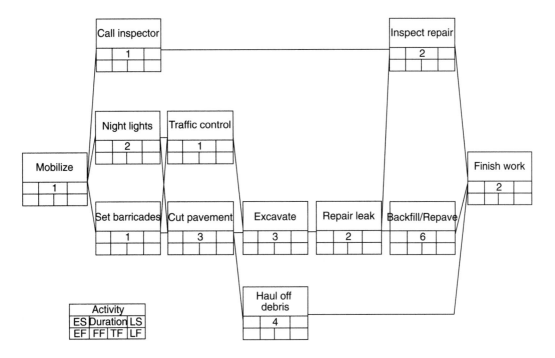

9. Schedule the precedence diagram for the following project. This problem is to be solved using a scheduling software program.

Activity List for a Board Fence Project

	Activity Description	Duration Days	Predecessors*
1	Mobilize	1	
2	Clearing and grubbing	3	1
3	Lay out fence alignment	1	2
4	Deliver posts to site	3	3 SS/1
5	Auger post holes	7	4 SS/2
6	Set posts in holes	7	5 SS/0
7	Backfill and plumb posts	10	6 SS/1
8	Install boards for fence	5	7 FF/1
9	Deliver boards for fence	3	8 SF/7
10	Paint board fence	12	8 SS/2
11	Plant shrubs along fence	7	10 SS/8
12	Final inspection	1	8, 10/2, 11
13	Demobilize	1	12

*Under Predecessors, the following notation is used:
x/#: Activity x must finish # days before this activity can begin.
xSS/#: The # of days after activity x starts that this activity can begin.
xFF/#: This activity cannot finish until # days after x is completed.
XSF/#: This activity must finish # days before x can start.

10. Schedule the precedence diagram for the following project. This problem is to be solved using a scheduling software program.

Activity List for Constructing a Planter

	Activity Description	Duration Days	Predecessors*
1	Mobilize	1	
2	Install security fence	1	1
3	Excavate planter foundation	2	1
4	Set forms	4	2, 3
5	Set rebar	1	4 SS/1
6	Place concrete	1	4, 5
7	Order concrete	1	6 SF/7
8	Strip forms	2	6/1
9	Construct masonry planter	12	8
10	Deliver bricks to site	1	9 SF/1
11	Place soil in planter	1	9
12	Remove security fence	2	11 FF/1
13	Cleanup	1	11
14	Demobilize	1	13

*Under Predecessors, the following notation is used:
x/#: Activity x must finish # days before this activity can begin.
xSS/#: The # of days after activity x starts that this activity can begin.
xFF/#: This activity cannot finish until # days after x is completed.
XSF/#: This activity must finish # days before x can start.

11. Schedule the precedence diagram for the following project. This problem is to be solved using a scheduling software program.

Activity List for a Reroofing Project

	Activity Description	Duration Days	Predecessors*
1	Mobilize	1	
2	Set up scaffolding	2	1
3	Construct security barricades	1	2 SS/1
4	Strip off old roofing	3	2, 3
5	Repair damaged decking	2	4 SS/1
6	Repair damaged cant strip	1	4, 5
7	Replace sheetmetal	1	5 SS/1
8	Install felt roofing	1	6, 7
9	Hot mop roof	2	8
10	Inspect roofing	1	9
11	Call for inspection	1	10 SF/3
12	Remove scaffolding	2	10
13	Haul off roofing materials	2	6, 7
14	Demobilize	1	12 FF/1, 13 FF/1

*Under Predecessors, the following notation is used:

x/#: Activity x must finish # days before this activity can begin.

xSS/#: The # of days after activity x starts that this activity can begin.

xFF/#: This activity cannot finish until # days after x is completed.

XSF/#: This activity must finish # days before x can start.

4

Determining Activity Durations

Once you plan your work, you must work your plan.

The development and use of schedules relies heavily on activity durations. Without carefully assigning durations to the various activities, the value of the schedules will be greatly diminished. It is therefore imperative that the durations of the various activities be developed with a certain amount of accuracy. The accuracy and usefulness of the schedule will be directly related to the accuracy that is inherent in the individual activity durations. There are several ways that these activity durations can be obtained or derived. There is always some degree of uncertainty that is present whenever project durations are developed. Despite this, a reasonable effort should provide results that will be acceptable. Some of these methods will be described.

ESTIMATING

To determine the probable cost of the project, construction estimates are prepared before a project is constructed. Thus, an estimate is, at best, a close approximation of the actual cost. There are many variations in the cost of materials, equipment, and labor from one locality to another and from one point in time to another. Therefore, the cost estimate of a project, even when compared to other similar projects, is unique to that project. To determine the cost accurately, the estimator should have an accurate estimate of the quantities of material, equipment hours, and labor hours required to complete a given project. The estimator then applies the proper unit costs to these items.

The soundness of the completed estimate depends on the following two factors: (1) the accuracy of the quantity take-off, and (2) the judicious selection of the unit costs and production rates to be used.

TYPES OF ESTIMATES

Construction estimates may be divided into at least two different types, depending on the purposes for which they are prepared: conceptual estimates and detailed estimates.

Conceptual Estimates

Conceptual estimates are generally used by the prospective owner of a project to determine the approximate cost of a project before making a final decision to construct it. In some instances, an agency, whether public or private, will need to know the approximate cost of a project before specific actions can be initiated. A public agency may require this information prior to holding a bond election. Private owners will need this information in order to know the extent of financing that will be required or the amount of capital that must exist for a project to be feasible. Conceptual estimates are rough approximations of the anticipated costs and are generally prepared by architects, engineers, or other consultants. The preparation of conceptual estimates requires a clear understanding of what an owner wants and a good "feel" for the probable costs.

Detailed Estimates

Detailed estimates are generally prepared by contractors prior to submitting bids on competitively bid contracts or when entering lump-sum or fixed-price contracts. A detailed cost estimate includes the costs of materials, labor, equipment, subcontracted work, overhead, and profit. To prepare a detailed estimate for a project, the estimator conducts a breakdown of the proposed project into the necessary operations required to complete the project. To some extent, the operations or items appear in the estimate in the order that they will be performed in the project. This reduces the danger of omitting the costs of one or more operations.

CONDUCTING A DETAILED ESTIMATE

The process of conducting a detailed estimate begins with a thorough analysis of the various physical quantities that must be incorporated in the final project. This information is obtained through a detailed quantity take-off. The quantity take-off is generally recorded on a standard sheet that includes information about the various items (numeric designation, description, and quantity). This quantity take-off is then used to develop a detailed cost estimate for the prospective project. Costs are generally captured separately for materials, equipment, labor, and subcontracted items. By separating the cost items, it is easier to check to verify that all cost categories have been included. During construction, this level of breakdown affords greater control of expenditures.

A project is made up of many activities. Separate take-offs are made for each job item, and the take-offs, in general, follow the same sequence as used to perform the activities. The primary elements of work required to complete each activity are tabulated and collected in an orderly and systematic manner. Since virtually all of the take-offs are double-checked, the estimator should use care in writing all numbers and notes clearly and legibly. (Of course, much of the estimating effort is performed with the use of computers.)

After the estimate is complete, it is a good practice to check the take-off to ensure that no job quantities have been overlooked and that no other obvious errors have been made. For unit price contracts, this verification includes a comparison with the quantities provided in the bid proposal. However, it is not wise to rely too heavily on these listed quantities. Unit price contracts may include work that must be done but for which there is no specific quantity pay item. For example, extensive scaffolding may be required to construct a project, but the owner will probably not allow scaffolding to be used as a pay item.

To develop a detailed estimate, well-defined costs must be linked to each of the items in the quantity take-off. These costs are easily defined for those activities that will be subcontracted. The costs of most materials can also be determined with some accuracy, as many suppliers quote specific prices for items that are purchased by a given date. The costs of labor and equipment, however, are not as simple to determine. A clear understanding is required of how many units of material can be put in place in a specified time period; i.e., the productivity of labor and equipment for each take-off item must be estimated with accuracy. These values of productivity may be derived from prior projects completed by a contractor. In some cases, the estimator may develop the productivity rates by discussing relevant factors about a work item with superintendents, project managers, and other experienced personnel. The estimator may also decide to obtain rough values of productivity from published sources. Whenever reference is made to another source for production values, the estimator should take care to make appropriate modifications to reflect the anticipated conditions.

Once the quantities have been priced, an estimate must be made for the job overhead and the home office overhead. The job site overhead includes all of those costs that cannot be attributed to a specific work activity, but that must be incurred to successfully complete the project. Examples include the cost of maintaining a job pickup truck, the salary of the superintendent, and the cost of an electric generator. Note that these are examples of costs that will continue for the duration of the project and that a longer project duration means that higher overhead costs will be incurred for these items. Some overhead items are "one-time" cost items. Examples of such cost items include the cost of a layout survey, security fence, signs, and mobilization. Of course, even these may not be one-time costs in all instances. Refer to Table 4.1 for examples of job site overhead items.

Home-office overhead must also be allocated to each project in some manner. Home-office overhead includes all costs incurred at the home office that cannot be attributed easily to one project. (See Table 4.2 for examples of this type of overhead.) There are a variety of philosophies on how these costs are to be charged to a project. Whatever means are employed, the home-office overhead charged to all of the company projects must be a reasonable approximation of the actual costs incurred at the

Table 4.1
EXAMPLES OF JOB SITE OVERHEAD

Salaries

Project Manager	Superintendent	Assistant Superintendent
Office Manager	Payroll Clerk	Safety Director
Project Engineer	Field Engineer	
Timekeeper	Security Watch	
Party Chief	Quality Control personnel	

Equipment Costs

Trucks	Automobiles	Cranes
Computers	Photocopiers	Hoists
Fax machines	Project pickup	Scaffolding
Mobilization ($/mile)	Air compressors	Water pumps
Water truck	Electric generators	
Welding machine	Forklift	

Temporary Facilities Costs

Job office trailer	Architect's field office	Janitor service
Storage sheds	Electric service	Water service
Tool bins	Security fence	Signs
Lights	Temporary toilets	Carpenter shed
Lavatories	File cabinets	

Temporary Protection and Safety Requirements

Fence	Canopies	Barricades
Safety nets	Security services	Vandalism/theft
Alarm system	Noise control	Fire extinguishers
Safety rails	Winter protection	Summer protection
Rain protection	Tree protection	Safety equipment

Engineering Support Services

Site survey	Project layout	Road layout
Soil borings	Field survey	

Reporting Expenses

CPM schedules	Progress reports	Certified payrolls
Photography		

Testing and Inspection

Concrete	Masonry	Steel
Load	Watertight testing	

Job Cleanup

Cutting and patching	General housekeeping	Disposal and dump charges
Dewatering	Dumpster fees	

Table 4.1
(continued)

Taxes and Insurance

Truck and auto	Public liability	Builder's risk
Special risk	Sales tax	Workers' Comp.
Performance bond	Social Security	

Communications Costs

Telephone/telegraph	Loudspeaker	Radios
Telephone service		

Permits

Building	Demolition	Sidewalk
Blasting	Water and sewer fees	

Expendables (Tools and Consumables Costing Less Than $500)

Hammers	Blades	Bits
Shovels	Bars	Cutters
Clamps	Fuels	Lubricants
Welding rods		

Supplies

Stationery	File folders	Plans
Specifications	As-built drawings	Postage
Drinking water	Ice	Dispensers
First aid supplies	Cups	

Other

Travel	Storage fees	Job sign
Ads (help wanted)	Data processing	Petty cash
Job parties	Temporary roads	Community education

home office. The allocation of home-office overhead begins with a separate estimate, at the corporate level, of the anticipated cost to maintain the home office. This cost determination is crucial to any system of overhead allocation. One method of allocation of home-office overhead to a project is to charge it in proportion to the total value of all projects undertaken within the period of one year. For example, assume the home-office overhead is $1,000,000 for one year and the total value of all contracts that year is $100,000,000. If the value of one project being bid is $20,000,000, then the home-office overhead charged to the project is $200,000. More sophisticated allocation methods can be devised that take into account the extent that the home office will be involved in a project, such as the number of worker hours required, or the duration of the project.

Table 4.2
EXAMPLES OF HOME-OFFICE OVERHEAD

Salaries

    Owners
    Officers
    Engineers
    Secretarial staff
    Receptionist

    President
    Estimator
    Accountant
    Clerical
    Roving or general superintendent

Equipment Costs

    Computers
    Photocopiers
    Company vehicles
    Idle equipment in the yard

    Typewriters
    Fax machines
    Vehicle operation (gasoline, oil, maintenance)
    Yard and maintenance shop

Building or Office Costs

    Rent, lease, or mortgage costs
    Furniture, fixtures
    Insurance

    Building maintenance
    Real estate taxes

Supplies

    Stationery
    General office supplies

    Company forms

Operating Expenses

    Electric
    Telephone
    Insurance: liability, etc.
    Radio
    Legal and consulting fees
    Custodial services
    Study courses
    Travel

    Water and sewer
    Postage
    Landscape maintenance
    Business taxes and licenses
    Sales promotion, entertainment
    Site investigation costs
    Contractor association dues
    Conventions

Other

    Lost plan deposits
    Charitable donations
    Political donations
    Company picnic
    Supervisory training expenses

    Interest expense/finance charges
    Goodwill expenditures
    Licensing fees
    Lost interest on retainage
    Manuals (software, estimating guides, etc.)

For many overhead items, it is essential that an accurate estimate be made of the project duration. Thus, it is important that some effort be expended on scheduling a project as the quantity estimate is performed. It is indeed foolish for a contractor to prepare an estimate without preparing a schedule. Since many of the costs of a project are a direct function of the amount of time required to complete the project, a reasonably accurate estimate should be made of the project duration. This can be accomplished by preparing a schedule with sufficient detail to give meaningful information.

The example on the following pages consists of a relatively simple evaluation of a basement to be constructed. The job items are as follows:

Mobilization
Excavation
Construct wall footing
Place granular fill
Finish grade (fill)
Erect form wall
Place concrete wall
Place concrete slab
Backfill
Cleanup

These job items are sufficiently detailed for a quantity take-off to be conducted. However, they do not reflect with sufficient accuracy the work tasks actually required to complete the project. It should be evident from the following activity listing that the estimator has assumed that a prescribed sequence of job activities will occur. For example, it is assumed that the slab will be placed after the concrete walls are completed, as no provision has been included for the costs of forms for the edge of the slab. Thus, it is obvious that the walls will be placed prior to the slab and that the concrete walls will function as concrete forms. Other sequences may also be identified.

For scheduling purposes, the following list of job activities gives a clearer idea of how the work will be done:

Mobilization
Excavation
Wall footing
   Install formwork
   Set rebar
   Place concrete
   Strip forms
Place granular fill
Finish grade
Concrete wall
   Form one side of form on the N and E walls
   Place rebar in N and E walls
   Form second side of form on the N and E walls

      Place concrete for N and E walls
      Strip forms from N and E walls
      Form one side of form on the S and W walls
      Place rebar in S and W walls
      Form second side of form on the S and W walls
      Place concrete for S and W walls
      Strip forms from S and W walls
  Concrete slab
      Set rebar
      Place concrete
  Backfill
  Cleanup

Note that the pricing of rebar will not vary whether the rebar is placed in a sequence of one wall at a time or if it is placed in all four walls at the same time. This decision will have an impact, however, on other costs. For example, if all four walls are formed at the same time, there is no possibility for reusing the forms on this project.

From the quantities that have been estimated, it is possible to determine an anticipated duration for each of the work items. To determine such durations requires that the productivity estimates are reasonably accurate. The following pages give examples of how the duration may be derived from the quantity estimate when productivity values are known. While activity durations can be established without a detailed estimate, it is not prudent to do so without at least considering the basic assumptions that were made about how the construction operations would be accomplished. Basically, the estimate should form the foundations on which the schedule is actually based. Ignoring the estimating assumptions may very well result in inaccuracies in the computed durations.

ESTIMATING DURATIONS

To a large extent, the method used to establish the duration of an activity will depend on the size, in terms of time consumption, of the activity and the amount of accuracy that is required. For example, the excavation of 375,000 cubic yards of material on a large dam would be considered a significant activity as considerable resources would be involved. Such an activity would warrant considerable investigation. Even some seemingly minor alterations in the work plan could result in considerable cost and time savings. While some schedulers with a significant amount of relevant experience might be able to make a good guess about the amount of time required to perform the work, most schedulers opt to analyze the operation in detail. The duration of such an excavation activity might take several days or even weeks, but this can only be determined with accuracy and with confidence with a close look at the operation and the quantities involved.

The determination of the duration of a small task might be made with confidence with only a limited evaluation. For example, suppose a schedule is being developed for a small, three-story office building. How might the duration be determined for the clearing and grubbing on the site? Assume the development lot is about an acre in area, with no trees, debris, or other significant vegetation on the site. How might the duration of the clearing and grubbing be determined? First of all, perhaps the activity will include the equipment mobilization, along with the actual clearing and grubbing activity. The scheduler might very well decide to allocate four days to the duration of the activity based on the assumption that it might take two days to get the equipment moved to the site. It is assumed that the actual clearing and grubbing will take slightly more than one day to complete. Thus, much of the duration has to do with preparation, and in this case, with mobilization. For an activity such as clearing and grubbing, the total cost will generally be fairly modest, but each site should be evaluated separately.

Suppose the activity on this same small building project is the installation of wallpaper on one accent wall in the reception area. Since the wall is a small flat area with no penetrations and a minimum of electric wall outlets, it is assumed that the work can be done in a matter of hours. The scheduler might very well allow a minimum of one day to do the work, both to allow time to prepare the work area and to fully mobilize for the work. Essentially no detailed analysis of this work would be performed.

On small commercial buildings and residential structures, most of the activities will take only a few days. For example, the roofing on a building can be estimated with considerable accuracy. The scheduler might determine that the new roofing system can be installed in two and one-half days. The first reaction might be to round up the computed value of the duration to about three or even four days. This added time might be needed to accommodate the mobilization of the roofing crew, include the installation of the roofing materials, and allow for cleanup/demobilization.

The scheduler could also think about productivity performances on past projects and thus make the duration estimates without actually performing any detailed analysis. For example, the drywall installation duration might be determined with minimal computations. This might be done if the scheduler can visualize the work actually being done at a given pace. Perhaps the drywall on a prior project was installed by a work crew at a rate of four rooms per day. If the proposed building is similar to the past project being referenced, a rough estimate might be made about the duration. If the new building has 30 rooms, the duration might logically be stated as being about eight days (30/4 = 7.5). An increase in crew size can result in shorter durations just as reductions in the crew size will result in longer durations.

On the short duration activities just described, a significant percentage of the time was devoted to allowances for mobilization. While mobilization may not be a major task for the members of some trades, they must still arrive on the job site at the appropriate time. Most building projects involve many subcontractors. It would not be realistic to expect each subcontractor to be on the job site at a precisely designated time. If a subcontractor is told to be on site at noon on a particular day but is unable to begin work because another crew has not yet completed its work, that subcontractor's crews are unlikely to wait while other subcontractors finish their tasks. What is even worse, that subcontractor's crews may not be assigned to show up the next day. It is costly for a subcontractor to be on a project at a stated time and then not be able to begin work. Thus, it is generally a good idea to allow some time to transition smoothly from one activity to another, especially when different specialty contractors are involved. This will allow time to contact the subcontractors to inform them of the specific needs of the project for their services.

But how is the duration determined if the scheduler has no experience with a particular work item? Probably the first step that most would take is to ask someone with experience. For example, the scheduler can talk with the estimator or the job superintendent. The job superintendent should have a very good idea of how the work plan is to be put into effect. The only note of caution here is that superintendents are often very optimistic about what they can accomplish in a given period of time. They often fail to include the transition time between activities. With experience, one can assess the need for modifying or adjusting the duration estimates provided by others.

If the firm has no individuals to contact concerning the duration of a specific activity, one might use reference guides that give productivity estimates for various tasks. Use reference guides with some degree of caution. These guides are generally conservative in their estimates, but they can be very helpful if no other resources are available. Note that each project has unique characteristics and that the reference guides cannot be expected to anticipate all the nuances that exist in project differences. One contractor stated that if a bid were prepared solely with the use of costs derived from reference guides, the firm would never win a contract. This may reflect the conservative nature of reference cost data, but this may not accurately reflect the true value of this additional resource.

Before using published reference materials, it is best to explore company files for relevant information. The historical data on past company projects can be a very valuable resource for the scheduler. Where detailed information is desired, this is perhaps the scheduler's best resource.

When historical data are used, the information of primary interest will be the productivity rates. That is, the scheduler will want to know the number of units of work accomplished in a given unit of time (generally hours). A couple of examples will help demonstrate how the scheduler can use this information to determine the duration. In the first example, assume the work task involves hand excavation on a relatively small project with no viable means of using heavy equipment. The volume of material to be excavated consists of nine cubic meters (11.8 cubic yards). The budgeted amount and the hourly costs are known, and from this, the minimum duration can be determined. If the duration is exceeded, the cost of the activity will also overrun the budget. The calculation of the duration is as follows:

Quantity of Excavation: 9 cubic meters (determined in the estimate)
Budgeted Amount: $950 (provided in the budget)
Hourly Cost Per Worker: $33.20 (includes general conditions and overhead ≈ 50%)
Cost Rate = Budgeted Cost / Quantity of Earth = $950 / 9 cubic meters = $105.56 /m^3
Production Rate = Cost Rate / Hourly Cost = $105.56 per m^3 / $33.20 per hour
= 3.18 hr. per m^3

$$\text{Excavation Time} = \text{Production Rate} \times \text{Quantity} = 3.18 \text{ hr. per m}^3 \times 9 \text{ m}^3 = 28.6 \text{ hr.}$$
If one worker is employed, duration = 3.6 days (use 4 days)
If two workers are employed, duration = 1.8 days (2 days)

In this example, the production rate was determined from the estimated quantity of material, the amount budgeted, and the cost of labor. Obviously, if the estimated values were in error, the production rates would also be in error. Presumably, the estimator used the production rates to derive the appropriate costs to be used in the estimate or budget.

The scheduler might use the production rates directly rather than determine the production rates that must be achieved to finish an activity within the budget. The following example assumes that the production rates for erecting 47 square meters (506 sq. ft.) of concrete formwork are well documented. In this example, the budgeted amount is not essential to deriving a duration.

Quantity of Forming: 47 square meters
Production Rate = 1.8 hr. per m^2 (obtained from historical data)
Forming Time = Production Rate $\times$ Quantity = 1.8 hr. per m^2 $\times$ 47 m^2 = 84.6 hr.
If one worker is employed: duration = 10.6 days (use 11 days)
If two workers are employed: duration = 5.3 days (5 or 6 days)
If three workers are employed: duration = 3.5 days (4 days)
If four workers are employed: duration = 2.6 days (3 days)

Note that durations are generally recorded in terms of whole days. This need not be done, especially with the use of computers, which can readily handle fractions of days. Rounding up to whole integer values does add some conservatism to the duration estimate and is often preferred. Some judgment must be exercised whenever activity durations are determined.

SCHEDULING ISSUES

While determining activity durations is important, it is also important to consider the actual scheduling of the activities. When using a schedule to control a project, a decision must be made about the use of float. Float is the leeway that is assigned to various activities in a network. This leeway may be effectively manipulated by a project manager. A very basic approach is to use either an "early start" schedule or a "late start" schedule. In an "early start" schedule, all of the available float of each activity remains controllable, while in a "late start" schedule, the float of each activity has been reduced to 0. In a "late start" schedule, all activities are critical.

An "early start" schedule has obvious advantages in that many activities are not critical. In a "late start" schedule, on the other hand, all flexibility has been removed from the network. In spite of the shortcomings of the "late start" schedule, contractors may be inclined to use them. This is particularly true if the owner is perceived as "owning" the float. The "early start" schedule has a disadvantage in that resources may not be consumed in a uniform fashion on the project. The same disadvantage exists on "late start" schedules. The obvious ideal solution probably lies in a compromise between the two types of schedules. By starting with an "early start" schedule, problems might be identified that can be resolved simply by making use of available float.

The "early start" schedule and the "late start" schedules may be used to predict the cash flow requirements on the project. In a cash-flow diagram, one would see that the area between the "early start" schedule and the "late start" schedule forms an envelope of money over a period of time. This "lazy S-curve" presents the full range of options open to the project manager. The S-curve could also contain actual cost data as compared to the predicted. This would show at a glance the magnitude of the various costs that are being incurred. It must be kept in mind that the expenditure of money does not necessarily imply that progress is being made or that a specific amount of profit has been earned. It is merely an indication of total project expenditures.

Once a schedule has been developed, the scheduling of certain activities should be examined. For example, major activities should be examined to determine the time of the year in which they are scheduled to occur. This helps management anticipate the need for certain precautionary measures, including those guarding against the effects of rain, snow, temperature extremes, high winds, and poor soil problems. To respond to this information, management may decide to winterize a portion of the project in order to stay on schedule. The schedule may also help the project manager anticipate problems caused by delayed shop drawing approvals, delays in material deliveries, the expiration of labor agreements, and deficiencies in the cash position on the project. A poor cash position may be handled by scheduling particularly expensive noncritical activities as late as possible to defer the large payment for the item. The project manager should also try to anticipate some rework and the loss of some time due to inspections. Efforts may be expended to minimize these adverse impacts.

FACTORS INFLUENCING CHOICE OF ACTIVITY SCHEDULES

Scheduling should satisfy various constraints that may be imposed by the contract or by specific needs of the firm. As noted earlier, the sequencing of activities depends on a variety of constraints or considerations. In developing the network logic, primary focus is placed on physical constraints. Other constraints (resource, environmental, management, safety, etc.) are not necessarily used to order the activities. Instead, these constraints are simply noted and left to be introduced at a later stage of the planning and scheduling process.

To the extent possible, activity start times are established such that the resulting schedule reflects these noted constraints. For example, if two activities require the same piece of equipment and have similar start times, it may be possible to set the scheduled start times of the two activities so that they do not overlap. Of course, the opportunities for accommodating these constraints are subject to available float and the implications of start times presented above. A more detailed discussion of scheduling procedures is presented in the chapter on resource allocation and leveling.

An additional consideration is that of materials delivery and storage. Staggering deliveries can reduce job site congestion and simplify coordination efforts on site. Delaying material deliveries can also take the pressure off when there is limited storage space. In most cases, it is wise to base the material inflow rate on the available storage space and the rate of installation. The schedule should reflect any significant concerns related to materials.

In other situations the scheduler may be faced with balancing the relative uncertainties inherent in any construction project. While one can anticipate these problems to occur during the life of a construction project, it is nonetheless desirable to develop as realistic a schedule as possible. Arrangements made for materials deliveries, subcontractor start dates, contract-specified milestone dates, and the like suggest that the initial schedule is important.

Selecting scheduled dates, therefore, may include a consideration of the likelihood that the assumptions of the network model will not hold true. An activity may, for example, be considered likely to experience problems that will extend its duration. In such a case this activity might be scheduled toward its early start to minimize the chance of project delay. Subsequent activities might be set to later scheduled starts so that arrangements made for the start of the activity would be less likely to be altered. An alternative, of course, would be to increase the duration of the initial activity to account for the amount of delay that might be expected.

Weather and the Schedule

Weather considerations or other contingencies may be incorporated into a schedule in the manner just described. The weather-dependent activity would be scheduled earlier and succeeding activities later. Alternatively, the durations of weather-dependent activities could be increased prior to proceeding with the network calculations. The amount by which the activity's duration would be increased would be based on the likelihood that a delay would occur as well as the extent of the delay. For example, if history indicated that rainfall could be expected 10% of the time (1 out of 10 days) during the period this activity would occur, then the duration would be increased by 10%. This activity weighting to accommodate adverse weather conditions is often a preferred approach. It should be noted that weather is a more serious consideration for the scheduler when the contract stipulates that the contract duration reflects the impact of adverse weather.

Another method of weather inclusion is to insert one or more specific activities into the network labeled as weather delays. These activities would be inserted into activity chains that contain weather-dependent activities. One drawback of this technique is that schedule monitoring is made more difficult because the extra activities make the schedule appear to be artificially tight (containing less float). If a weather activity is added to the network, should this appear early in the project or late in the project? The decision of the actual placement of the weather activity is not a trivial or arbitrary one. A stronger case can probably be made for placing the weather activity at the end of the network. In this way, if no adverse weather is encountered, the schedule remains intact and the project can be completed with fewer revisions of the network. If the weather activity is early in the schedule, essentially all activities will be scheduled later than they would normally occur. That is, the accuracy of the schedule will probably suffer if the weather activity is inserted early and no adverse weather occurs; i.e., early start times for activities will be scheduled as if it is certain that the adverse weather will occur early in the project.

One other weather-related scheduling consideration is whether the contract is administered on the basis of working days or calendar days and how weather affects the definition of working days. (This issue of how weather is handled in contracts is addressed further in the next chapter.)

Uncertainty in Duration Estimates

When most network schedules are developed, the activity durations are assumed to be fixed at a particular duration estimate. While this approach is a standard practice, there is a fundamental flaw with this method of activity duration estimation. The flaw is not fatal and is certainly defendable in that any other approach would make the solution of the network very unwieldy. Nonetheless, it is important to have a good appreciation of the true meaning of a duration estimate, which requires at least a rudimentary understanding of probability and statistics.

Table 4.3 gives information on the probabilities associated with certain estimates as defined in terms of standard deviation. The values in the table show the amount of the area under the bell-shaped curve when a value is selected a given number of standard deviations to the right of the mean. When the mean is selected, the number of standard deviations to the right of the mean is 0.0 and the area to the left of this point is 0.5, or 50% of the total area. If the area is sought for a point that is one standard deviation to the right of the mean, the area is determined to be 84.13% of the total area (see Table 4.3). Thus, it can be said that with a duration estimate of one standard deviation beyond the mean, there is an 84.13% chance of completion in the allotted time. With two standard deviations, the estimate of the area to the left is found to be 97.72% of the total area. Examine Table 4.3 and develop a clear understanding of the way the above percentages were derived.

Let's first consider what is meant by the value that is assigned to an activity duration. This will be demonstrated by considering the time durations for two different subcontracted activities on a building project. One of the activities is related to the masonry work and the other relates to laying carpet. Both subcontractors were asked to estimate the time to allow for their work. The estimate from the masonry contractor was 28 days and the estimate from the carpet installer was 23 days. When asked how they arrived at these durations, both said that they had data on seven prior projects that were similar to the one being constructed. The masonry contractor considered such factors as the number of bricks, the number of window openings, the number of doorways, the number of corners in the structure, and various other features. The carpet installer considered room regularity, size of rooms, access to the work areas, and

Table 4.3
AREA UNDER A STANDARD NORMAL CURVE

x = The number of standard deviations to the right of the mean

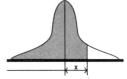

The area under the curve (as shown in the table) always includes the portion containing the mean.

X	0	1	2	3	4	5	6	7	8	9
0.0	.5000	.5040	.5080	.5120	.5160	.5199	.5239	.5279	.5319	.5359
0.1	.5398	.5438	.5478	.5517	.5557	.5596	.5636	.5675	.5714	.5754
0.2	.5793	.5832	.5871	.5910	.5948	.5987	.6026	.6064	.6103	.6141
0.3	.6179	.6217	.6255	.6293	.6331	.6368	.6406	.6443	.6480	.6517
0.4	.6554	.6591	.6628	.6664	.6700	.6736	.6772	.6808	.6844	.6879
0.5	.6915	.6950	.6985	.7019	.7054	.7088	.7123	.7157	.7190	.7224
0.6	.7258	.7291	.7324	.7357	.7389	.7422	.7454	.7486	.7518	.7549
0.7	.7580	.7612	.7642	.7673	.7704	.7734	.7764	.7794	.7823	.7852
0.8	.7881	.7910	.7939	.7967	.7996	.8023	.8051	.8078	.8106	.8133
0.9	.8159	.8186	.8212	.8238	.8264	.8289	.8315	.8340	.8365	.8389
1.0	.8413	.8438	.8461	.8485	.8508	.8531	.8554	.8577	.8599	.8621
1.1	.8643	.8665	.8686	.8708	.8729	.8749	.8770	.8790	.8810	.8830
1.2	.8849	.8869	.8888	.8907	.8925	.8944	.8962	.8980	.8997	.9015
1.3	.9032	.9049	.9066	.9082	.9099	.9115	.9131	.9147	.9162	.9177
1.4	.9192	.9207	.9222	.9236	.9251	.9265	.9279	.9292	.9306	.9319
1.5	.9332	.9345	.9357	.9370	.9382	.9394	.9406	.9418	.9429	.9441
1.6	.9452	.9463	.9474	.9484	.9495	.9505	.9515	.9525	.9535	.9545
1.7	.9554	.9564	.9573	.9582	.9591	.9599	.9608	.9616	.9625	.9633
1.8	.9641	.9649	.9656	.9664	.9671	.9678	.9686	.9693	.9699	.9706
1.9	.9713	.9719	.9726	.9732	.9738	.9744	.9750	.9756	.9761	.9767
2.0	.9772	.9778	.9783	.9788	.9793	.9798	.9803	.9808	.9812	.9817
2.1	.9821	.9826	.9830	.9834	.9838	.9842	.9846	.9850	.9854	.9857
2.2	.9861	.9864	.9868	.9871	.9875	.9878	.9881	.9884	.9887	.9890
2.3	.9893	.9649	.9898	.9901	.9904	.9906	.9909	.9911	.9913	.9916
2.4	.9919	.9920	.9922	.9925	.9927	.9929	.9931	.9932	.9934	.9936
2.5	.9938	.9940	.9941	.9943	.9945	.9946	.9948	.9949	.9951	.9952
2.6	.9953	.9955	.9956	.9957	.9959	.9960	.9961	.9962	.9963	.9964
2.7	.9965	.9966	.9967	.9968	.9969	.9970	.9971	.9972	.9973	.9974
2.8	.9974	.9975	.9976	.9977	.9977	.9978	.9979	.9979	.9980	.9981
2.9	.9981	.9982	.9982	.9983	.9984	.9984	.9985	.9985	.9986	.9986
3.0	.9987	.9987	.9987	.9988	.9988	.9989	.9989	.9989	.9990	.9990

other features. The specific project data that they used to make their estimates are shown as follows:

	Masonry Data		Carpet Data	
Project #	Duration (Dur)	(Dur-Mean)^2	Duration (Dur)	(Dur-Mean)^2
1	14	36	18	4
2	17	9	21	1
3	27	49	19	1
4	16	16	22	4
5	18	4	17	9
6	25	25	22	4
7	23	9	21	1
	Sum$_1$ = 140	Sum$_2$ = 148	Sum$_1$ = 140	Sum$_2$ = 24
	Mean$_1$ = 20	(Sum$_2$/n−1)$^{1/2}$ = 4.97	Mean$_1$ = 20	(Sum$_2$/n−1)$^{1/2}$ = 2
	n = 7	Std. Dev. = 5 (approx.)	n = 7	Std. Dev. = 2

Note that the average duration of performing the work on previous projects was 20 days for both specialty contractors. Yet the average value is not the duration they quoted. To use the average duration would mean that the estimated duration would be too small in many instances. So how did these firms come up with the duration estimates of 28 days and 23 days? This is where the standard deviation about past project data is helpful. In this example, let's assume that the subcontractors would want to give an estimate that will not be understated in 95% of the cases, meaning that we want a high level of confidence that the estimate is not too small. If the estimate of 20 days had been given, the contractors would have a level of confidence of 50% that sufficient time had been allowed to perform the work (see Table 4.3). If the masonry contractor's estimate had been 25 days (essentially one standard deviation beyond the mean), the confidence would increase to 84%. To be 95% confident that the duration will not be too small, the estimate must be 1.645 standard deviations above the mean, or about 28 days for the masonry work (20 + 1.645 × 5 = 28.2). Similarly, the carpet installer used the same logic and came up with a duration of 23 days (20 + 1.645 × 2 = 23.3). The carpet installer had less scatter in the past project durations, as is evidenced by the smaller standard deviation of 2. The desired estimate should bear a realistic value of the duration that does not generally understate the duration. This is where statistics can be valuable.

The most relevant information needed to make an estimate with confidence is the mean and the standard deviation. The standard deviation implies the amount of variability that exists in a collection of numbers. In our case, the average was 20 days for both contractors and the standard deviations were five days and two days, respectively. In our example, we assumed that the various estimates are essentially normally distributed, similar to a bell curve. If either subcontractor's estimate had been 20 days, it should be apparent that there is only a 50% chance of achieving this duration. These are not desirable odds. On the other hand, if 98% confidence was desired, the duration estimates would be about 30 and 24, respectively, for the masonry and carpet work.

Consider another similar problem. The amount of time to set the flatwork (sidewalk) forming on a project is to be determined. The data on past projects indicate that the duration has a mean of 130 worker hours and a standard deviation of nine hours. What estimate should be used if we are to have a 95% confidence that the estimated duration will not be exceeded? Referring to Table 4.3, find the number of standard deviations that include 95% of the area under the curve.

From the table, it can be interpolated that the number of standard deviations is 1.645 for a confidence of 95%. Thus, the estimate that we would be well advised to use is 144.8 or 145 hours (130 + 1.645 [9 hours]).

While contractors may not generally go though this rigorous type of statistical approach, the essence of their logic is essentially the same. Contractors must compute a realistic duration or have a very good intuitive sense of the appropriate estimate to be used.

When the durations for the various activities are considered from the perspective of probability and statistics, it should be clear that the durations attached to many activities represent best guesses. A contractor will view an activity with a duration of eight days as being one that will be completed in the vicinity of eight days, but it is recognized that the duration could be more or less. Owners may view durations a bit differently. The owner on a project may view the duration of eight days as implying that the duration will not exceed eight days. While these viewpoints may not seem to be drastically different, they could be the source of much conflict on a project. For example, the issue of float ownership becomes a serious matter on some projects. If the owner owns the float, an activity with a duration of eight days had best not exceed that duration. If the contractor exceeds that duration, there may be a cost associated with it. Contractors may very well argue that, since the durations are "fuzzy," it is inappropriate for the owner to seize the float. The contractor likes to utilize float without the owner's interference.

FINAL COMMENTS

The duration estimates for activities can be determined by various methods. At the root of many of these estimates is the quantity take-off. Another key piece of information is the production rate associated with the installation of those quantities. The quantity take-off is project specific while the production rate is relatively constant for that particular type of activity. Despite the computations associated with the activity durations, it must be recognized that these are essentially best guesses and should not be regarded as absolute values. Some deviation from the estimated durations is to be expected. Allowances may have to be made for the impact of adverse weather conditions and for other factors, such as the accommodation of small tasks that are not specifically included in the schedule.

REVIEW PROBLEMS

1. Scaffolding is to be constructed on all faces of a six-story building in order for major masonry rehabilitation work to be performed. The exposed surface area of the exterior walls of the building has been estimated to be 432,000 sq. ft. Historical records of the company show that for this simple type of structure, the production

rate for erecting scaffolding is 0.002 hours per square foot of building surface area. If ten workers (working 8-hour days) are assigned to do the entire scaffolding erection, what duration (in working days) should be used for this activity? Round up your answer to the nearest whole integer day.

2. Tiles (6" by 6") can be installed at a rate of 50 per worker hour. The floor of a large lobby area (80' by 40') is to be covered with these tiles. If three workers are assigned to this task, what duration (to the nearest whole day) should be assigned to this activity?

3. Historic records show that the mean duration of an activity is expected to be eight days with a standard deviation of 2.4 days. What duration should be used if there is to be a 95% confidence that the duration will not be exceeded? What is the probability that the duration will actually exceed ten days?

4. The task of framing a building has been estimated to take an average of 25 days with a standard deviation of four days. What duration should be used if there is to be a 90% confidence that the duration will not be exceeded? What is the probability that the duration will actually exceed 23 days?

5. A masonry facade consisting of 3,800 square feet is to be constructed for a building. The total cost per worker hour is estimated to be $31.50 and the total estimated cost of the labor for this task is $10,500. Assuming eight-hour work days and a crew of six workers, how many days should be allowed to complete this task? What is the production rate that the crew must attain to keep the project on schedule and within the budget?

6. Based on historical weather data, a particular project is expected to be impacted by six adverse weather days. Using one of the procedures described in this chapter, make the appropriate allowance for adverse weather and solve the network for the basic scheduling information. For each activity, show the computed value of early start, early finish, late start, late finish, total float, and free float.

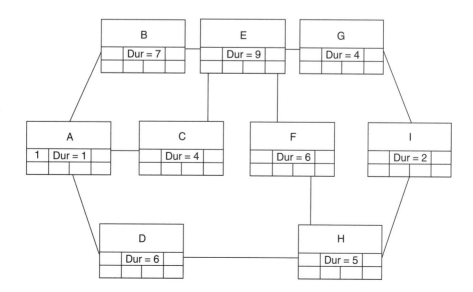

5

Time in Contract Provisions

If it weren't for the last minute, nothing would ever get done.

Although the contractor has specific needs for planning the overall sequence of activities for a project, decisions about project sequences and the contractor's approach in undertaking a project are impacted by the contractual language. A variety of provisions undoubtedly influence the contractor as he or she plans a project. First of all, the contract includes provisions that relate directly to the project's duration. Still other provisions address the timing of the owner's cash disbursements to the contractor. Some provisions may even stipulate the scheduling technique that must be used on a project. It is important that the contractor evaluate all contract provisions. Some may present inherent risks for the contractor; others may be regarded as reducing the contractor's risk. The degree of risk should logically be reflected in the amount of the bid.

This chapter provides various provisions as examples of how risk can be alleviated or imposed on a contractor. These provisions come from actual contract documents; however, some modifications were made when appropriate. For example, an effort was made to eliminate the use of masculine pronouns. While not all contracts contained masculine pronouns, this is still a reasonably widespread practice. Provisions were also rewritten so that they use only the terms *Owner* and *Contractor*. Contracts that have been customized for particular owner agencies tend to use terms that specifically identify the owner. The provisions were modified to the extent that they use only the term *Owner* while the source document may have used such terms as *Agency, Authority, City, Department, District,* or *Highway Administration.* Similarly, only the term *Owner's Representative* has been used in the provisions, while the actual contract documents used such terms as *Chief Engineer, Engineer, Director, District Engineer, Resident Engineer, Architect, Consultant,* and *Owner's Agent.* The provisions have been italicized to distinguish them from text.

TIME IS OF THE ESSENCE

It is imperative to most owners that all parties to the contract consider time as an elemental aspect of the contract. This emphasis on time gives greater validity to the inclusion of liquidated damages provisions, and it also allows owners to pursue claims for breach of contract when contractors do not fulfill the time requirements stipulated in the contract. The issue of time is emphasized by such provisions as the following:

Time shall be strictly of the essence of the contract. The Contractor shall promptly begin the work under the contract and all portions of the project made the subject of the contract shall be begun and so prosecuted with necessary plant, equipment, procedures, and overtime that they shall be completed and ready for full use in the time stated in the special conditions.

REQUIREMENTS FOR PROJECT COORDINATION

To ensure a smooth progression of work on a project, the owner will often stipulate that the major parties involved discuss the project's overall organization at a meeting prior to the commencement of any construction work. The following provision demonstrates the basic intent of such meetings:

Within ten days following award of the contract, a preconstruction conference will be held in the Owner's office. The Contractor shall be represented at the meeting by the job superintendent or foreman. All aspects of the job will be discussed and representatives of various agencies with interest in the project will be present to answer questions.

The provision for a preconstruction conference may also be more specific. The following provision stipulates that specific tasks be performed by the contractor in preparation for the conference:

Before commencing with any construction work, the Contractor shall meet with the Owner's Representative for a preconstruction conference. At this time the Contractor shall submit....
A progress schedule showing the order in which the Contractor proposes to carry out the work and the contemplated dates on which the Contractor and any Subcontractors will start and finish each of the salient features of the work, including any scheduled period of shutdown. The schedule shall also indicate any anticipated periods of multiple-shift work.
A list showing anticipated dates for procurement of materials and equipment, the ordering of articles of special manufacture, the furnishing of plans, drawings, and other data required under this contract and for any other events such as inspection of structural steel fabrication.
A list showing all proposed subcontractors and material suppliers....

When it comes to cooperation, the issue cannot be unilateral: both parties must be responsible. Subcontractors need to know when their services are needed, and the subcontractors must respond at the appropriate time so the construction effort progresses smoothly. This mutual need is addressed in the following subcontract provision:

SCHEDULE. Contractor shall verbally notify Subcontractor at least seven (7) days prior to the date Subcontractor is to commence work and shall provide immediate notification of any changes affecting the start date. Subcontractor shall commence work immediately on the date given with notice to proceed. Subcontractor shall provide adequate supplies of materials and necessary equipment together with competent full-time on-site supervision and a competent and adequately sized work force working diligently to complete the Work in a timely manner in accordance with the project schedule as periodically revised as well as the direction of the Contractor's Project Superintendent. The updated schedule shall always be available to the Subcontractor in the Contractor's on-site project office. Subcontractor shall cooperate with the Contractor and other Subcontractors at all times to facilitate the scheduled completion date set forth in the Main Contract, including attendance at coordination meetings on the site.

COOPERATION

Depending on the nature of the construction project, the project site may be used by parties other than the contractor. This use may be related to other contractors who have contracts with the owner or to employees of the owner. If such dual occupancy or use of the facility is anticipated, it is prudent to include a provision that stipulates that the contractor is to cooperate with these other parties. Naturally, it is also important for the owner to try to convey the specific nature of the work being performed by the other parties. This would best be conveyed in the special conditions. The following is a provision that addresses the general nature of the need for the contractor to cooperate with other parties on the site:

The Owner, or other contractors performing work on behalf of the Owner, shall be at liberty to enter upon the site of the work with workers and materials to do work, and the Contractor shall afford any such workers all reasonable facilities and cooperation to the satisfaction of the Owner's Representative. . . . Any conflict arising between the Contractor and any other contractor employed by the Owner, or between the Contractor and the workers of the Owner with regard to their work, shall be submitted to the Owner's Representative for a decision on the matter. If the work of the Contractor is delayed because of any acts or omissions of any other contractor or of the Owner, the Contractor may claim for extensions of time and/or compensation in accordance with the contract.

PROGRESS SCHEDULE

The owner of a project has a specific need for the completed facility. The time in which this project is ready for occupancy by the owner is of considerable interest. Decisions will be made based on the anticipated delivery date of the completed project. Because of the importance of the delivery date, owners want to make their own judgments about project completion dates, so they often independently monitor projects to evaluate adherence with the schedule. One way to have a reasonably reliable schedule is to monitor the project with a schedule to which both the owner and the contractor have agreed. In some cases owners prepare the schedules and have the contractors make suggestions for modifications. It is more typical for the contractor to prepare the schedule for the owner's approval. The requirements for the "project schedule" may vary considerably from project to project or from owner to owner. Many contract documents do not include requirements for a project schedule. Other documents vary in the level of detail that is required. The following provisions provide a basic sampling of different project schedule provisions:

When required by the special provisions, the Contractor shall submit to the Owner's Representative a practicable progress schedule within 20 working days of approval of the contract, and within ten working days of Owner's Representative's written request at any other time. The form of the schedule shall be the Contractor's choice, unless the Owner's Representative furnishes a specific form to be used. If the Owner's Representative furnishes a form, the Contractor may be requested, on or before the last day of each month, to indicate the status of work actually completed during the preceding estimate period.

The schedule shall show the order in which the Contractor proposes to carry out the work, the dates on which the several salient features of the work will start (including procurement of materials, plant, and equipment), and the contemplated dates for completing the said salient features. The progress schedule submitted shall be consistent in all respects with the time and order of work requirements of the contract. Subsequent to the time that submittal of a progress schedule is required in accordance with these specifications, no progress payments will be made for any work until a satisfactory schedule has been submitted to the Owner's Representative.

While the above provision may provide some latitude for the contractor, it does stipulate that no progress payments will be authorized until a satisfactory schedule has been submitted. The following provision is more generic and essentially includes no sanctions for the contractor for failing to submit the schedule in a specific format. Indeed, the contractor could submit a bar chart or even a list of activities in order to comply with the provision.

Prior to beginning construction operations, the Contractor shall submit to the Owner's Representative a chart or brief outlining the manner of prosecution of the work that is intended to be followed in order to complete the contract within the allotted time. The Contractor shall begin the work to be performed under the contract within 30 days after the date of the authorization to begin work and shall continuously prosecute same with such diligence as will enable the Contractor to complete the work within the time limit specified.

The following provision is more specific than the preceding one. It establishes the schedule as a monitoring and communication tool to be used by the contractor and owner during the course of the project.

After the award of the contract and prior to starting work, the Contractor shall submit to the Owner's Representative a satisfactory progress schedule or critical path schedule which shall show the proposed sequence of work, and how the Contractor proposes to complete the various items of work within the number of working days set up in the contract or on or before the completion date specified in the contract.

This schedule shall be used as a basis for establishing the controlling item of construction operations and for checking the progress of the work. The controlling item shall be defined as the item which must be completed either partially or completely to permit continuation of progress. It shall be the responsibility of the Contractor to show the intended rate of production for each controlling item listed on the schedule during the period such item is controlling.

The Contractor shall confer with the Owner's Representative at regular intervals in regard to the prosecution of the work in accordance with the progress schedule or critical path schedule.

Another provision with similar intent is as follows:

The Contractor shall submit a progress schedule on an approved form within 15 days after the execution of the contract showing how the work is to be prosecuted. If the Contractor's operations are materially affected by changes in the plan or in the amount of the work or if there is a failure to comply with the approved schedule, the Contractor shall submit a revised progress schedule, if requested by the Owner's Representative, which schedule shall show how the balance of the work is to be prosecuted. The Contractor shall submit the revised progress schedule within ten days after the date of the request. The Contractor shall incorporate into every progress schedule submitted, any contract requirements regarding the order of performance of each portion of the work. All practicable means shall be used to make the progress of the work conform to that shown on the progress schedule that is in effect. No payment will be made to the Contractor while delinquent in the submission of a progress schedule. Should the prosecution of the work, for any reason, be discontinued, the Contractor shall notify the Owner's Representative at least 24 hours in advance of resuming operations.

Note that the following provision, which is similar to the one above, specifically outlines items or activities that are to be shown in the project schedule.

A contract schedule, reflecting all engineering, equipment and material procurement, fabrication and shipping, mobilization, and construction activities required for the orderly performance and completion of work, shall be prepared by the Contractor and submitted for Owner's Representative's approval within 30 calendar days from and after Contractor's receipt of written Notice of Award and before work is commenced. Approval of the contract schedule by the Owner's Representative shall be a condition precedent to making any payment to the Contractor. Contractor shall submit separate written justification for any changes that are made to the schedule submitted. The schedule shall be a time-scaled plan, CPM network, or bar chart for implementing and completing work and shall include but not be limited to (a) all activities that have a duration of four weeks or longer, (b) dates of order and delivery of all major items of equipment and/or material when supplied by the Contractor, (c) milestone dates, (d) restraints, (e) interfaces with other contractors or subcontractors, and (f) proposed weighted values for each category of work, reflecting that category as a percentage of the total scope of work.

The following provision is more specific in its requirements. The contract may require that the schedule be in the form of a critical path schedule or a "similar method." (It is not clear what a similar method is.) This provision includes a requirement for updating the schedule and for submitting periodic reports. Note that payments will be delayed for failure to submit an approved schedule. Additionally, the owner is specifically relieved of any responsibility that the contractor may wish to confer on the owner through the schedule.

Immediately after being awarded a contract, the Contractor shall submit a progress schedule to the Owner's Representative. This schedule and any supplemental schedule shall show: (1) completion of all work within the specified contract time, (2) the proposed order of work, and (3) projected starting and completion times for major phases of the work and for the total project. If the contract requires, the schedule shall be developed by the critical path method or a similar method. The Contractor shall provide sufficient material, equipment, and labor to meet the completion times in this schedule.

As the work proceeds, the Contractor shall submit supplemental progress schedules. These shall reflect any changes in the proposed order of the work, any construction delays, or other conditions that may affect the progress of the work. The original and all supplemental progress schedules shall not conflict with any time and order-of-work requirement in the contract.

No progress payments will be issued to the Contractor until a progress schedule has been submitted and approved. If the Owner's Representative deems that the original and any necessary supplemental progress schedule is not satisfactory, the Owner may withhold progress payments until a satisfactory schedule has been submitted by the Contractor and approved by the Owner's Representative.

The Owner's Representative's approval of any schedule shall not transfer any of the Contractor's responsibility to the Owner. The Contractor alone shall remain responsible for adjusting forces, equipment, and work schedules to ensure completion of the work within the time(s) specified in the contract.

The contract may dictate the level of detail to be embodied in the schedule. This may apply to the duration of the activity as:

Duration of any activity shall not exceed 30 days except 2% of the activities may exceed this limit without prior written approval of the Owner.

The constraints may also extend to the amount to be paid to the contractor for each activity. For example, one public agency contract had the following provision:

The value for any one activity shall not exceed $50,000 except 2% of the total Contract Sum may exceed this limit without the Owner's prior written approval.

One contract placed limitations on the float by stating:

No more than 30% of the monetary value of the Project activities shall have float of less than seven days and no more than 20% of the monetary value of the Project shall have zero float without prior written authorization of the Owner.

The last provision is apparently included to ensure that the contractor does not submit a late start schedule, one that would result in no float for any activities.

The following provision specifically requires the use of the critical path method and that a computer be employed to perform the computations. Requirements are stipulated for quarterly reporting and updating. Updating may be more frequent. The receipt of each progress payment may be jeopardized by failing to submit a satisfactory schedule or to properly update the schedule.

After execution of the contract and before the first progress payment is prepared, the Contractor shall submit, for approval by the Owner's Representative, a graphic and tabular construction schedule prepared by a critical path method of analysis. The construction schedule shall be prepared from estimates of the required duration and sequence of each item of work and function to be performed.

Tabulation and analysis of the work schedule shall be performed by computer using a commercially available critical path program. The graphic and tabulated construction schedule shall be updated and resubmitted quarterly, before every third monthly progress payment. The quarterly schedule submittals shall be prepared in the same format as the initial schedule submittal. Progress payments may be withheld until an acceptable schedule is submitted.

The Contractor shall supplement the quarterly construction schedule with monthly narrative-type reports covering progress on all major structures, highlighting any areas or items that are over one month behind on the latest schedule. Substantial changes may be grounds for an immediate updating of the construction schedule at the Owner's Representative's discretion.

If the Contractor fails to adhere to the accepted construction schedule as modified by any extension of time, the Owner may at any time withhold the Contractor's progress payment or any portion thereof. When the contractor regains adherence to the accepted construction schedule, amounts so withheld will be released and paid to the Contractor.

Another contract stipulated that updating of the schedule would occur monthly. It stated:

The Contractor shall submit monthly updates of the schedule and reports with the monthly request for payment. This update shall report the actual construction progress by updating the mathematical analysis of the Owner-approved percentage completion of each activity.

The provision also stated that the most recent submittal was to highlight the adjustments that were made in the schedule. It stated that the contractor was to submit a *separate listing of all revisions and/or data entries made since the last report.* This listing was to include information on logic changes, duration changes, and the actual start and finish dates.

The following provision shows that the owner will prepare a schedule, but the contractor is encouraged to submit a schedule for approval. If the contractor does not submit such a schedule, the owner's schedule will become the controlling document for the project.

The Owner will furnish a form that will show the distribution of the contract. This form will show the total contract time allowed for completion of all work on the project, a list of the various operations to be performed on the project, and a schedule of time estimates during which the Owner suggests each operation can be performed.

At the preconstruction meeting, the Contractor shall present for approval by the Owner's Representative a detailed construction schedule showing completion of all work at or before the time allowed by the contract. This schedule shall show all sequencing and all other aspects of how work on the project will be scheduled and performed.

If the schedule is accepted without change in writing by the Owner's Representative, it will be considered the official schedule for all purposes, including, but not limited to, the calculation of liquidated damages and the computation of time used in proving all claims filed. If the schedule is not accepted in writing or if no schedule is presented for approval at the preconstruction meeting, the schedule contained in the contract will be the official schedule for all purposes, as stated above. The decision by the Owner's Representative is final and binding.

The project schedule should be prepared with diligence and care. In the following provision, it is clear that progress will be monitored carefully with the project schedule. Should the contractor fail to stay on schedule, the owner has the contractual power to order an increase in the workforce and the addition of other resources, all at no added cost to the owner.

Contractor shall submit for Owner's Representative's approval, a progress schedule showing rate of progress expected for the various phases of work. The schedule shall show percentage of work completed at any time, anticipated monthly payments by Owner, as well as significant dates (such as completion of excavation, concrete, foundation work, underground lines, superstructure, roughing-in, enclosure, hanging of fixtures, etc.), which shall serve as checkpoints to determine compliance with the approved schedule. There will be no payment of any periodic estimate until the list of subcontractors and suppliers; progress and payment schedule; and the breakdown of the Contractor's proposal have been approved.

... Should the Contractor fail in any respect to adhere to the approved schedule, the Owner's Representative may require that additional workers, plant, and equipment be placed on the work or require that hours in addition to regular hours be worked until progress is as scheduled, with no additional cost to Owner.

While most provisions stipulate a project schedule, the following provisions address short-interval schedules. These are required on a weekly basis. An additional clause of interest is that any changes in the work activities that require additional staff members of the owner on the project may not be implemented for ten days. In other words, sudden changes in the work that will impact inspection requirements may be delayed.

At a mutually convenient location and time, the Contractor shall meet weekly with the Owner's Representative to discuss construction activities; however, some meetings may be waived if mutually agreed to, due to weather conditions, work progress, or for other reasons. At these meetings, the Contractor shall provide the Owner's Representative with a detailed, written schedule of major construction activities and phases of work for the forthcoming two-week period. This written schedule shall detail the start and anticipated completion dates of major phases of work as well as indicate the status of major ongoing activities. Minutes of the weekly meetings will be kept by the Owner's Representative and a copy given to the Contractor. Failure to provide an accurate, appropriate schedule may be grounds for the suspension of the work.

Schedule changes requiring an increase in the Owner's personnel will not be put into effect for ten days after the submission of weekly schedules detailing such activities, or until the Owner's Representative has made arrangements for additional personnel, whichever is the shorter time.

Should construction progress differ significantly from the progress schedule presented at the preconstruction conference, the Owner's Representative may request that the Contractor submit a revised progress schedule and anticipated completion dates of the major phases of work remaining and the anticipated completion date of the work.

The following provision concerns the requirement for a short-interval schedule. This provision was extracted from a document that also contained provisions for a CPM project schedule.

A construction two-week schedule shall be prepared in a format approved by the Owner's Representative and submitted for approval on a weekly basis by 11:30 A.M. on Thursday of each week. Each shall be a detailed plan of the activities and resources, for the coming two weeks, to meet the 90-day schedule. Input to these schedules shall be by use of the same activity codes as established for the pay items. The Contractor shall adjust this schedule as directed by the Owner's Representative to meet changing project requirements.

The 90-day schedule mentioned is an intermediate form of schedule that shows more detail than the project schedule. On larger projects the 90-day schedule would have particular utility for project management personnel.

OWNERSHIP OF FLOAT

The subject of float is increasingly addressed in construction contracts. The specific issue associated with float concerns the party that has entitlement to the float. The specific language regarding the ownership of float should be carefully examined in every contract. With no contract language on the subject, the contractor will probably assume that the float associated with any activity is available for the sole use of the contractor. The rationale behind this assumption is that the schedule is an approximation of how the project will be constructed and that the float is needed to make adjustments in the schedule. Some owners take exception to this line of thinking and regard any float to be completely at the disposal of the owner or that it is available to the party that uses it first. The impact on the schedule can be significant. For example, if the owner makes a change on a portion of the project that is not critical (e.g., activities contain float), the owner will be reluctant to grant any time extensions to the contractor. The contractor will want a time extension, as the change that consumes the available float will result in limited flexibility to the contractor. With these added restrictions, the contractor has fewer options and will want to receive compensation for the added constraints.

It could be argued that if the owner takes possession of the float, then the owner has, in fact, directly interfered with the contractor's control of the methods, means, techniques, sequences, and procedures. If a court might agree with this assessment, the owner could find it a costly decision to take ownership of any float. Fondahl (1975) suggested that it would be interesting to evaluate the legal effect of a specific activity—added near the end of the network and on the critical path—that had the label "estimated time allowance for non-excusable delays" (p. 18).

The issue of float ownership is a very real concern in many contracts. In one early (1970s) contract, a provision essentially stated the following:

The positive float existing in the progress schedule at any time is not considered to be "owned" by either the Owner or the Contractor. The amount of positive float and its relative position within the overall progress schedule will be factors to be considered during negotiations to resolve time extensions.

While this is one of the earlier provisions on float ownership to be found in contracts, it is not overtly onerous as on the surface. Note that the provision indicates that the status of the project must be taken into account to determine if the owner might have the right of such ownership. The inference seems to be that the owner cannot have float ownership if certain project criteria are not met.

The following provision, taken from a U.S. federal construction contract, addresses the ownership of float but limits the provision to addressing total float only. The total float is essentially owned by the first party that requires it, a distinct advantage to owners. Note that extensions of time would generally not be granted under this provision as long as only total float is being utilized by the delay:

Float or slack is defined as the amount of time between the early start date and the late start date, or the early finish date and the late finish date, of any of the activities in the schedule. Float is not time for the exclusive use or benefit of either the Owner or the Contractor. Extensions of time for performance required under the contract clauses entitled Changes, Differing Site Conditions, Termination for Default, Damages for Delay, Time Extensions, or Suspension of Work will be granted only to the extent that equitable time adjustments for the activity or activities affected exceed the total float or slack.

Another contract provision, taken from a contract for the construction of a public school, defines float as being owned by both the owner and the contractor. This is a harsher provision in that it defines total float to include the difference between the contractually mandated completion date and contractor's scheduled (conceivably earlier) completion date. Thus, changes mandated by the owner could delay the contractor's completion of the project, but no compensation is allowed as long as the contractor essentially completes the project by the contractually established date. Most U.S. government contracts restrict float-related contract provisions to the float as determined from the contractor's schedule. The following provisions grant considerable control of float to the owner:

Any float time, including time between the Contractor's anticipated completion date and the end of the contract time, shall be deemed to be for the joint use of the Owner and the Contractor provided, however, that, in the event of any conflicting need for or overlapping use of any float time, the Owner shall have priority for the use of such float time, and the Contractor shall not be entitled to any adjustment in the Contract Time, the Progress Schedule, or the Contract Sum, or to any additional payment of any sort by reason of the Owner's use of any float time.

Contractors would prefer to work on projects where the float ownership remains with the contractor. Clearly, with certain contract provisions this right is lost. One mechanism employed by some contractors to avoid the adverse impact of such provisions is to submit late start schedules to the owner. This has been done successfully by several contractors. Should the owner contest such a schedule? Should the owner mandate an early start schedule? The possibility is very real that such owner intervention could result in a legal interpretation that the owner has assumed control over the methods, means, techniques, sequences, and procedures.

NOTICE TO PROCEED

The notice to proceed is the document that typically marks the beginning of the actual construction duration. Most documents make reference to the notice to proceed. The following are examples of this.

Example A

Within five days after the execution of the contract by the Owner, written notice to proceed will be given by the Owner to the Contractor. The Contractor may expect the Owner to execute the contract within five days of the Owner's receipt of the Contractor's executed contract. Notwithstanding any other provision of the contract, the Owner shall not be obligated to accept or to pay for any work furnished by the Contractor prior to delivery of notice to proceed whether or not the Owner has knowledge of the furnishing of such work. The Contractor shall not be allowed on the site of the work until the contract bonds and evidence of insurance comply with all contract requirements.

Example B

The Contractor agrees to commence work on the date the "Notice to Proceed" is received from the Owner, and the entire work shall be completed within the number of consecutive calendar days stated in the Special Conditions. While time is of the essence of each and every portion of this contract, liquidated damages for delay in completion of the work as such are not prescribed.

Before the Contractor will be permitted to begin work, there must be in the hands of the Owner the following properly executed instruments: contract, performance bond, and certificate or policies of required insurance. Within ten days after Contractor's receipt of the Notice to Proceed, Contractor shall submit to the Owner for approval a progress and payment schedule, and a complete breakdown of the Contractor's proposal. No pay of any periodic estimate will be made until Contractor has received approval of these submitted materials.

Example C

The Contractor shall commence with work within ten calendar days after receipt of the Notice to Proceed and will be completed by (specified date) unless the period for completion is extended in accordance with the methods provided therefor in the contract and further agrees to pay the Owner as liquidated damages the sum of $_____ for each day thereafter, excluding Sundays and Holidays.

Note that liquidated damages may be stipulated in the provisions or they may be specifically omitted from the contract.

TIME OF COMPLETION

The contract duration is typically stated in terms of the total number of working days or the total number of calendar days allotted for completion. Sometimes, specified completion dates are given. Italicized text in this chapter has been taken from various construction contracts.

The Contractor shall commence work under this contract on the date the "Notice to Proceed" is received from the Owner, and shall fully complete the project within_____ consecutive calendar days.

This provision states the total amount of time permitted for project completion and also stipulates the day on which the days will begin to be counted. Note that in the above provision the days start to be counted on the date the notice to proceed is received. Some contracts state that the counting will begin on the date specified in the notice to proceed or in a given number of days after the notice to proceed is received.

UNITS OF TIME: WORKING DAYS OR CALENDAR DAYS

When setting durations, some consideration must be given to the units of time used. Most typical construction schedules use days. Shorter time periods might be preferred on some projects. For example, in turn-around work, where a contractor is to make significant modifications to a facility while normal operations are shut down for a short period of time, the durations might be set in terms of shifts or even hours. If the schedule is conceptual in nature, the durations might very well be stated in longer time units such as weeks, months, or even years. The time units must be matched to the detail required of the project needs.

While many different units of time might be used, most construction schedules are defined in terms of days, but even this is not a simple matter. What is a day? There is a big difference between working days and calendar days. Many job tasks will automatically be given in working days, representing all days with the exclusion of weekends and holidays. For example, if wall framing is estimated to take six days, there is little doubt that these are all working days. A working day is a logical unit to use on most jobs, but what about those durations that are not restricted to weekdays? There certainly are a variety of such activities. The curing of concrete will take place over the weekends whether or not any workers are on site. Even the material requisitions will include materials that will be en route over weekends. How can these differences be handled? While some computer programs address these differences, using a single unit may avoid confusion. Note that some computer programs offer considerable sophistication in defining durations. For example, some programs accept such information as electricians working 40 hours one week and working 32 hours every other week (four days in alternate weeks). Some crafts might recognize holidays that are not recognized by others. Some crews may be scheduled to work six-day workweeks while others work the five traditional five-day workweeks. These are all nuances that can be addressed by several of the current scheduling software programs.

Why are some schedules planned in working days and others in calendar days? The question is not trivial. In general, if a project is vulnerable to the weather or if weather can dramatically impact work progress, scheduling with working days is most common. Projects designed by engineers (e.g., dams, highways, bridges, canals, airports, and utility line installations) tend to be those in which site work is a significant component of the project and susceptible to adverse weather. Projects that are not as susceptible to adverse weather are generally scheduled in terms of calendar days. These projects tend to be those designed by architects (e.g., schools, office buildings, retail stores, warehouses, and high-rise buildings). To a considerable extent, the use of working days or calendar days may be guided by the contract. Public owners tend to use working days to define project duration while private owners tend to use calendar days. Thus, calendar days are the most common.

Project durations may be defined by a specific completion date in which the substantial completion of the project is linked to a specific calendar date or milestone date. Such dates are preferred by owners who have specific dates on which their projects are to be functional. For example, a retail store, in an attempt to capture significant revenues from Christmas shoppers, should open prior to Thanksgiving. An apartment building near a university will ideally be completed prior to the beginning of the fall term. A school building should be completed prior to the start of classes in the fall. Completion dates on some buildings are dictated by the ending date on the lease of a currently occupied building. When completion dates are established in the contract, owners tend to be reluctant to grant time extensions for delays. The stipulation of a specific completion date for a project would appear to fix the project duration. While project completion dates do establish when a project will be completed, they do not generally establish exact starting dates. The contractor must recognize that project start is linked to the notice to proceed but that is not a date that is known on bid day; the contract must be awarded and then signed by both parties before the notice to proceed is issued.

Many projects requiring calendar days can be converted to working days. For example, if concrete is to cure for 28 days, this is obviously stated in calendar days. This could be converted to working days with little difficulty. The curing of the concrete could be stated as requiring 20 working days (5/7 of 28 days). If the lead time on a material delivery is stated as being 40 calendar days, it could similarly be shown on the schedule as requiring 29 working days (5/7 of 40). These conversions are not difficult to make and they do not compromise schedule accuracy. However, this conversion may present problems if the project work schedule subsequently is changed to a six-day workweek or if the project work schedule changes to four-10s (four 10-hour days per week).

Working-day schedules are perhaps the more complex ones with which to plan activities. If the work progresses as scheduled, there is essentially no difference between the use of working days and calendar days. If work does not progress as scheduled, the working-day schedules may present additional challenges to the contractor. Assume that a project scheduled in working days is about 50% complete but is slightly behind schedule. In an attempt to get the project back on track, the contractor considers working on Saturdays. To the contractor's dismay, the owner on this project will count any day that is worked as a working day, including weekends and holidays. The owner justifies this practice on the grounds that the owner will have to staff the project on all days that are worked. Thus, the contractor cannot get caught up on such a project by working weekends. Under such circumstances, the contractor may instead be well advised to work overtime. The contractor must evaluate the owner's practices prior to making key changes in the work schedule. The following provision, found in a federal construction contract, defines the workday and further places restrictions on overtime work.

Unless otherwise specified, the Contractor will be permitted to do the work between the hours of 7:45 A.M. and 4:30 P.M., Monday through Friday. Federal holidays (New Year's Day, Martin Luther King, Jr.'s Birthday, Presidents' Day, Memorial Day, Independence Day, Labor Day, Columbus Day, Veterans Day, Thanksgiving Day, Christmas Day) that fall within the workweek will not be considered workdays. Prior to the Contractor performing any work during hours other than those specified, the Contractor shall submit an overtime request to the Owner's Representative for review and approval. Overtime requests shall be submitted no less than 24 hours prior to the time the Contractor desires to work.

As mentioned, working days are commonly used on projects that are particularly susceptible to adverse weather. While building-type projects are susceptible to adverse weather, this impact is generally restricted to the early phases of the construction work. When working days are used to define project duration, the contractual duration of the project generally includes an allowance for delays resulting from normally anticipated weather. Thus, time extensions are not generally granted if the delays are caused by normally anticipated weather. A problem often faced by contractors is that many contracts do not define "normally anticipated weather." In those instances in which normally anticipated weather is defined, reference is often made to climatological data that is collected nearby. If no effort is made to contractually define the weather, the contractor may assume considerable risk. For example, the local climatological data may show that for a given metropolitan area there are, on average, eight rain days in March. The inference is that nine rain days during March would entitle the contractor to a one-day time extension, and ten rain days during the month would entitle the contractor to a two-day time extension. While this may appear equitable, the contractor may benefit substantially if all of the rain days occurred on the weekends. Similarly, the contractor is adversely impacted if all of the rain days occur on weekdays. The following provision is an example of one in which the weather data source is specifically mentioned.

U.S. Weather Bureau meteorological data for the general area are included as an attachment to this contract. Complete weather records and reports may be obtained from the local U.S. Weather Bureau Office nearest to the work site. The Contractor shall evaluate unique hazards that are likely to arise from weather conditions during the construction period.

At first, it might appear that a working day is any weekday other than holidays. This definition is too simple for most contracts. Remember, the use of working days to define project duration essentially assumes that weather may adversely impact the project duration. Under these circumstances, the definition of *working day* may need to be further defined. For example, working days may be more broadly defined to exclude weekends, holidays, and those days on which no work can be performed. What constitutes a day on which no work can be performed? This, too, should be clearly addressed prior to entering the construction contract. The practices vary considerably on this definition, but the spirit of the definition tends to be consistent. Some owners will define a nonworking day as one in which weather conditions reduce the work efficiency by more than 50% on critical work tasks. Others define the workday in terms of the percent of the workforce that cannot be put to work or is absent as a result of the weather. Still others define the workday in terms of the percent of the day that can actually be worked. The actual percentages used may vary by owner on these measures, but the intent is to objectively define each working day.

It is obviously important for the contractor to be clear on how to address weather delays. This, too, is often addressed in the contract. Practices vary considerably on this issue. Some contracts will stipulate that adverse impacts of the weather be reported at each occurrence, weekly, monthly, or at the end of the contract. With such diverse requirements, the contractor must be careful to examine the requirements in detail to ensure that legitimate time extensions are not denied.

LIQUIDATED DAMAGES—DAMAGES FOR LATE COMPLETION

The use of liquidated damages provisions in construction contracts is common. There is little difference between the liquidated damages provisions encountered in the industry. The following provision is typical:

Damages for avoidable delay as set forth in this contract shall be in the amount of $_____ per day.

Some exceptions do exist. The following paragraph not only does not include a liquidated damages provision, but it specifically states that such liquidated damages are not prescribed. Instead, the contractor will be liable for the actual damages incurred by the owner. This type of provision will probably result in more litigation as the contractor may take issue with the manner in which the owner decides to compute the actual damages.

Example D

Time is expressly declared to be of the essence in completion of the work covered by these specifications, and the Contractor shall be liable for actual damages for delay in completion of work, but liquidated damages as such for delay in completion of work are not prescribed.

Example E

For each and every day that any portion of the work remains unfinished after the time fixed for completion in the contract documents as modified by any extension of time, damage will be sustained by the Owner. Because of the difficulty in computing the actual material loss and disadvantages to the Owner, it is determined in advance and agreed by the parties thereto that the Contractor will pay the Owner the amount of damages set forth as liquidated damages, an amount representing a reasonable forecast of the actual damages that the Owner will suffer by the failure of the Contractor to complete the work within the stipulated time. The execution of the agreement shall constitute acknowledgment and agreement by the Contractor that the Owner will actually suffer damages in the amount herein fixed for each and every day during which the completion of the work is avoidably delayed beyond the stipulated completion date.

The following liquidated damages provision is similar to the above; however, this provision specifically states that the liquidated damages are not to be regarded as a penalty. Instead, the amount reflects a best estimate of the actual losses expected for late completion of the project. Additionally, provided the contractor has made a concerted effort to complete the project within the contract time, a limit is placed on the total amount of liquidated damages to be charged against the contractor for late completion. Such limitations are not common.

For each and every day that any portion of the project remains unfinished after expiration of the contract time, the Contractor shall pay the Owner, not as a penalty but as liquidated damages, the amount of $_____ per day or part of a day. Because of the difficulty in computing the actual damages that will result, the amount of liquidated damages as set forth is hereby estimated, agreed upon, and determined in advance by the parties hereto as a reasonable forecast of the actual damages that the Owner will suffer by the failure of the Contractor to complete the work within the agreed contract time. The Owner may retain from any monies due the Contractor after the time fixed in the contract documents for completion of the work such amount as may be necessary to pay said liquidated damages. Should such amount due to the Contractor not be sufficient to pay such damages, the contractor shall immediately pay the deficiency to the Owner. By executing the contract, the Contractor hereby acknowledges a full understanding of the liquidated damages provision and has estimated and ascertained and agrees that the Owner will actually suffer damages in the amount herein fixed for each and every day that the project completion is delayed beyond the contract time.

The Owner shall not assess liquidated damages under this contract in an amount in excess of 20 percent of the contract price unless the Contractor's failure to complete the project within the contract time is due to willful and intentional neglect.

WEATHER

Construction work is often sensitive to weather conditions. This is particularly true of earthwork projects, civil works, or portions of building projects in which site work (clearing, paving, excavation, etc.) is performed. The following provision shows that the owner not only acknowledges this but that the contractor is not to compromise performance as a consequence of adverse weather.

During unfavorable weather and other conditions, the Contractor shall pursue only such portions of the work as shall not be damaged thereby. No portions of the work whose satisfactory quality or efficiency will be affected by any unfavorable conditions shall be constructed while those conditions occur, unless by special means or precautions acceptable to the Owner's Representative, the Contractor shall be able to overcome them.

USE OF COMPLETED PORTIONS OF THE WORK

Some projects are undertaken in operating facilities in which the owner or the owner's employees will have a continuing presence. In other projects the owner may wish to begin using portions of a project that have been completed. If such use is anticipated, a specific contractual provision to that effect is required. The contractor must then anticipate the impact that such use or occupancy by the owner's personnel will have on the construction operations. The following examples are provisions that must be assessed carefully by the contractor:

Example F

The Owner shall have the right to take possession of and use any completed or partially completed portions of the work, even if the time for completing the work or such portion may not have expired, but such taking possession and use shall not be deemed an acceptance of any part of the work. If such prior possession or use increases the cost of the work, the Contractor shall be entitled to claim for extra compensation within five days of each occurrence. The amount of extra compensation shall be in accordance with the terms of this contract. The Contractor shall not, however, be entitled to claim extra compensation for portions of the work that are specifically required by the contract to be placed into use or operation before completion of all work under this contract.

Example G

Contractor agrees that Owner, upon advance notification to Contractor, in writing, will be permitted to occupy and use any completed or partially completed portions of the project notwithstanding time for completion of the entire work. If such prior occupancy increases the cost of the work or delays its completion, provided that the same occur prior to the completion date fixed by the "Notice to Proceed," the Contractor shall be entitled to extra compensation or extension of time, or both. However, such claims shall be made in writing prior to such partial occupancy and must be substantiated with supporting data satisfactory to the Owner or Owner's Representative, and no such claim shall be honored in the event the time for completion has expired prior to such partial occupancy.

Note that both of the above provisions stipulate that the contractor shall be entitled to added compensation if warranted. Sufficient caveats exist in each provision so that the contractor cannot dismiss these provisions. In Example F, the contractor must assess the possible cost impact of an owner's occupancy of a portion of a project within five days of such occupancy. In Example G, the contractor must make a claim that is considered satisfactory to the owner and it must be made prior to occupancy or use by the owner. Fairness from the owner may be questioned by a contractor in a claim situation.

SUBSTANTIAL COMPLETION

The project duration is typically marked or ended with the substantial completion of a project. The project is considered substantially complete when it reaches the point that the owner can begin to occupy and utilize the facility for its intended purpose. At that point the only contractor tasks that remain on the project itself (disregarding demobilization, etc.) consist of addressing the punch list items. These items are minor aspects of the project that require the contractor's attention. On building projects, these typically consist of cosmetic corrections. Examples of punch list items include paint on the carpet, missing cover plates on an electrical outlet, smudges on the wall, caulking on a window pane, or a nick in the trim. The following is a general provision regarding substantial completion:

A project shall be deemed to have reached substantial completion on that date, as certified by the Owner's Representative, when the construction of the project or a specified part thereof is sufficiently completed, in accordance with the contract documents, so that the project or specified part can be utilized for the purposes for which it is intended.

NOTICE OF DELAYS

With time being of the essence in most construction contracts and with the project duration clearly defined, it is also important to define those conditions under which the duration might be extended. This is relevant particularly in the area of delays. The following outlines procedures to be followed when delays occur:

Whenever the Contractor foresees any delay in the prosecution of the work, and in any event immediately upon the occurrence of any delay that the Contractor regards as unavoidable, the Contractor shall notify the Owner's Representative in writing of the probability of the occurrence of such delay and its cause in order that the Owner's Representative may take immediate steps to prevent, if possible, the occurrence or continuance of the delay or, if this cannot be done, may determine whether the delay is to be considered avoidable or unavoidable, how long it continues, and to what extent the prosecution and completion of the work are to be delayed thereby. It will be assumed that any and all delays that have occurred in the prosecution and completion of the work have been avoidable delays, except such delays as shall have been called to the attention of the Owner's Representative at the time of their occurrence and found to have been unavoidable. The Contractor shall make no claims that any delay not called to the attention of the Owner's Representative at the time of its occurrence has been an unavoidable delay.

AVOIDABLE DELAYS

While most owners grant time extensions for legitimate delays in the work, some contracts specifically outline delays that are not to be considered grounds for time extensions. The following stipulates examples of avoidable delays for which time extensions are not to be granted:

Avoidable delays in the prosecution or completion of the work shall include all delays which, in the opinion of the Owner's Representative, could have been avoided by: the exercise of care, prudence, foresight, and diligence on the part of the Contractor or the Subcontractors. Delays in the prosecution of parts of the work that may in themselves be unavoidable but do not necessarily prevent or delay the prosecution of other parts of the work or the completion of the whole work within the time herein specified; reasonable loss of time resulting from the necessity of submitting samples of materials and drawings to the Owner's Representative and from making of tests of materials, measurements, and inspections; loss of time or interference with construction schedules due to rejection of unacceptable materials or products; and reasonable interference of other contractors employed by the Owner that do not necessarily prevent the completion of the whole work within the time agreed upon shall constitute avoidable delays within the meaning of this contract. Delays in delivery of equipment or material purchased by the Contractor or any subcontractors or suppliers shall be considered avoidable delays. The Contractor shall be fully responsible for the timely ordering, scheduling, expediting, delivering, and installing of all equipment and materials.

UNAVOIDABLE DELAYS

Time extensions are typically granted for unavoidable delays. Such extensions are not always automatically granted. The contractor must carefully document information related to delays and request extensions in accordance with the contract provisions. Following is one such provision that specifically addresses the issue of weather delays not constituting grounds for time extensions:

Unavoidable delays in the prosecution or completion of the work shall include all delays that in the opinion of the Owner's Representative, result from causes beyond the control of the Contractor or the subcontractors and suppliers and that could not have been avoided by the exercise of care, prudence, foresight, and diligence on the part of the Contractor or the subcontractors and suppliers. Delay in completion due to contract modifications ordered by the Owner's Representative and unforeseeable delays in the completion of work of other contractors employed by the Owner will be considered unavoidable delays insofar as they interfere with the Contractor's completion of the work. Delays due to adverse weather conditions, except for Acts of God, that prevent the Contractor from proceeding with the controlling item on the accepted critical path schedule will not be regarded as unavoidable delays. The contract time specified allows for delays due to normal adverse weather conditions.

EXTENSION OF TIME (AVOIDABLE DELAYS)

While time extensions are typically restricted to unavoidable delays, the following makes allowance for a possible exception. Note that while the time extension may be granted, the contractor may still be asked to pay for any added costs that the owner may incur as a result of the time extension.

In case the work is not completed in the time specified, including such extensions of time as may have been granted for unavoidable delays, the Contractor will be assessed damages for delay in accordance with the liquidated damages provisions of this contract. The Owner, however, shall have the right to grant an extension of time for avoidable delays if it is deemed in the Owner's best interest to do so. During such extension of time, the Contractor may be charged for engineering and inspection services and other costs, but will not be assessed liquidated damages.

EXTENSION OF TIME (UNAVOIDABLE DELAYS)

The occurrence of unavoidable delays will typically have a less adverse impact on the contractor as time extensions are often automatic when certain procedures are followed. Note that of the following provisions, in Example H the delay will result in a time extension only if "the unavoidable delay involves the controlling item on the critical path."

From this it may be inferred that the owner will essentially take possession of any float that activities may have. If activities have float, they are not generally regarded as "controlling activities." These provisions must be carefully evaluated to determine the extent that risk is placed on the contractor.

Example H

For delays that the Owner's Representative considers to be unavoidable, the Contractor shall, pursuant to a written request, be allowed an extension of time beyond the time herein set forth, proportional to such delay or delays, in which to complete the contract. During such extension of time, neither extra compensation for engineering and inspection shall be charged to the Contractor. Time extensions will be granted only if the unavoidable delay involves the controlling item on the critical path. Time extensions granted as a result of a change order shall not be duplicated or otherwise deemed to cause further extensions within the terms of this provision.

Example I

All time limits stated in the contract documents are of the essence of the contract. However, if the Contractor is delayed at any time in the progress of the work by any act or neglect of the Owner or the Owner's Representative or by changes ordered in the work, or by strikes, lockouts, fire, or other causes, not limited to the kinds mentioned, beyond the Contractor's reasonable control, or by delay authorized by the Owner, then the time of completion may be extended for a reasonable time and to such date as certified by the Owner's Representative; provided further that the Contractor shall, within seven days from beginning of any such delay, notify the Owner's Representative in writing of the cause of the delay, and the Owner's Representative shall ascertain the facts and extent of the delay and notify the Contractor within a reasonable time of the decision in the matter.

Example J

If the Contractor is delayed at any time in the progress of the work by an act or neglect of the Owner or the Owner's Representative, or by any separate contractor employed by the Owner, or by changes ordered in the work, or by strike, lockouts, fire, unusual delay in transportation, unavoidable causalities, or any causes beyond the Contractor's control, or by delay authorized by the Owner's Representative, or by any cause that the Owner's Representative shall decide to justify the delay, the time of completion shall be extended for such reasonable time as the Owner's Representative may decide.

Some documents go to considerable lengths to spell out exactly which causes of delays constitute grounds for extension of the contract time. Many causes are for time extensions only and the delay in the project does not give any grounds for the contractor to claim for damages or for additional costs, expenses, overhead, profit, or other compensation. Examples of such causes for time extensions are as follows:

- Floods, fire, strikes, lockouts, labor disputes, pickets, war, acts of the public enemy, acts of God

- Change orders
- Acts of performance or delays in performance caused by persons other than the Contractor and other than persons acting for and on behalf of the Contractor
- Causes beyond the control of the Contractor, the delays from which could not have been avoided through the exercise of reasonable care, prudence, foresight, and diligence on the Contractor's part and that of the subcontractors

SUBMITTALS

Submittals constitute a major component of the procurement of materials and equipment incorporated in most construction projects, although submittals may be required on items other than materials and equipment. Without proper approval of submittals on materials, the contractor is well advised to defer the delivery of such materials. This may cause a delay in the delivery of materials, which must be anticipated well in advance. Consequently, contractors should prepare documents for submittal early in the project. Following are examples of provisions that address submittals:

Example K

The Contractor shall furnish all drawings, specifications, descriptive data, certificates, samples, tests, methods, schedules, and manufacturer's instructions as specifically required in the specifications, and all other information as may be reasonably required to demonstrate fully that the materials and equipment to be furnished and the methods of work comply with the provisions and intent of the specifications and drawings... The Contractor shall ensure coordination of submittals among all related crafts. Information shall be submitted in time to allow one month to review and return to the Contractor without interfering with the accepted construction schedule... If the review and checking indicates limited corrections are required, copies will be returned marked "MAKE CORRECTIONS NOTED" and the Contractor may begin immediately to incorporate the material and equipment covered by the corrected submittal into the work... If the review and checking indicates insufficient or incorrect data have been submitted, copies will be returned marked "AMEND AND RESUBMIT." No work may begin on incorporating the material and equipment covered by this submittal into the work until the submittal is revised, resubmitted, and returned marked either "NO EXCEPTIONS TAKEN" or "MAKE CORRECTIONS NOTED."... Similarly, no work may begin on submittals marked "REJECTED" and such work cannot begin on incorporating the material and equipment covered by this submittal into the work until the submittal is revised, resubmitted, and returned marked either "NO EXCEPTIONS TAKEN" or "MAKE CORRECTIONS NOTED."

Example L

Contractor shall submit samples of materials at least 20 days before materials are required to be ordered for scheduled delivery to site. Materials delivered prior to receipt of Owner's Representative's approval are subject to rejection and immediate removal from site. Contractor shall schedule submittal of shop drawings to Owner's Representative so that no delays will result in delivery of materials and equipment, advising Owner's Representative of priority for checking drawings, but a minimum of two weeks shall be provided for this purpose. All shop drawings must be submitted prior to the receipt of the third partial payment request. After the second payment has been made, no further payments will be made until all shop drawings have been submitted.

PROGRESS PAYMENTS

It is common for payments in the construction industry to be made on a periodic basis, most commonly, on a monthly basis. This payment arrangement alleviates financial hardship that the contractor would assume if payment was not made until project completion. These payment conditions must be carefully established in order to avoid any misunderstandings. The following is a typical payment provision:

The Contractor shall, within ten days of receipt of notice to proceed, submit a complete breakdown of the contract price showing the value assigned to each part of the work including an allowance for profit and overhead. Upon acceptance of the breakdown of the contract price by the Owner's Representative, it shall be used as one of the bases for all requests for payment. The Contractor shall also submit a schedule of payments that are anticipated to be earned during the course of the work.

While front-end loading of billings may be expected and even permitted by some owners, others prefer that the contractor request payment for work that bears a reasonable comparison to the true value of work put in place. While the contractor may still wish to front-end load on billings to some degree, provisions such as the following send the contractor a clear notice to avoid this:

An unbalanced breakdown estimate, providing for overpayment of the Contractor on items of work that would be performed first under the lump sum item of contract, will not be accepted and shall be revised and resubmitted until acceptable to the Owner's Representative.
  The Owner's Representative will, by the 15th day of the month, prepare a request for payment for materials furnished and work completed during the previous payment period. The first request for payment will be the value of the work done, purchase value of delivered equipment and material, and the value of work done on undelivered manufactured items and equipment since the Contractor shall have begun the performance of the contract. Every subsequent request for payment, except the final one, shall be the value of the work done, purchase value of delivered equipment and material, and the value of work done on undelivered manufactured items and equipment since the last request was made.
  Payment for mobilization, bond, and insurance costs will be made when the monthly partial payment estimate of the amount earned, not including the amount earned for mobilization, bonds, and insurance, is five percent or more of the original contract amount. Progress payment for mobilization, bond, and insurance costs will be limited to five percent of the original contract amount. If the amount bid for mobilization,

bonds, and insurance costs exceeds five percent of the original contract amount, the remaining amount will be paid at the time the final payment is made.

Delivered equipment and material shall be suitable for permanent incorporation in the work and suitably and safely stored at the site of the work or in a bonded warehouse in the vicinity of the work. Payment requests for delivered equipment and material must be accompanied with certified invoices from the supplier when requested by the Owner's Representative. Payment for delivered equipment and material shall not exceed 50 percent of the purchase value until the Contractor has delivered to the Owner's Representative five sets of acceptable manufacturer's operating and maintenance instructions covering that item of equipment or equipment assembly provided in this contract. The operating and maintenance instructions, also referred to as operating and maintenance manuals, shall include the following:

1. *An itemized list of all data provided.*
2. *Name and location of the manufacturer, the manufacturer's local representative, the nearest supplier, and spare parts warehouse.*
3. *Accepted submittal information applicable to operation and maintenance.*
4. *Recommendation for installation, adjustment, start-up, calibration, and troubleshooting procedures.*
5. *Recommendation for lubrication and an estimate of yearly quantity needed.*
6. *Recommendation for step-by-step procedures for all modes of operation.*
7. *Complete internal and connection wiring diagrams.*
8. *Recommendation for preventive maintenance procedures and schedule.*
9. *Complete parts lists, by generic title and identification number, with exploded views of each assembly.*
10. *Recommended spare parts.*
11. *Disassembly, overhaul, and reassembly instructions.*
12. *Completion, as applicable, of operating and maintenance instruction transmittal forms and summary sheets required by this contract.*

No later than the first day of each calendar month, Contractor shall submit a request of partial payment to the Owner by mailing or delivering same to the Owner's Representative.

All statements shall be subject to approval of the Owner's Representative, after which the Owner will prepare its standard payment form and submit same to the Contractor for signature. Payment of voucher will be made by the Owner within 15 days after voucher has been returned properly executed.

In making such partial (monthly) payments, there shall be retained ten percent of the estimated amounts until final completion and acceptance of all work covered by the contract. The Owner's Representative, with the Owner's approval, may at any time after 50 percent of the work has been completed, if satisfactory progress is being made and with written consent of surety, recommend that any of the remaining partial payments be paid in full. Any amount previously retained shall be held by the Owner until completion of the contract.

The payment provisions must be carefully evaluated. The contractor will want assurances that monthly payments will be made. In addition, there may be considerable interest in the payment policy for materials delivered but not installed and in the payment policy concerning mobilization, insurance, and bonds. The retainage to be withheld will also be of considerable interest to the contractor.

PAYMENT FOR MATERIALS

While the periodic payment provision may address the payment of materials delivered but not installed, some documents may address this as a separate issue. Following is an example of such a provision:

Ordinarily no allowance will be made in estimates for materials delivered on site of work and not incorporated in the work; however, items considered by Owner's Representative to be major items of considerable magnitude, if suitably stored in a bonded warehouse or on the site, will be allowed in estimates on the basis of 90 percent of invoices.

FINAL PAYMENT

The final payment is typically considered the payment consisting of the release of all of the withheld retainage. The last periodic payment will probably precede the final payment by a considerable duration, particularly for some owners. If the timing of the final payment is clearly outlined, the contractor can more accurately predict the cash-flow requirements of a project. Following is a provision that stipulates a reasonably clear plan for the timing of the final payment:

The Owner will make final payment to the Contractor in the manner provided by law following the expiration of 35 days after the acceptance of the work and filing of the notice of completion by the Owner. Such final payment shall include the entire sum so found to be due hereunder, after deducting therefrom all previous payments and such other lawful amounts as the terms of the contract prescribe.

SUSPENSION

For a variety of reasons, the owner may find it necessary to halt work on a construction project. Since the actual reasons may not be anticipated when the contract documents are drafted, the owner typically will want to include a broad provision that preserves the right of authorizing a suspension in the work. The following are examples of such provisions:

Example M

The Owner's Representative may at any time suspend work, or any part thereof, by giving notice to the Contractor in writing. The work shall be resumed by the Contractor within ten days after receiving written notice from the Owner's Representative to do so.

If all of the work under the Contract is suspended and if the Owner's Representative does not give notice in writing to the Contractor to resume work at a date within 90 days of the date of the written notice to suspend, then the contract shall be assumed to have been terminated and the Contractor shall be entitled to appropriate compensation. Reasonable time allowances required by the Owner to schedule equipment or process shutdowns or tie-ins, as specified in the Technical Specifications, will not be considered suspension of the work.

Example N

When conditions at the site of the proposed work are considered by the Owner to be unsatisfactory for prosecution of the work, the Contractor may be ordered, in writing, to suspend the work or any part thereof until reasonable conditions exist. When such suspension is not due in the opinion of the Owner to fault or negligence of the Contractor, time allowed for completion of such suspended work will be extended by a period of time equal to that lost due to the delay occasioned by the ordered suspension.

Example O

The Owner's Representative may order parts of the work suspended should the weather or season be such that any part of the work cannot be done properly and with due regard to durability, finish, or appearance. The Contractor may be required to protect the several parts of the exposed work from damage by the elements or other causes.

Note that the suspension provisions may only grant equitable time extensions to the contractor or they may also stipulate compensation to the contractor. These are not to be regarded as subtle differences in the provisions.

TERMINATION BY CONTRACTOR

Project termination is a harsh reality that a contractor occasionally faces. While this may seem a remote possibility when a project is being bid, the contractor should at least review the provisions related to termination. In particular, the contractor may wish to evaluate the conditions under which he or she can terminate a contract. Following is an example of a termination-by-contractor provision:

If work shall be stopped under an order of any court or other public authority for a period of three months through no act or fault of the Contractor or of anyone employed by the Contractor, the Contractor may on seven days' written notice to the Owner and the Owner's Representative stop work or terminate this contract and recover from the Owner payment for all work executed, any losses sustained on any plant or material, and a reasonable profit. If the Owner's Representative shall fail to issue any certificate for payment within ten days after it is due, or if the Owner shall fail to pay the Contractor within 15 days after its maturity, then the Contractor may on seven days' written notice to the Owner stop work and give written notice of the intention to terminate this contract. If the

Owner shall fail to pay the Contractor within seven days after receipt of such notice, then the contractor may terminate the contract and recover from the Owner payment for all work executed, any losses sustained upon any plant or materials, and a reasonable profit.

FINAL COMMENTS

The construction contract dictates much about the way a project is to be scheduled. The contract might contain a variety of requirements that are directly associated with the preparation of the project schedule. Additional provisions generally outline the manner in which various scheduling matters will be handled. The judicious contractor evaluates the various provisions in terms of the degree of risk that is associated with each. Many contractual provisions directly address the timing of the events that commonly take place on construction projects. These must all be given careful consideration prior to entering into the contract agreement.

REVIEW QUESTIONS

1. What is the primary concern of contractors when contract provisions address the issue of the ownership of float?
2. What defines the start date and what defines the finish date of a contract, with the bracketed time being the project duration?
3. How can it be argued that the inclusion of a liquidated damages provision for late completion of a project simplifies matters for the contracting parties?
4. To the contractor, what is the significance of the difference between an avoidable delay and an unavoidable delay?
5. What types of projects tend to be scheduled in terms of working days? Why?
6. In terms of counting days, which type of measure presents the greatest potential for conflict, working days or calendar days?
7. What kinds of controls have owners included in their construction contracts to ensure the receipt of realistic schedules of values or schedules of payments from their contractors?
8. What distinguishes the last periodic payment from the final payment made to the contractor on most construction projects?
9. What are the primary grounds on which the contractor can typically terminate the contract?
10. How are owners often assured of receiving realistic schedules from the contractors?

6

Resource Allocation and Resource Leveling

Time is nature's way of keeping everything from happening all at once.

The realization that resources are being managed through the use of a network has been an underlying theme of the preceding chapters. Although time as a primary "resource" has been the focus thus far, the discussion will now focus specifically on the analysis and manipulation of more tangible resources, such as equipment and labor, particularly in situations when limitations are imposed.

When a network is developed for a project, the logic of the network presumably reflects the sequences of the activities as they must take place, without regard to limitations imposed by the availability of resources. In practice this is not typically the case. Invariably, the logic initially incorporated in a network already reflects the fact that some resources must be efficiently utilized. For example, on a multistory building the network may show that the insulation installation on the second floor cannot take place until the insulation has been installed on the first. From the perspective of pure logic, the two activities are independent, yet few would question the sequencing arrangement. The logic reflects the fact that only one insulation crew will be used on the building, thereby necessitating the sequential arrangement of the activities by floor. As discussed earlier, the inclusion of many such constraints in the network may unnecessarily extend the duration and cost of the project.

THE MANAGEMENT OF RESOURCES

With no regard for limitations imposed by resources, most activities are intuitively scheduled on an early start schedule. That is, the float of all noncritical activities is preserved by scheduling them on their earliest start dates. However, limited resources do exist and they must be managed. The types of resources that are subject to consideration

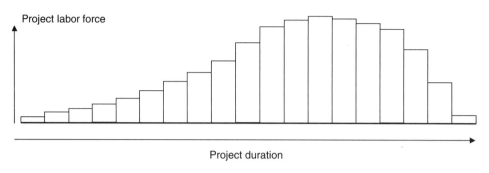

Figure 6.1 Typical Labor Utilization During a Project

are those that impose the greatest limitations on the schedule. These limitations arise because the resources may be limited in availability. In many cases additional resources may be available; however, these added resources are available only at such a premium that it is only practical to treat them as a limited resource. For example, on a project to which a 200-ton crane is assigned, if the activities are sequenced in such a way that a second crane would be required for a few days, it would appear impractical to do so. Consequently, the project schedule would be subject to modification so that a single crane could perform all activities.

Another resource that is commonly managed is personnel assigned to a project. The productivity of crews on a project—whether they consist of carpenters, pipe fitters, or ironworkers—will generally be highest if a constant crew size is maintained for the project duration. Additional crews hired for a short duration require additional orientation and their productivity will not be as high as that of a seasoned crew.

An example of the utilization of labor on a project is depicted in Figure 6.1. From this histogram, no apparent problem is noted. This is because of the many different trades and specialty firms involved in the project.

Material is another resource that must be managed. In some instances this resource is truly limited. For example, a schedule may stipulate that 2,000 cubic yards of concrete are to be placed per day. If the batch plant on site can deliver only 1,500 cubic yards per day, the concrete that is available each day is limited. During the "oil embargo" of the mid-1970s, fuel was a limited resource. All of the equipment owned by large companies could not be operated on a full-time basis. Thus, it was imperative for some firms to carefully manage their equipment operations.

WHEN RESOURCES ARE LIMITED (RESOURCE ALLOCATION)

Where resource management is concerned, there are two broad scenarios. One method, termed here *resource allocation*, concerns the allocation of resources that are limited, while the other method, *resource leveling*, concerns the efficient use of the required resources when the project duration cannot be altered. This text examines two methods of resource allocation: the series method (also the Brooks method) and the parallel method. (Other variations of these methods do exist, depending on the assumptions or additional constraints to be imposed on the resource allocation.)

Resource allocation, as described here, uses the precedence diagram as the basis for the solution. The next step is to clearly define the limitations of the availability of a specified resource or resources. Any number of limited resources may be defined; however, the manual solution of a network becomes quite cumbersome with more than two resources. The assumption of limited resources may extend the project duration beyond the completion time that would be predicted if no resource constraints were imposed. If the duration is extended excessively, management decisions may be required to resolve the conflict between the limitation of resources and the project duration.

Once all resource constraints are known, the series method resource allocation solution can be developed. Under this method an activity is scheduled to start as soon as its predecessors have been completed, provided that the utilization of available resources is not exceeded. Thus, if an activity does not utilize a limited resource, the activity is scheduled to start as soon as its predecessors have been completed. Once an activity has been started, it is not interrupted. If two different activities requiring the use of the same limited resource can be scheduled (assuming the resources needed would exceed the available resources if both activities are scheduled concurrently), the activity with the *earliest late start date* is given priority. If the two activities are tied for the earliest late start date, priority is given to the activity with the least *total float*. If a tie still exists, preference is given to the activity with the *largest number of resources*. Then, if a tie still exists, preference is given to the *input order*. Other rules might be defined, but the above order of priorities generally gives an acceptable solution.

The series method relies on the assumption that once an activity has been started, it cannot be interrupted. This is a practical assumption in many cases, simply because productivity may be unduly compromised by a break in an activity.

The parallel method is similar to the series method with one basic difference: the parallel method permits activities to be interrupted. The rules for scheduling activities with limited resources might be summarized as follows:

- Schedule activities to start as soon as their predecessors have been completed.
- If more than one activity using a specific limited resource can be scheduled, priority is given to the activity with the *earliest late start.*
- If the activities are tied for the early late start date, give priority to the activity with *least total float.*
- If the activities are tied for total float, give priority to the activity with the *largest number of resources.*
- If the activities are tied in the number of resources, give priority to the *activity that has already started* (top priority in the series method).
- If no activity has been selected with the above rules, simply start the activity that occurs first in the *input order.* Any other rule could be used here. After the other rules have resulted in ties, there is probably no great consequence associated with selecting one activity over another. Input order is simply one way of making a

choice. Additional rules could relate to the equipment being used, the numeric value assigned to the activity, or some other criteria.

The Manual Solution for Resource Allocation

With the widespread use of microcomputers on construction projects, it is reasonable to assume that most resource allocation on construction projects is done with the use of computers. An examination of the manual approach for resource allocation will give a clearer impression of how this is done. To demonstrate the manual solution, a simple network will be used. Although this will appear to be quite a manageable task, projects entailing many activities and many resources pose a considerable challenge in a manual solution.

The sample illustration uses a simple network with two different types of limited resources. The first step is to do a manual "forward pass" and then to do a "backward pass." The primary purpose is to determine the late start dates for each activity. The network is still to be used in the solution, but most of the solution itself will be done on a tabular form.

On the tabular form, first list all of the activities, along with each activity's required resources, duration, late start, and total float (early start may also be shown). Use a "status" column to indicate which activities are eligible to be scheduled and those that have been fully scheduled. The resource allocation can then begin. The solution will appear as a bar chart, with the resources being tallied at the bottom of the bar chart for each day.

After examining the network, indicate in the status column which activity or activities can be started. This can be indicated by simply drawing a diagonal line in the status column next to each activity that can be scheduled. The next step is to actually schedule those activities. Naturally, any activity that does not utilize any of the limited resources will be scheduled as soon as it becomes eligible. The activities utilizing the limited resources will require further examination, particularly if the available resources are exceeded when all eligible activities are scheduled simultaneously. Obviously, not all activities can be scheduled if the available resources are exceeded. Thus, a rational and orderly approach must be used to determine which activities are to be scheduled. The decision for determining which activities are to be scheduled should be based on the rules or priorities given earlier.

With the parallel method, it is essential to review each activity that is eligible for each day in the project. With the series method, a similar approach is used, but once an activity is selected for scheduling, it must not be interrupted. A sample network illustrating the series solution method is shown in Figure 6.2. Computations will show the steps to follow as the solution evolves.

The information needed to solve the series solutions is shown in Figure 6.2. The tabular information obtained from the forward and backward passes is shown in Figure 6.3.

Solve the schedule by the series method of resource allocation. The resource limits not to be exceeded are as follows: 5 M (masons) 2H (helpers)

Resource limits are not to be exceeded under any circumstances.

In the first step, only Activity A is eligible for scheduling. This is indicated by the slash placed next to Activity A in the eligibility column headed by a "/". Activity A is then

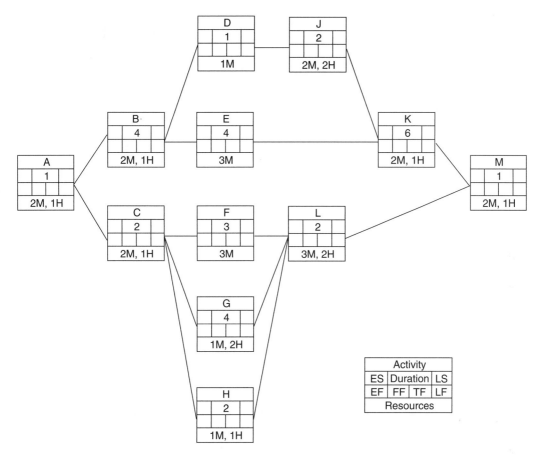

Figure 6.2 Network Used in Series Solution Example

scheduled on Day 1 on the chart. Since Activity A requires two masons and one helper per day and since the duration is one day, the resources are applied to the first day. The status column in Figure 6.4 is shaded to indicate that Activity A has been scheduled.

At the completion of Activity A, Activities B and C are eligible for scheduling. Since the total number of available resources of five masons and two helpers is not exceeded, both activities are scheduled on Days 2 and 3 (see Figure 6.5). After Day 3, Activity C will be completed while Activity B still has two remaining days. The eligibility column of Activity C is shaded.

To schedule Days 4 and 5, Activity B has not been completed so it will be scheduled as the highest priority. When activity C was completed in the last step, Activities F, G, and H became eligible for scheduling. Of these, activity G has the earliest late start, so it is considered first for scheduling on Day 4. Examination of the resources needed to complete Activity G are such that the number of helpers required along with Activity B would be three, more than the two that are available. Thus, Activity G cannot

Chapter 6

Act	Resource		Dur	TF	LS	1	2	3	4	5	6	7	8	9	10	11	12	13	14	15	16	17	18	19	20	21	22
A	2M	1H	1	0	1																						
B	2M	1H	4	0	2																						
C	2M	1H	2	6	8																						
D	1M	0H	1	1	7																						
E	3M	0H	4	0	6																						
F	3M	0H	3	7	11																						
G	1M	2H	4	6	10																						
H	1M	1H	2	8	12																						
J	2M	2H	2	1	8																						
K	2M	1H	6	0	10																						
L	3M	2H	2	6	14																						
M	2M	1H	1	0	16																						

Figure 6.3   Resource Allocation by the Series Method

Act	Resource		Dur	TF	LS	/	1	2	3	4	5	6	7	8	9	10	11	12	13	14	15	16	17	18	19	20	21	22
A	2M	1H	1	0	1		2/1																					
B	2M	1H	4	0	2																							
C	2M	1H	2	6	8																							

Figure 6.4   Allocation of Resources for Day 1

Act	Resource		Dur	TF	LS	/	1	2	3	4	5	6	7	8	9	10	11	12	13
A	2M	1H	1	0	1		2/1												
B	2M	1H	4	0	2			2/1	2/1										
C	2M	1H	2	6	8			2/1	2/1										

Figure 6.5   Allocation of Resources for Days 2 and 3

Resource Allocation and Resource Leveling 123

Act	Resource		Dur	TF	LS	/	1	2	3	4	5	6	7	8	9	10
A	2M	1H	1	0	1	/	2/1									
B	2M	1H	4	0	2	/		2/1	2/1	2/1	2/1					
C	2M	1H	2	6	8	/		2/1	2/1							
D	1M	0H	1	1	7											
E	3M	0H	4	0	6											
F	3M	0H	3	7	11	/				3/0	3/0					
G	1M	2H	4	6	10	/										
H	1M	1H	2	8	12	/										
J	2M	2H	2	1	8											
K	2M	1H	6	0	10											
L	3M	2H	2	6	14											
M	2M	1H	1	0	16											

Figure 6.6 Allocation of Resources for Days 4 and 5

be scheduled. The next priority is then Activity F, which can be scheduled without exceeding the five masons and two helpers (see Figure 6.6). Activity H cannot be scheduled because of the resource limits.

At the end of Day 5, Activity B will be completed. The eligible activities now also include Activities D and E. On Day 6, the first activity scheduled is Activity F (previously started). Activity E has the earliest late start but this would exceed the resources available. Activity D is then considered and found to be eligible for scheduling. This is followed by Activity G, which is also scheduled for Day 6 (see Figure 6.7).

The same procedure is followed for each day until all activities have been scheduled. The completed schedule is shown in Figure 6.8. The project duration is now 20 days, four days longer than if the resource limits had been ignored.

Figure 6.9 shows the network used to apply resource limits with the series solution. This same network will now be used to show how the parallel solution is derived.

Solve the schedule by the parallel method of resource allocation (see Figure 6.10). The resource limits are as follows: 5M (masons) 2H (helpers)

Resource limits are not to be exceeded under any circumstances.

Figure 6.7 — Allocation of Resources for Day 6

Act	Resource		Dur	TF	LS	/	1	2	3	4	5	6	7	8	9	10
A	2M	1H	1	0	1	/	2/1									
B	2M	1H	4	0	2	/		2/1	2/1	2/1	2/1					
C	2M	1H	2	6	8	/		2/1	2/1							
D	1M	0H	1	1	7	/						1/0				
E	3M	0H	4	0	6	/										
F	3M	0H	3	7	11	/				3/0	3/0	3/0				
G	1M	2H	4	6	10	/						1/2				
H	1M	1H	2	8	12	/										
J	2M	2H	2	1	8	/										
K	2M	1H	6	0	10	/										
L	3M	2H	2	6	14	/										
M	2M	1H	1	0	16	/										

Figure 6.7 Allocation of Resources for Day 6

Figure 6.8 — Schedule of Activities for the Project

Act	Resource		Dur	TF	LS	/	1	2	3	4	5	6	7	8	9	10	11	12	13	14	15	16	17	18	19	20	21	22
A	2M	1H	1	0	1	/	■																					
B	2M	1H	4	0	2	/		■	■	■	■																	
C	2M	1H	2	6	8	/								■	■													
D	1M	0H	1	1	7	/							■															
E	3M	0H	4	0	6	/						■	■	■	■													
F	3M	0H	3	7	11	/				■	■	■																
G	1M	2H	4	6	10	/						■	■	■	■													
H	1M	1H	2	8	12	/										■	■											
J	2M	2H	2	1	8	/								■	■													
K	2M	1H	6	0	10	/										■	■	■	■	■	■							
L	3M	2H	2	6	14	/															■	■						
M	2M	1H	1	0	16	/																	■					
Masons Scheduled							2	4	4	5	5	5	4	4	4	5	2	3	3	2	2	2	2	3	3	2		
Helpers Scheduled							1	2	2	1	1	2	2	2	2	2	2	2	2	1	1	1	1	2	2	1		

Figure 6.8 Schedule of Activities for the Project

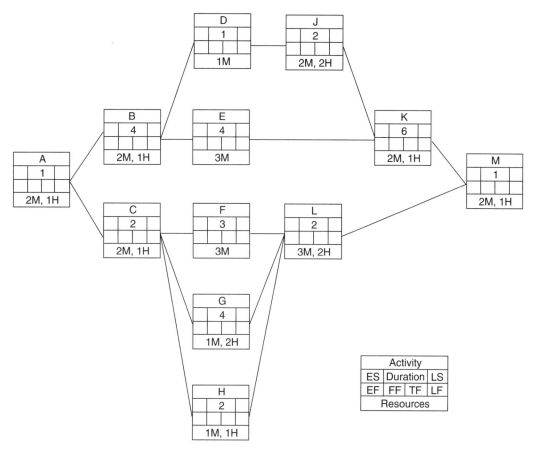

Figure 6.9 Parallel Solution Example

In the first step, only Activity A is eligible for scheduling. As with the series method, this is indicated by the slash placed next to Activity A. Activity A is then scheduled on Day 1 on the chart (Figure 6.11).

At the completion of Activity A, Activities B and C are eligible for scheduling. Since the total number of resources of five masons and two helpers is not exceeded, both activities are scheduled on Days 2 and 3. After Day 3, Activity C will be completed while Activity B still has two remaining days (Figure 6.12).

To schedule Days 4 and 5, Activity B has not been completed and Activities F, G, and H are also eligible for scheduling. Of these, Activity B has the earliest late start so it is considered first for scheduling on Day 4. Activity G is considered for scheduling, but this cannot be done as the resources needed to complete Activity G are such that the number of helpers (three) required along with Activity B would exceed the two that are available. Thus, Activity G cannot be scheduled. The next priority is then Activity F, which can be scheduled without exceeding the five masons and

Act	Resource		Dur	TF	LS	/	1	2	3	4	5	6	7	8	9	10	11	12	13	14	15	16	17	18	19	20	21	22
A	2M	1H	1	0	1																							
B	2M	1H	4	0	2																							
C	2M	1H	2	6	8																							
D	1M	0H	1	1	7																							
E	3M	0H	4	0	6																							
F	3M	0H	3	7	11																							
G	1M	2H	4	6	10																							
H	1M	1H	2	8	12																							
J	2M	2H	2	1	8																							
K	2M	1H	6	0	10																							
L	3M	2H	2	6	14																							
M	2M	1H	1	0	16																							

Figure 6.10   Resource Allocation by the Parallel Method

Act	Resource		Dur	TF	LS	/	1	2	3	4	5	6	7	8	9	10	11	12	13	14	15	16	17	18	19	20	21	22
A	2M	1H	1	0	1		2/1																					
B	2M	1H	4	0	2																							
C	2M	1H	2	6	8																							

Figure 6.11   Allocation of Resources for Day 1

Act	Resource		Dur	TF	LS	/	1	2	3	4	5	6	7	8	9	10	11	12	13	14	15	16	17	18	19	20	21	22
A	2M	1H	1	0	1		2/1																					
B	2M	1H	4	0	2			2/1	2/1																			
C	2M	1H	2	6	8			2/1	2/1																			

Figure 6.12   Allocation of Resources for Days 2 and 3

Act	Resource		Dur	TF	LS	/	1	2	3	4	5	6	7	8	9	10
A	2M	1H	1	0	1	/	2/1									
B	2M	1H	4	0	2	/		2/1	2/1	2/1	2/1					
C	2M	1H	2	6	8	/		2/1	2/1							
D	1M	0H	1	1	7	/										
E	3M	0H	4	0	6	/										
F	3M	0H	3	7	11	/				3/0	3/0					
G	1M	2H	4	6	10	/										
H	1M	1H	2	8	12	/										
J	2M	2H	2	1	8	/										
K	2M	1H	6	0	10											
L	3M	2H	2	6	14											
M	2M	1H	1	0	16											

Figure 6.13 Allocation of Resources for Days 4 and 5

two helpers (see Figure 6.13). Because of resource limitations, Activity H cannot be scheduled at this time.

At the end of Day 5, Activity B will be completed. The eligible activities now also include Activities D and E. On Day 6, of the eligible activities (D, E, F, G, and H), the first activity scheduled is Activity E (earliest late start). Activity D is then considered eligible for scheduling and is scheduled to start on Day 6. This is followed by Activity G, which is also scheduled for Day 6 (see Figure 6.14). Note that Activity F has a lower priority owing to its late start date and it cannot be scheduled at this stage without exceeding the resource limits. Thus, Activity F will be discontinued after Day 5.

The same procedure is followed until all activities have been scheduled. The completed schedule is shown in Figure 6.15. Because of the criteria that permit an activity to be interrupted, note how Activities F and G are now scheduled differently than they were with the series method. Such interruptions of lower-priority activities often result in reduced project durations. Note that the series solution of the same network also resulted in a project duration of 20 days, but that different activities were scheduled on certain days.

Figure 6.14 Allocation of Resources for Day 6

Act	Resource		Dur	TF	LS	/	1	2	3	4	5	6	7	8	9	10
A	2M	1H	1	0	1		2/1									
B	2M	1H	4	0	2			2/1	2/1	2/1	2/1					
C	2M	1H	2	6	8			2/1	2/1							
D	1M	0H	1	1	7							1/0				
E	3M	0H	4	0	6							3/0				
F	3M	0H	3	7	11					3/0	3/0					
G	1M	2H	4	6	10							1/2				
H	1M	1H	2	8	12											
J	2M	2H	2	1	8											
K	2M	1H	6	0	10											
L	3M	2H	2	6	14											
M	2M	1H	1	0	16											

Figure 6.15 Schedule of Activities for the Project

Act	Resource		Dur	TF	LS	/	1	2	3	4	5	6	7	8	9	10	11	12	13	14	15	16	17	18	19	20	21	22
A	2M	1H	1	0	1		■																					
B	2M	1H	4	0	2			■	■	■	■																	
C	2M	1H	2	6	8					■	■																	
D	1M	0H	1	1	7							■																
E	3M	0H	4	0	6							■	■	■	■													
F	3M	0H	3	7	11					■	■	■																
G	1M	2H	4	6	10							■		■						■	■							
H	1M	1H	2	8	12									■	■													
J	2M	2H	2	1	8								■	■														
K	2M	1H	6	0	10											■	■	■	■	■	■							
L	3M	2H	2	6	14																		■					
M	2M	1H	1	0	16																			■				
Masons Schedule							2	4	4	5	5	5	5	5	4	5	3	3	2	2	2	1	1	3	3	2		
Helpers Schedule							1	2	2	1	1	2	2	2	2	1	2	2	1	1	1	2	2	2	2	1		

THE BROOKS METHOD OF RESOURCE ALLOCATION

The series and parallel methods of resource allocation constitute good graphical procedures by which resources can be allocated to a project. A related, though slightly different approach, is the Brooks method. The Brooks method is not a graphical procedure but rather a tabular one. The user begins with a network, similar to those used in the series or parallel methods. Standard computations are made of the early and late occurrence times of each activity. Of particular interest in the Brooks method is the late start date. Priorities are assigned to the various activities on the basis of their late start dates, with the highest priority given to the activity with the smallest or earliest late start date. (Note: The literature about the Brooks method assigns priority on the basis of the total number of days available from the late start date to project completion, a priority designation called the ACTIM value. The highest priority is given to the activity with the highest or largest number of days from the late start date to the project completion. This procedure results in the same priority listing as when the highest priority is given to the activity with the earliest late start date.) Once the activities in a network have been prioritized based on their late start dates, the activities are rearranged in a table in the order of their priority.

Once the priorities are determined, the resource allocation can begin. The number of resources available must be constantly monitored as the allocations are being made. In addition, it is imperative that the scheduler maintains a constant vigil to ascertain which activities are available for scheduling. Priority is always given to the available activity with the highest ACTIM priority. It is possible to impose constraints concerning the ability to interrupt activities or not to interrupt activities. In fact, it would be possible to have both these conditions imposed on different activities in a network. This would be a more realistic approach, as some activities cannot be interrupted while others may be interrupted without seriously compromising productivity.

In addition to ACTIM, this method also uses terms as *TNow* (the current time under consideration or the particular day being scheduled) and *Act. ready* (those activities whose immediate predecessors have already been scheduled, making them eligible for scheduling). In assigning start dates to the various activities, the primary focus will be on the TNow values and the activities that are ready for scheduling (Act. ready). TNow starts with an assigned value of 1, and the Act. ready for the first assignment date consists of the first activity only, assuming the network begins with a single activity. As the activities are assigned to the various start dates, more activities (succeeding those already completed) will be included in the Act. ready cell in the table. The activity to be assigned first in the next TNow cell is the one that has the highest ACTIM priority, unless an uninterruptable activity has already started in a previous TNow assignment. As the assignments are made, the scheduler must use care to stay within the resource limits that exist on the project. As long as careful accounting is used to keep track of the activities, the approach is relatively simple.

(The network shown in Figure 6.16 will be used to demonstrate resource allocation by the Brooks method. In the example, it will be assumed that no activities can be interrupted once started.) The resource limits are as follows: 5M (masons) 2H (helpers)

Resource limits are not to be exceeded under any circumstances. The resource allocation information will be captured in Figure 6.17.

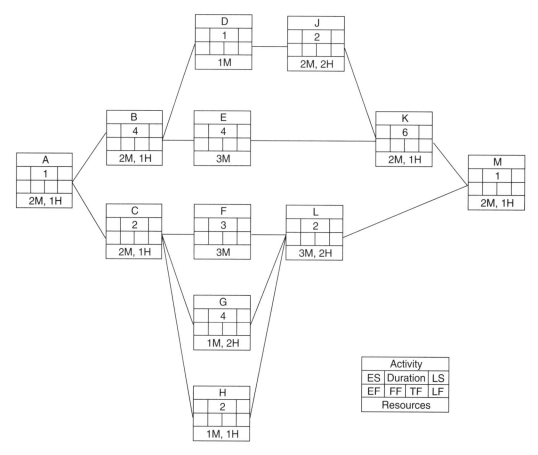

Figure 6.16   Brooks Example

The Brooks method solution begins with determining the ACTIM values, the difference between the completion date (when no resource constraints exist), and the late start of each respective activity. If two activities have the same ACTIM value, the highest priority is given to the activity with the least total float, similar to the series and parallel methods.

Once the ACTIM values are determined, the scheduling of the activities can begin. On Day 1, the only activity that is ready for scheduling is Activity A. Notice that Figure 6.18 captures all of the information required to perform the Brooks method of resource allocation. The second row of the figure is simply a listing of the ACTIM values. The activities are ordered by their ACTIM values to simplify the procedure. The middle portion records information about each activity. All that remains to be determined for these activities are the start and finish times for each activity. This is computed as the process is completed. The bottom portion simply keeps track of the resources actually being utilized at each stage of the schedule.

Once Activity A is scheduled, Activities B and C are ready for scheduling. Of these, Activity B is scheduled first (higher ACTIM priority), but the resource needs of

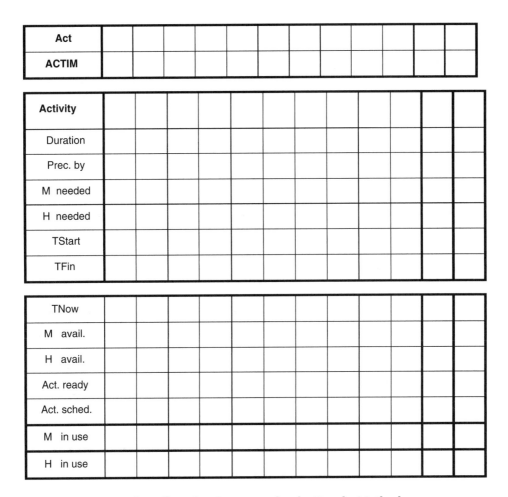

Figure 6.17 Form for Allocating Resources by the Brooks Method

the activities are such that Activity C can also be scheduled. Since the finish date (TFin) for Activity A is listed as 2, the start date for Activities B and C is 2. The durations of these activities are then added to the start times to determine the finish dates. Once this is done, the resource utilization information is recorded (see Figure 6.19).

The TNow date of interest at this point is Day 4, the day that Activity C is completed. On Day 4, Activity C has been completed and Activities F, G, and H are eligible for scheduling. Of course, Activity B is already started, so that will be the first activity scheduled. Of the remaining eligible activities (F, G, and H), Activity G is considered first, but it is not scheduled because the resource limits would be exceeded. Activity F is the next activity to consider, based on the ACTIM values, and it can be scheduled (see Figure 6.20). Resource limits would be exceeded if Activity H were also scheduled.

The next day to schedule is Day 6, the day Activity B is completed. The eligible activities now also include D and E, since Activity B is completed. Activity F has already been begun, so it will be scheduled first. The next activity to consider is Activity E, but

Act	A	B	E	D	J	C	K	G	F	H	L	M
ACTIM	16	15	11	10	9	9	7	7	6	5	3	1

Activity	A	B	E	D	J	C	K	G	F	H	L	M
Duration	1	4	4	1	2	2	6	4	3	2	2	1
Prec. by		A	B	B	D	A	E,J	C	C	C	FGH	K,L
M needed	2	2	3	1	2	2	2	1	3	1	3	2
H needed	1	1	0	0	2	1	1	2	0	1	2	1
TStart	1											
TFin	2											

TNow	1											
M avail.	5											
H avail.	2											
Act. ready	A											
Act. sched.	A											
M in use	2											
H in use	1											

Figure 6.18 Scheduling of the First Step

scheduling it would result in exceeding the resource limits. Activity D is then considered and is found to be acceptable for scheduling. The next activity to consider is Activity G, and it too can be scheduled (see Figure 6.21). The limits on resources would be exceeded if Activity H was scheduled.

The next day to consider for scheduling is Day 7, the day Activities D and F are completed. The eligible activities now include E, G, H, and J. Since Activity G has already been started, it is scheduled first. Activity E is considered next and can be scheduled as shown in Figure 6.22. Because of resource limitations, Activities H and J cannot be scheduled.

This procedure is followed until all activities have been scheduled (see Figure 6.23). A comparison with the series solution will reveal that the same resource distribution is obtained.

Resource Allocation and Resource Leveling

Activity	A	B	E	D	J	C	K	G	F	H	L	M
Duration	1	4	4	1	2	2	6	4	3	2	2	1
Prec. by		A	B	B	D	A	E,J	C	C	C	F,G,H	K,L
M needed	2	2	3	1	2	2	2	1	3	1	3	2
H needed	1	1	0	0	2	1	1	2	0	1	2	1
TStart	1	2				2						
TFin	2	6				4						

TNow	1	2										
M avail.	5	5	5	5	5	5	5	5	5	5	5	5
H avail.	2	2	2	2	2	2	2	2	2	2	2	2
Act. ready	A	B,C										
Act. sched.	A	B,C										
M in use	2	4										
H in use	1	2										

Figure 6.19 Scheduling of the Second Step

Activity	A	B	E	D	J	C	K	G	F	H	L	M
Duration	1	4	4	1	2	2	6	4	3	2	2	1
Prec. by		A	B	B	D	A	E,J	C	C	C	F,G,H	K,L
M needed	2	2	3	1	2	2	2	1	3	1	3	2
H needed	1	1	0	0	2	1	1	2	0	1	2	1
TStart	1	2				2			4			
TFin	2	6				4			7			

Figure 6.20 Scheduling of the Third Step

134 Chapter 6

TNow	1	2	4								
M avail.	5	5	5	5	5	5	5	5	5	5	5	5
H avail.	2	2	2	2	2	2	2	2	2	2	2	2
Act. ready	A	B,C	B,F,G,H								
Act. sched.	A	B,C	B,F								
M in use	2	4	5								
H in use	1	2	1								

Figure 6.20 *(continued)*

Activity	A	B	E	D	J	C	K	G	F	H	L	M
Duration	1	4	4	1	2	2	6	4	3	2	2	1
Prec. by		A	B	B	D	A	E,J	C	C	C	F,G,H	K,L
M needed	2	2	3	1	2	2	2	1	3	1	3	2
H needed	1	1	0	0	2	1	1	2	0	1	2	1
TStart	1	2		6		2		6	4			
TFin	2	6		7		4		10	7			

TNow	1	2	4	6								
M avail.	5	5	5	5	5	5	5	5	5	5	5	5
H avail.	2	2	2	2	2	2	2	2	2	2	2	2
Act. ready	A	B,C	B,F,G,H	D,E,F,G,H								
Act. sched.	A	B,C	B,F	F,D,G								
M in use	2	4	5	5								
H in use	1	2	1	2								

Figure 6.21 Scheduling of the Fourth Step

Resource Allocation and Resource Leveling   135

Activity	A	B	E	D	J	C	K	G	F	H	L	M
Duration	1	4	4	1	2	2	6	4	3	2	2	1
Prec. by		A	B	B	D	A	E,J	C	C	C	F,G,H	K,L
M needed	2	2	3	1	2	2	2	1	3	1	3	2
H needed	1	1	0	0	2	1	1	2	0	1	2	1
TStart	1	2	7	6		2		6	4			
TFin	2	6	11	7		4		10	7			

TNow	1	2	4	6								
M avail.	5	5	5	5	5	5	5	5	5	5	5	5
H avail.	2	2	2	2	2	2	2	2	2	2	2	2
Act. ready	A	B,C	B,F,G,H	D,E,F,G,H	E,G,H,J							
Act. sched.	A	B,C	B,F	F,D,G	G,E							
M in use	2	4	5	5	4							
H in use	1	2	1	2	2							

Figure 6.22 Scheduling of the Fifth Step

Activity	A	B	E	D	J	C	K	G	F	H	L	M
Duration	1	4	4	1	2	2	6	4	3	2	2	1
Prec. by		A	B	B	D	A	E,J	C	C	C	F,G,H	K,L
M needed	2	2	3	1	2	2	2	1	3	1	3	2
H needed	1	1	0	0	2	1	1	2	0	1	2	1
TStart	1	2	7	6	10	2	12	6	4	12	18	20
TFin	2	6	11	7	12	4	18	10	7	14	20	21

Figure 6.23 Scheduling of All Project Activities

TNow	1	2	4	6	7	10	11	12	14	18	20
M avail.	5	5	5	5	5	5	5	5	5	5	5
H avail.	2	2	2	2	2	2	2	2	2	2	2
Act. ready	A	B,C	B,F,G G,H	D,E,F, G,H	E,G, H,J	E,H,J	H,J,K	H,K	K,L	L	M
Act. sched.	A	B,C	B,F	F,D,G	G,E	E,J	J	K,H	K	L	M
M in use	2	4	5	5	4	5	2	3	2	3	2
H in use	1	2	1	2	2	2	2	2	1	2	1

Figure 6.23 *(continued)*

WHEN PROJECT DURATION IS FIXED (RESOURCE LEVELING)

Resources may not always be limited in absolute terms. It may simply be more costly to "double up" on resources at certain times in a project. This added cost may be justified if the construction contract places a tight constraint on the project completion date. The issue of liquidated damages is not always the only cost that must be considered for justifying a time overrun on a project. A company's reputation may be riding on the timely completion of a project. When the project duration is set without regard to limitations in the availability of resources, the construction manager or scheduler is not in a position to extend the project duration. When these constraints exist, resource leveling is an effective scheduling strategy.

Resource leveling, while not constrained by limited resources, tends to reduce the maximum demands for a given resource on any given day. This is desirable, particularly where personnel are concerned, because it helps avoid or minimize the need for hiring short-term workers. On some projects and in some locations, it may be difficult to locate skilled workers. New workers must be given an orientation session and some may require additional training. Newly hired workers are also less efficient in performing the work. Considerable paperwork is associated with each worker who is hired and dismissed. In addition, the dismissal of workers can be potentially bad for morale on a project. These shortcomings can be reduced with resource leveling.

Because of the changing needs of a project, it is not uncommon for the job demands for a particular craft or trade to fluctuate widely (see Figure 6.24). This type of fluctuation in the labor requirements for a single craft can occur, but this can be costly to the employer. When the project needs for a craft drop to zero, it is not realistic to dismiss all of the workers, especially if additional activities remain that rely on that craft. Hopefully, there are other projects to which some workers can be assigned on a temporary basis. If not, the employer will be inclined to retain the better employees, regardless of the project needs for that craft.

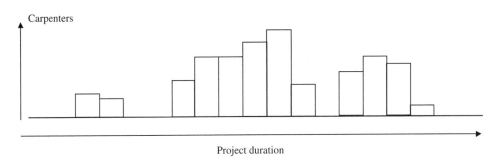

Figure 6.24 Utilization of Carpenters on a Project

A more prudent approach is to reorganize the project activities in such a manner that the resource requirements are more uniform for the project duration. For example, by simple logic modifications, the carpentry requirements on the project mentioned earlier might be leveled as shown in Figure 6.25.

Resource leveling will hold the duration of the project constant. Without altering the project time, resource leveling is an effective means of smoothing the utilization of resources. In other words, resource leveling is an effective means of reducing the number of resources required on "spike" days and it also fills in resources on low resource utilization days. In addition, the day-to-day fluctuation of resource needs will be reduced. The method of resource leveling that will be described uses a minimum-moment algorithm.[1] Essentially the resource requirements on a project are smoothed or leveled by making use of the available free float. The activities are first arranged by an early start schedule. With resource leveling, one can systematically evaluate the impact of using any float associated with each activity.

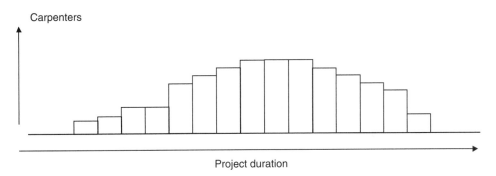

Figure 6.25 Distribution of Carpenter Utilization After Leveling

---

[1] A detailed discussion of the minimum-moment algorithm is presented in Robert B. Harris, *Precedence and Arrow Networking Techniques for Construction* (New York: John Wiley & Sons, Inc., 1978), 277.

The minimum-moment algorithm that will be used for the resource leveling assumes that once an activity has been started, it cannot be interrupted. Another assumption is that resource consumption is constant over the duration of an activity. The network logic is not questioned when resource leveling is done; however, this should be done if the final solution is not considered acceptable. This method will not alter the critical activities but will focus on the merits of shifting any noncritical activities by reducing their float. In the final solution, resource buildups should be minimized.

A number of different approaches can be used for resource leveling. If the project duration is held constant, the different methods will have the same objective: to use up available float if smoothing of resources will occur.

THE MANUAL SOLUTION FOR RESOURCE LEVELING

As with the manual solution for resource allocation, the manual solution for resource leveling is typically limited to simple networks. Although computers can solve these problems much more readily, the objective of this exercise is simply to present the basic principles of resource leveling calculations.

It is assumed that *activities cannot be interrupted*; once an activity has started, it must be completed. Note also that *only one resource is leveled at a time*. It is assumed by leveling one resource that other resources will similarly be leveled to some extent. It would also be very difficult to address the resource leveling process manually on a project for which several different resources were to be leveled simultaneously. Since the duration of the project is fixed, the critical activities are not manipulated, meaning only the start times of activities with positive free float can be adjusted.

The sample begins with a bar chart showing all activities of the network in their early start positions. The resource being leveled is tallied on one line (iteration zero). This will show the extent of the resource utilization and whether there is an unacceptable buildup of resource needs during the project. If an identical number of resources were required each day, no leveling would be required. This is generally not the case. The smoothing of most resources occurs by using the minimum-moment approach. This smoothing effect will be visible with each iteration as the resources will be tallied each time an activity start date is altered.

After the early-start bar chart is developed and the resources have been tallied, the resource-leveling process begins. With the minimum-moment algorithm procedure, the process begins at the end of the project duration and works systematically in a "backward" fashion to the beginning of the project. For each day in the bar chart, all the activities that could be scheduled to occur on the day in question are considered. For each activity that is considered, the optimal number of float days to be utilized is established. This is done by calculating an improvement factor for each potential change in the start date. For example, if an activity has three days of free float, an improvement factor is determined when the start date uses up one day of free float, when two days of free float are used, and when all the float days are used.

The calculation of the improvement factor forms the basis of the leveling decisions. The general equation for determining the improvement factor (IF) can be stated as follows:

$$\text{IF}(A,N) = R \times (R_v - R_o - R \times (N_r))$$

where

A = activity designation
N = number of free float days consumed
R = the number of resources used by the activity per day
R_v = number of resource days currently assigned to those days that will be vacated when the activity start date is changed
R_o = number of resource days currently assigned on those days that will be occupied when the activity start date is changed
N_r = the smaller value of the number of days of free float consumed and the duration of the activity

The improvement factor must be 0 or some positive value in order for a benefit to be derived by reassigning the start date of an activity. The largest improvement factor determines the number of free float days to use. If improvement factors are calculated for several activities, the governing value is the activity with the largest improvement. If two activities are tied with the same improvement factor, priority reassigning of the start date is given to the activity with the most resources per day. If a tie still exists, the activity that will use up the largest number of free float days is selected. If still tied, the activity with the latest start date is selected. If a tie still exists, priority is given to the activities on the basis of input order.

When an activity start date is changed to a later date, the free float available to that activity is reduced. However, the activity still has flexibility in that the start date could now be earlier. This is known as *back float*. Suppose an activity has five days of free float, and through the leveling process its start date is delayed three days. This activity will now have two days of free float and three days of back float.

When resource leveling calculations are made, an accounting should be kept of the number of days of back float for each activity. Just as calculations were made to determine if the free float of activities should be consumed, similar calculations must be made to determine if the back float of activities should be used. Note that when a "backward pass" is made, an activity start date is reassigned if the improvement factor is zero. A value of zero means that no smoothing is generated by reassigning the start date. Since this is not detrimental to resource utilization results, the reassignment is made to "make room" for other activities. That is, by shifting the start dates, free float is given to other activities that might otherwise not have free float. When the entire project has been evaluated with the forward pass that considered the back float, the resource leveling process is completed.

A simple network will demonstrate the process of performing a manual resource leveling solution. The network is shown in Figure 6.26 along with a table that shows the resource (R) utilization per day along the duration of the project (Figure 6.27). Note that resources are needed for each day, but the total number of resources required

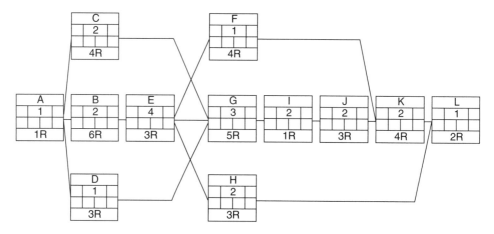

Figure 6.26 Sample Network to Demonstrate Resource Leveling

each day varies considerably. The activities in Figure 6.27 are shown in their early start positions and the free float of the activities is shown as dashed lines. The extent that the resources will be successfully leveled will depend on the network logic and the resources required of the individual activities.

The following process is by no means the only way that the procedure can be followed. Variations of the process might be justified, but circumstances will not always favor one method over another. The process of leveling begins with a backward pass where the project is evaluated one day at a time. The purpose of this backward pass is to evaluate the merits of utilizing some of the free float to create a more efficient resource utilization. In conducting this backward pass, first consideration is given to Day 17. Since there is no activity that has any free float on that day, there is no opportunity to make any changes in resource utilization. On Day 16, there is a free float day associated with Activity H, so we will now consider moving Activity H to determine if the resources can be leveled. When we examine Activity H, it is obvious that it actually has several days of free float. We will consider assigning Activity H to each of the days of free float to find the optimal allocation. The ideal location will be determined from the improvement

	1	2	3	4	5	6	7	8	9	10	11	12	13	14	15	16	17
		C 4	C 4					F 4									
	A 1	B 6	B 6	E 3	E 3	E 3	E 3	G 5	G 5	G 5	I 1	I 1	J 3	J 3	K 4	K 4	L 2
		D 3						H 3	H 3								

	1	2	3	4	5	6	7	8	9	10	11	12	13	14	15	16	17
Cycle 0	1	13	10	3	3	3	3	12	8	5	1	1	3	3	4	4	2

Figure 6.27 Summary of Resource Utilization with No Resource Leveling

factor. The improvement factor is determined for moving activity H one day, two days, three days, and so on. Improvement factors will be determined for seven possible positions (7 days of free float) to which Activity H could be assigned. The improvement factors are shown as follows:

$$IF\,(H, N) = R \times (R_v - R_o - R \times (N_1))$$
$$IF\,(H,1) = 3 \times (12 - 5 - 3 \times (1)) = 12$$
$$IF\,(H,2) = 3 \times (20 - 6 - 3 \times (2)) = 24$$
$$IF\,(H,3) = 3 \times (20 - 2 - 3 \times (2)) = 36 \Leftarrow$$
$$IF\,(H,4) = 3 \times (20 - 4 - 3 \times (2)) = 30$$
$$IF\,(H,5) = 3 \times (20 - 6 - 3 \times (2)) = 24$$
$$IF\,(H,6) = 3 \times (20 - 7 - 3 \times (2)) = 21$$
$$IF\,(H,7) = 3 \times (20 - 8 - 3 \times (2)) = 18$$

From the improvement factors, it is clear that the utilization of any of the free float will improve the resource utilization of Activity H. The maximum improvement is realized when three days of free float are used. The movement of Activity H and the status of resource use are shown in Figure 6.28.

After Activity H has been moved, the backward pass can continue. The next day to consider is Day 15. Free float does occur on this day, but this belongs to Activity H, which has already been moved to its ideal location. The backward pass continues to Day 14, on which Activity F has free float. As no other activities have free float on Day 14 (Activity H was moved in Cycle 1), improvement factors will be computed for each possible location of Activity F.

$$IF\,(F, N) = R \times (R_v - R_o - R \times (N_1))$$
$$IF\,(F,1) = 4 \times (9 - 5 - 4 \times (1)) = 0$$
$$IF\,(F,2) = 4 \times (9 - 5 - 4 \times (1)) = 0$$
$$IF\,(F,3) = 4 \times (9 - 4 - 4 \times (1)) = 4$$
$$IF\,(F,4) = 4 \times (9 - 4 - 4 \times (1)) = 4$$
$$IF\,(F,5) = 4 \times (9 - 3 - 4 \times (1)) = 8$$
$$IF\,(F,6) = 4 \times (9 - 3 - 4 \times (1)) = 8 \Leftarrow$$

	1	2	3	4	5	6	7	8	9	10	11	12	13	14	15	16	17
		C 4	C 4					F 4									
	A 1	B 6	B 6	E 3	E 3	E 3	E 3	G 5	G 5	G 5	I 1	I 1	J 3	J 3	K 4	K 4	L 2
		D 3									H 3	H 3					

	1	2	3	4	5	6	7	8	9	10	11	12	13	14	15	16	17
Cycle 0	1	13	10	3	3	3	3	12	8	5	1	1	3	3	4	4	2
								−3	−3		+3	+3					
Cycle 1	1	13	10	3	3	3	3	9	5	5	4	4	3	3	4	4	2

Figure 6.28 Summary of Resource Utilization After Cycle 1

	1	2	3	4	5	6	7	8	9	10	11	12	13	14	15	16	17
		C 4	C 4											F 4			
	A 1	B 6	B 6	E 3	E 3	E 3	E 3	G 5	G 5	G 5	I 1	I 1	J 3	J 3	K 4	K 4	L 2
		D 3									H 3	H 3					

	1	2	3	4	5	6	7	8	9	10	11	12	13	14	15	16	17
Cycle 1	1	13	10	3	3	3	3	9	5	5	4	4	3	3	4	4	2
								−4						+4			
Cycle 2	1	13	10	3	3	3	3	5	5	5	4	4	3	7	4	4	2

Figure 6.29 Summary of Resource Utilization After Cycle 2

There is a clear advantage in moving Activity F to utilize its free float. In this cycle (Cycle 2), the improvement is the same whether we move Activity F five or six days. When ties occur, the normal procedure is to move the maximum number of days, primarily to allow additional space for other activities. Although no other activities could benefit from this move of five versus six days, to be consistent we will move the maximum of six days. The summary of the resource use after Cycle 2 is shown in Figure 6.29.

The backward pass continues. Free float days are noted on many days (Days 13, 12, 11, etc.), but many of these represent free float of Activities H and F, and in some cases it is really the back float of Activities H and F. It is only when the backward pass gets to Day 7 that the free float of Activities C and D is encountered. Activities C and D cannot be scheduled to occur later than Day 7 (during the back float days created by moving Activities F and H) because they must be finished before Activity G is started. Since both these activities have free float on Day 7, improvement factors will be calculated for each. Note that Activity C has four days of free float and Activity D has five days. Their improvement factors are determined as follows:

$$IF(C, N) = R \times (R_v - R_o - R \times (N_1))$$
$$IF(C,1) = 4 \times (13 - 3 - 4 \times (1)) = 24$$
$$IF(C,2) = 4 \times (23 - 6 - 4 \times (2)) = 36$$
$$IF(C,3) = 4 \times (23 - 6 - 4 \times (2)) = 36$$
$$IF(C,4) = 4 \times (23 - 6 - 4 \times (2)) = 36 \Leftarrow$$

$$IF(D, N) = R \times (R_v - R_o - R \times (N_1))$$
$$IF(D,1) = 3 \times (13 - 10 - 3 \times (1)) = 0$$
$$IF(D,2) = 3 \times (13 - 3 - 3 \times (1)) = 21$$
$$IF(D,3) = 3 \times (13 - 3 - 3 \times (1)) = 21$$
$$IF(D,4) = 3 \times (13 - 3 - 3 \times (1)) = 21$$
$$IF(D,5) = 3 \times (13 - 3 - 3 \times (1)) = 21$$

From the improvement factors, it is evident that the greatest improvement occurs when Activity C is moved forward four days (when ties occur, always move the greatest number of days). At the end of Cycle 3, the resource utilization is as shown in Figure 6.30.

Resource Allocation and Resource Leveling

	1	2	3	4	5	6	7	8	9	10	11	12	13	14	15	16	17
						C 4	C 4							F 4			
	A 1	B 6	B 6	E 3	E 3	E 3	E 3	G 5	G 5	G 5	I 1	I 1	J 3	J 3	K 4	K 4	L 2
			D 3								H 3	H 3					

	1	2	3	4	5	6	7	8	9	10	11	12	13	14	15	16	17
Cycle 2	1	13	10	3	3	3	3	5	5	5	4	4	3	7	4	4	2
		−4	−4			+4	+4										
Cycle 3	1	9	6	3	3	7	7	5	5	5	4	4	3	7	4	4	2

Figure 6.30 Summary of Resource Utilization After Cycle 3

As we are about to continue the backward pass, it should be apparent that Day 7 should still be considered. Activity D does have free float, so it should be considered next. Note that Activity D could have been moved in Cycle 3 but did not have an improvement factor as large as Activity C. The improvement factors for moving Activity D are now recomputed:

$$\text{IF}(D, N) = R \times (R_v - R_o - R \times (N_1))$$
$$\text{IF}(D,1) = 3 \times (9 - 6 - 3 \times (1)) = 0$$
$$\text{IF}(D,2) = 3 \times (9 - 3 - 3 \times (1)) = 9$$
$$\text{IF}(D,3) = 3 \times (9 - 3 - 3 \times (1)) = 9 \Leftarrow$$
$$\text{IF}(D,4) = 3 \times (9 - 7 - 3 \times (1)) = -3$$
$$\text{IF}(D,5) = 3 \times (9 - 7 - 3 \times (1)) = -3$$

Activity D should be moved three days. Note how the improvement factors for Activity D have changed after first moving Activity C. The procedure is followed as before to yield a summary of resource utilization, as shown in Figure 6.31.

	1	2	3	4	5	6	7	8	9	10	11	12	13	14	15	16	17
						C 4	C 4							F 4			
	A 1	B 6	B 6	E 3	E 3	E 3	E 3	G 5	G 5	G 5	I 1	I 1	J 3	J 3	K 4	K 4	L 2
					D 3						H 3	H 3					

	1	2	3	4	5	6	7	8	9	10	11	12	13	14	15	16	17
Cycle 3	1	9	6	3	3	7	7	5	5	5	4	4	3	7	4	4	2
		−3			+3												
Cycle 4	1	6	6	3	6	7	7	5	5	5	4	4	3	7	4	4	2

Figure 6.31 Summary of Resource Utilization After Cycle 4

The essence of the solution is now as shown in Figure 6.31. Remember that when activities use up free float, back float is created. The back float may become free float for some preceding activities. In the process of performing the computations, whenever there was a positive improvement factor or a tie in the improvement factor for different positions, the activities were always moved the maximum amount of time. This was done to create free float for other activities so as to increase the flexibility of scheduling. The activities must now be examined to determine if any activities should be moved to use up some of the back float. This is done by starting on Day 1 and conducting a forward pass. As back float is encountered, improvement factors are computed. In this process, Day 2 is noted to contain back float for Activities C and D. Therefore, improvement factors are computed for both activities as follows:

$$IF (C, N) = R \times (R_v - R_o - R \times (N_1))$$
$$IF (C,1) = 4 \times (7 - 6 - 4 \times (1)) = -12$$
$$IF (C,2) = 4 \times (14 - 9 - 4 \times (2)) = -12$$
$$IF (C,3) = 4 \times (14 - 9 - 4 \times (2)) = -12$$
$$IF (C,4) = 4 \times (14 - 12 - 4 \times (2)) = -24$$

$$IF (D, N) = R \times (R_v - R_o - R \times (N_1))$$
$$IF (D,1) = 3 \times (6 - 3 - 3 \times (1)) = 0 \Leftarrow$$
$$IF (D,2) = 3 \times (6 - 6 - 3 \times (1)) = -9$$
$$IF (D,3) = 3 \times (6 - 6 - 3 \times (1)) = -9$$

All improvement factors are negative except one. The improvement factor of Activity D is zero when it is moved back one day. Note that this means that there is no improvement, but it does schedule the activity earlier so as to give greater flexibility to the overall network. The summary of resource use is shown in Figure 6.32.

As the forward pass continues, improvement factors are computed for Activities F and H.

$$IF (H, N) = R \times (R_v (R_o - R \times (N_1))$$
$$IF (H,1) = 3 \times (4 - 5 - 3 \times (1)) = -12$$
$$IF (H,2) = 3 \times (8 - 10 - 3 \times (2)) = -24$$
$$IF (H,3) = 3 \times (8 - 10 - 3 \times (2)) = -24$$

$$IF (F, N) = R \times (R_v - R_o - R \times (N_1))$$
$$IF (F,1) = 4 \times (7 - 3 - 4 \times (1)) = 0 \Leftarrow$$
$$IF (F,2) = 4 \times (7 - 4 - 4 \times (1)) = -4$$
$$IF (F,3) = 4 \times (7 - 4 - 4 \times (1)) = -4$$
$$IF (F,4) = 4 \times (7 - 5 - 4 \times (1)) = -8$$
$$IF (F,5) = 4 \times (7 - 5 - 4 \times (1)) = -8$$
$$IF (F,6) = 4 \times (7 - 5 - 4 \times (1)) = -8$$

From the improvement factors, it is determined that activity F can be moved back one day. The final results are shown in Figure 6.33. Note how the use of resources

Resource Allocation and Resource Leveling

1	2	3	4	5	6	7	8	9	10	11	12	13	14	15	16	17
					C 4	C 4							F 4			
A 1	B 6	B 6	E 3	E 3	E 3	E 3	G 5	G 5	G 5	I 1	I 1	J 3	J 3	K 4	K 4	L 2
			D 3							H 3	H 3					

	1	2	3	4	5	6	7	8	9	10	11	12	13	14	15	16	17
Cycle 4	1	6	6	3	6	7	7	5	5	5	4	4	3	7	4	4	2
				+3	−3												
Cycle 5	1	6	6	6	3	7	7	5	5	5	4	4	3	7	4	4	2

Figure 6.32 Summary of Resource Utilization After Cycle 5

1	2	3	4	5	6	7	8	9	10	11	12	13	14	15	16	17
					C 4	C 4							F 4			
A 1	B 6	B 6	E 3	E 3	E 3	E 3	G 5	G 5	G 5	I 1	I 1	J 3	J 3	K 4	K 4	L 2
			D 3							H 3	H 3					

	1	2	3	4	5	6	7	8	9	10	11	12	13	14	15	16	17
Cycle 0	1	13	10	3	3	3	3	12	8	5	1	1	3	3	4	4	2
								−3	−3		+3	+3					
Cycle 1	1	13	10	3	3	3	3	9	5	5	4	4	3	3	4	4	2
								−4						+4			
Cycle 2	1	13	10	3	3	3	3	5	5	5	4	4	3	7	4	4	2
		−4	−4		+4	+4											
Cycle 3	1	9	6	3	3	7	7	5	5	5	4	4	3	7	4	4	2
		−3			+3												
Cycle 4	1	6	6	3	6	7	7	5	5	5	4	4	3	7	4	4	2
				+3	−3												
Cycle 5	1	6	6	6	3	7	7	5	5	5	4	4	3	7	4	4	2
													+4	−4			
Cycle 6	1	6	6	6	3	7	7	5	5	5	4	4	7	3	4	4	2

Figure 6.33 Summary of Resource Utilization After Cycle 6

per day has been significantly leveled between Cycles 0 and 6. This is purely a mechanical solution and does not take into account unique project characteristics. Thus, this is the manual solution but further study of the network and the consideration of changes in scheduling logic should be explored.

A final comment seems warranted on this method of resource leveling: this approach will generally give an acceptable solution in that the "spikes" and "dips" in resource needs will be smoothed. However, the solution may not be an optimal one. It must be recognized that the solution is developed in an iterative fashion. This means that the approach simply evaluates the impact of considering the merits of reassigning the start dates of those activities under consideration. The overall impact of a decision is not considered. Although the decision may not be optimal, it will generally be sufficiently close to being optimal as to not warrant further concern.

The Sports Facility Project

The sports facility that was briefly described in Chapters 1 and 3 will be used to demonstrate how resources might be leveled by computer. For this project, it is assumed that all of the work is being performed by one contractor, essentially the only employer on site. It is assumed that all the workers can perform various tasks on this project. The activity descriptions, durations, predecessors, and resource (labor) requirements are shown (see Figure 6.34). Figure 6.35 is a computer-generated labor utilization chart for the project. It shows the resource utilization for this project when every activity is scheduled to begin at its early start date. Figure 6.36 is another resource utilization chart that shows how the start dates for some activities might be altered to reduce the peak worker requirements. Note how the peak resource demands have been reduced in Figure 6.36.

Before the resources are leveled, the resource utilization for this project will be as shown in Figure 6.35. The schedule shows each activity at its early start time. Since no work occurs on Saturdays and Sundays on this project, the resource utilization on weekends is shown to be zero. Note the peaks in resource utilization. Ideally, the contractor will want to minimize the need to lay off workers and rehire them. Any excess hirings necessitated by high resource demands are costly and attempts will generally be made to avoid them.

Activity	Dur	Predecessors	Workers required
Mobilize	2	-	4
Stock materials	4	Mobilize	5
Clearing and grubbing	6	Mobilize	2
Grading for road	7	Mobilize	4
Finish grade	5	Stock materials, clearing and grubbing	8
Prefab bleachers	16	Stock materials	3
Landscape	12	Stock materials, clearing and grubbing	4
Pave roadway	8	Grading for road	5
Place tennis court	10	Finish grade	8
Erect/paint bleachers	7	Prefab bleachers, clearing and grubbing	8
Curbing	5	Landscape, pave roadway	5
Final inspection/cleanup	3	Place tennis court, erect/paint bleachers, curbing	4

Figure 6.34 Labor Requirements for the Sports Facility Project

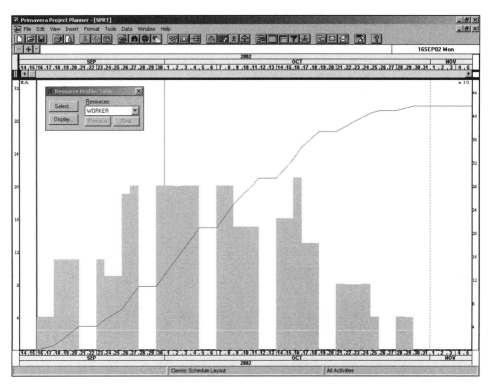

Figure 6.35 Resource Utilization of Labor on the Sports Facility Project with Early Start Schedule (Primavera Project Planner®) (*Reprinted with permission from Primavera Systems, Inc.*)

By utilizing some of the free float that is available, it is possible to level the demands placed on the number of workers required to construct the project. A simple computer manipulation of the activities, by using the available free float, results in a somewhat leveled resource schedule (see Figure 6.36). On a larger project, the leveling potential is generally increased.

FINAL COMMENTS

The efficient control of resources is fundamental to successful project scheduling. While much attention in scheduling is paid to the issue of time, it must be kept in mind that time in activity durations is driven by the resources that are required compared to those resources that are available to accomplish the task. Resources may consist of materials, labor, equipment, subcontractors, and money, in addition to time itself. While

148   Chapter 6

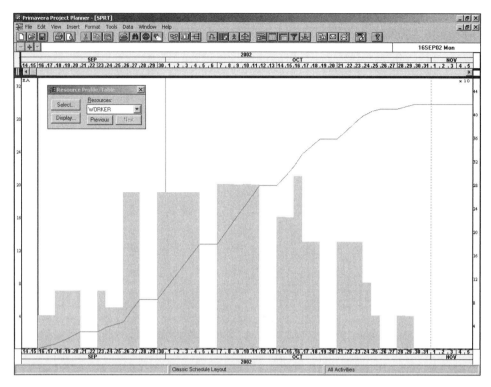

Figure 6.36   Resource Utilization of Labor on the Sports Facility Project After Leveling (Primavera Project Planner®) (*Reprinted with permission from Primavera Systems, Inc.*)

the normal scheduling of work activities may acknowledge the limitations of some obvious resource constraints, this is not a systematic approach. Where resources are particularly costly or scarce, it behooves the project scheduling team to specifically address those limitations.

Computer software can be used to address resource constraints. It is important for the user of such software to recognize the different algorithms that might be used by the software. Many different approaches can be used to determine the means by which priority is given to selected activities. These should be carefully examined before actually using the software. The software's approach should be consistent with the nature of the constraints on the project.

Resource limitations should never be ignored, as this will result in an unrealistic schedule. Acknowledging legitimate resource constraints in a schedule often makes the difference between an idealistic schedule and a realistic one. Only realistic schedules are of value on construction sites; otherwise they quickly lose their credibility. An untrustworthy schedule will not be an effective management tool.

REVIEW PROBLEMS

1. Solve the following by the parallel method of resource allocation. The resource limits are as follows: 5M and 2H.

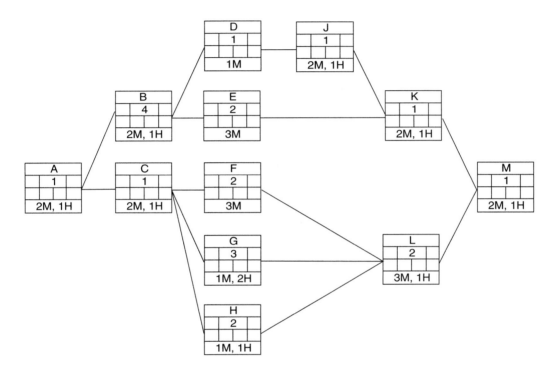

2. Solve this resource scheduling problem with the series method. Resource limits: 4C, 3L.

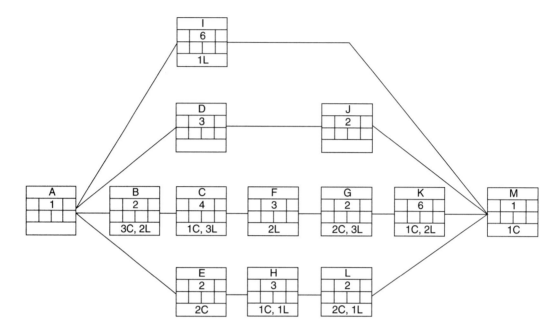

3. Solve the following network by the Brooks method.

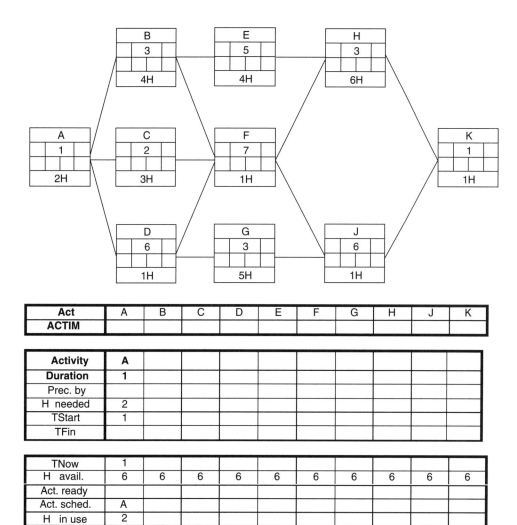

152   Chapter 6

4. Solve the following network by the Brooks method.

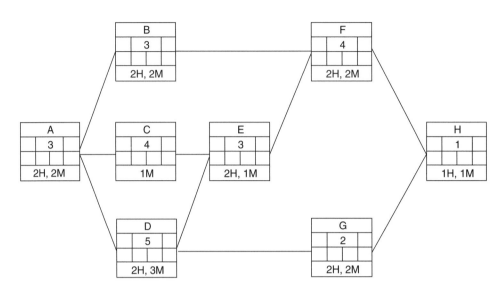

Act	A	B	C	D	E	F	G	H
ACTIM								

Activity	A	D	C	B	E	F	G	H
Duration	3	5	4	3	3	4	2	1
Prec. by		A	A	A	C,D	B,E	D	F, G
H needed	2	2	0	2	2	2	2	1
M needed	2	3	1	2	1	2	2	1
TStart								
TFin								

TNow	1							
H avail.	4	4	4	4	4	4	4	4
M avail.	3	3	3	3	3	3	3	3
Act. ready	A							
Act. sched.								
H in use	2							
M in use	2							

5. Solve the following network by the Brooks method.

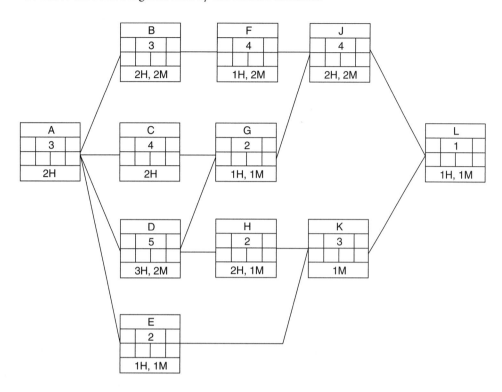

Act	A	B	C	D	E	F	G	H	J	K	L
ACTIM (LS)	1	4	5	4	10	7	9	10	11	12	15

Activity	A	D	B	C	F	G	H	E	J	K	L
Duration	3	5	3	4	4	2	2	2	4	3	1
Prec. by		A	A	A	B	C,D	D	A	F,G	H,E	J,K
H needed	2	3	2	2	1	1	2	1	2	0	1
M needed	0	2	2	0	2	1	1	1	2	1	1
TStart	1										
TFin											

TNow	1										
H avail.	4	4	4	4	4	4	4	4	4	4	4
M avail.	3	3	3	3	3	3	3	3	3	3	3
Act. ready	A										
Act. sched.											
H in use											
M in use											

154   Chapter 6

6. Solve the following resource (*R*) leveling problem.

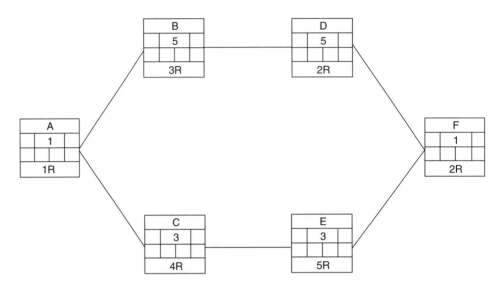

Bar chart

Resource leveling

7. Solve the following resource (*R*) leveling problem.

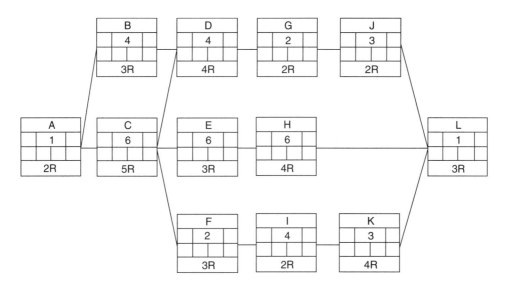

Bar chart

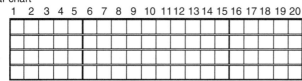

Resource leveling

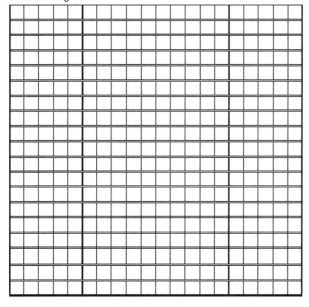

Chapter 6

8. Level the resources (*R*) for the following network:

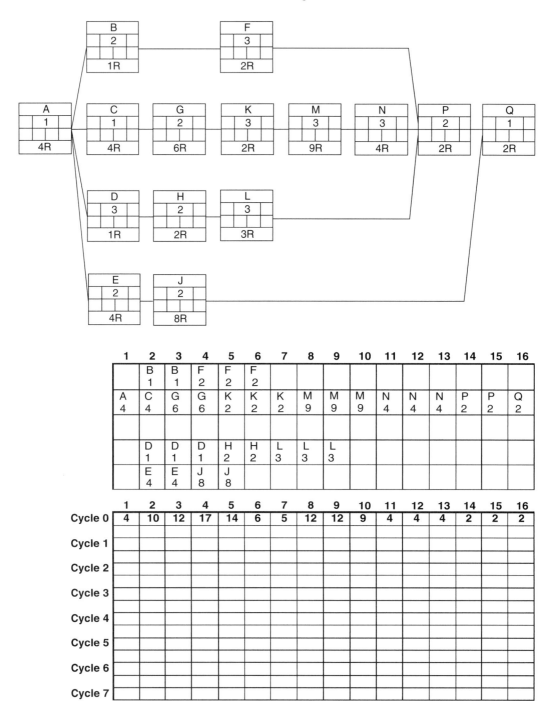

Resource Allocation and Resource Leveling 157

9. Level the resources (*R*) for the following network:

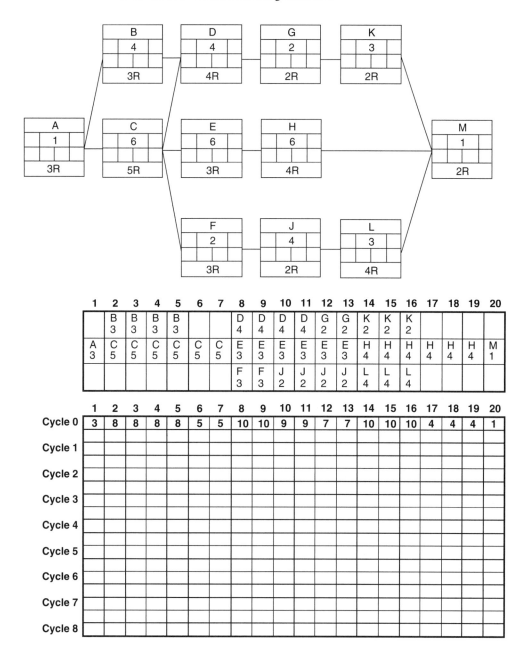

158   Chapter 6

10. Level the resources (*R*) for the following network:

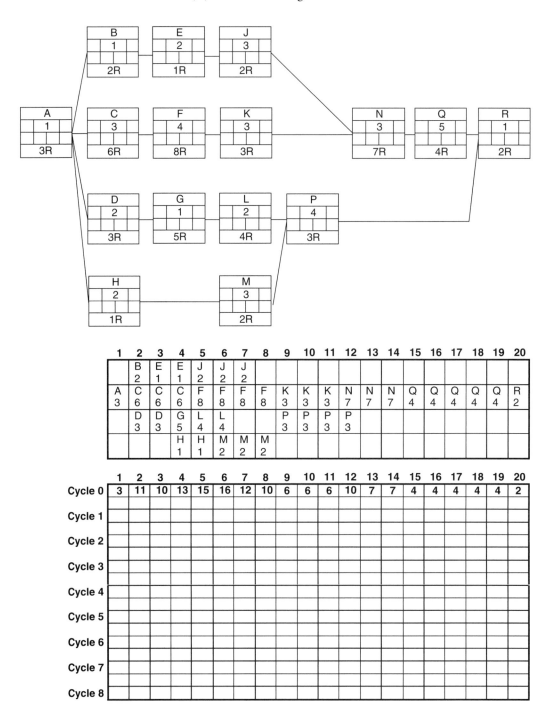

11. Level the resources (*R*) for the following network:

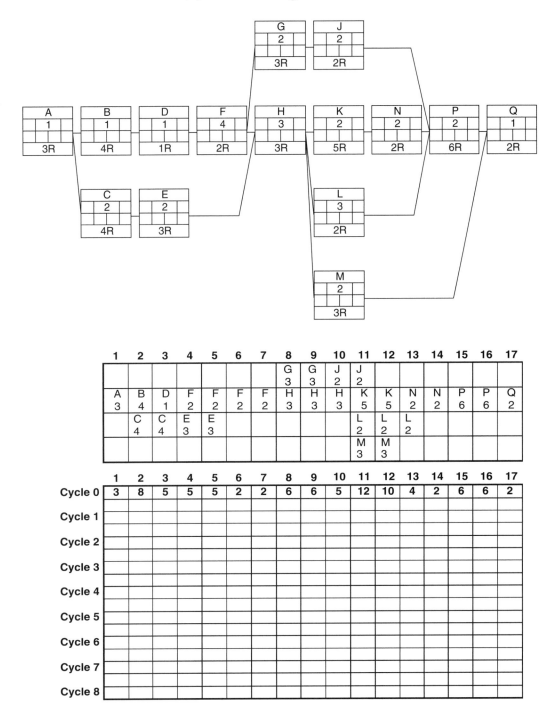

160 Chapter 6

12. Level the resources (R) for the following network:

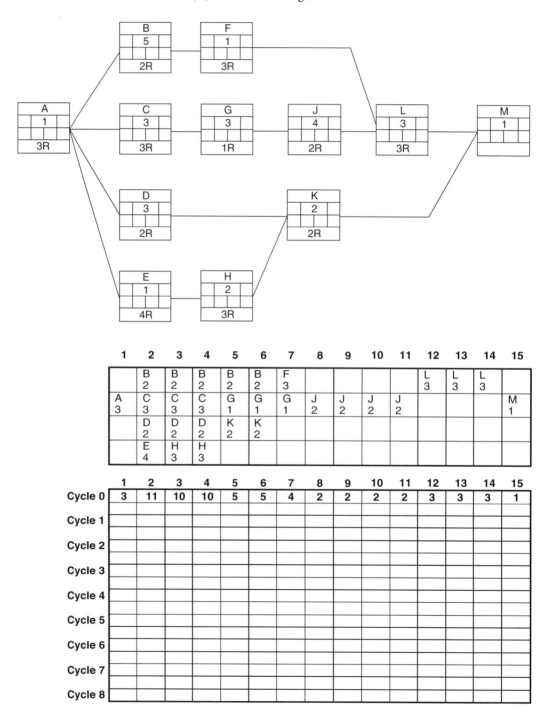

13. Level the resources (R) for the following network:

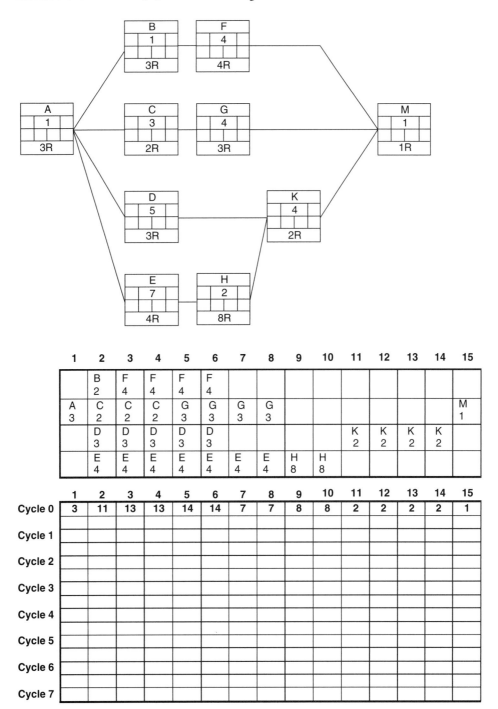

162 Chapter 6

14. Level the resources (R) for the following network:

Show exactly which activity is being moved at each cycle and how many days are being moved. Show all cycles required to utilize the free float and show all cycles required to utilize the back float.

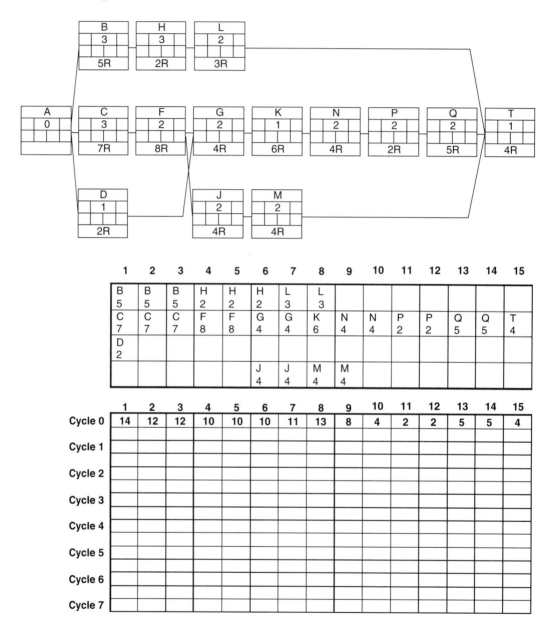

Resource Allocation and Resource Leveling

Resource allocation by series method

Resource allocation by parallel method

7

Money and Network Schedules

Time is the coin of our lives. We must take care how we spend it.
—Carl Sandburg

As with any businesspersons, contractors are primarily concerned with making money in their businesses. The decisions they make in preparing for and running each of their projects are inevitably related to costs and income. Therefore, it is important to explore the role that critical path method techniques have in managing money.

In short, CPM provides a means for relating time and money. Previously we dealt with resources in conjunction with our network analysis. Clearly the application of resources to a project (materials, manpower, and machines) relates directly to another resource, money. The value of the resources applied to each activity represents a substantial component of the cost of carrying out the project. Thus, knowing when and how many resources are being used permits the contractor to examine the temporal distribution of costs and, ultimately, income. Furthermore, CPM provides a tool to analyze such distributions and evaluate the monetary impacts of alternative schedules.

The following discussion is presented in two sections. The first deals with a contractor's cash flow. Central to this discussion are the time value of money and the timing of the payments to the contractor over the duration of the construction project. The second section examines the relationship between activity times (durations) and their costs, and presents a methodology for using CPM to perform time/cost trade-off analyses. These analyses can lead to shortened project durations and cost savings or be used to evaluate whether to speed up a project in order to recover from delays.

CASH FLOW

Cash flow is a very meaningful term to most contractors, who must be sensitive to the issue of cash flow if they are to survive. Many construction firms that were undertaking profitable projects have failed in business simply because of severe cash-flow problems. Essentially, these problems exist when funds are not available to meet current financial obligations. Thus, it is possible for a project that is profitable "on paper" to generate insufficient funds to meet current financial obligations. This is well known among contractors, who tend to pay close attention to this topic. The primary objective of most firms is to make a profit, but the maintenance of a positive cash flow is also of crucial importance.

A cash-flow analysis is an investigation of a project in which the focus is on the flow of money. This includes a separate analysis of the expenditures of money during the construction phase of a project and the revenues or income that will be generated as the project is being completed. A positive cash flow is when the cumulative revenues exceed the cumulative expenditures, meaning that surplus cash is available. On large projects, this excess cash can be invested in various short-term opportunities. A negative cash flow can also occur, and this means that more money is being spent than received. This will mean that the firm will need to take the appropriate steps to borrow from within the firm or from financial institutions to cover the shortfall in funds.

The Time Value of Money

One might look at a contractor's operation in very simple terms. For example, suppose a contractor carefully prepared an estimate of the cost to construct a specific project. This included an acceptable allowance for profit and the successful submittal of the lowest bid to an owner, resulting in the award of the contract. If the contractor estimated the cost of the project exactly, then the amount of profit realized will be exactly as originally allowed. This appears to be the ideal that is always sought. However, is the value of the profit actually received the same as was originally anticipated? Time changes the value of the profit realized. Because of inflation and the delay in the actual receipt of all revenues, the profit, once realized, will no longer permit the contractor to purchase the same value of goods as when the bid was originally submitted. Contractors are paid only after construction work has been performed. This lag or delay in the receipt of revenues is often a considerable time after the costs of construction are actually paid.

Interest Rates

If inflation was zero and the interest rate was zero, it would be a simple matter to assess profitability. Even if inflation is at an extremely low level, interest rates are rarely at a level that they can be ignored. For a contractor, the important question is, "What interest rate should be used?" Two different contractors might justifiably use quite different interest rates. For example, one contractor might use a rate of 11 percent for funds needed to cover negative cash flow during a project. This value of 11 percent might be based on the actual cost of borrowing the money needed to "carry" the firm during the cash deficit. Another firm might use an interest rate of eight percent, based on the interest rate realized on the company's investment of surplus funds. During a negative cash flow, the company can simply use these surplus funds to "cover" the deficit. For this second company, a rate of eight percent would also seem appropriate when a positive cash flow is realized. For the first company, however, a rate other than 11 percent would probably be used, unless the company can earn 11 percent on its excess funds. Contractors must be careful in selecting the appropriate interest rate to use for the analysis of positive and negative cash flows, respectively. Overall, interest rates for deficit funds flow should reflect the cost of borrowing the money or, if no money is borrowed, the interest rate should reflect the value of opportunities that are foregone for internal funds that are used to meet the financial obligations.

Contractor Cash Disbursements

The cash-flow analysis of a project entails an examination of the disbursement of funds and the receipt of funds. Their timing has a direct impact on the amount of cash available or the deficit to be realized at differing times during the construction of a project. The disbursement of funds that are typical on many projects will be discussed first.

Labor: The cost of labor on a project can be substantial. A rough estimate puts labor at approximately one-third of the total cost of a construction project. However, this is largely dependent on the type of project and the amount of work subcontracted. The timing of the disbursement of funds for labor is largely dictated by law. Hourly wages are paid weekly. The primary difference between companies might be simply the day of the week on which the wages are actually paid; otherwise, there are few differences. The wage rates and any overtime pay should be included in the estimate, as this information must be known prior to submitting the bid. The timing of the payment of these wages is of particular concern in the cash-flow analysis.

Equipment: Equipment used on a project will generally fall into one of two categories: equipment owned by the company versus equipment leased or rented by the company. For cash-flow analysis, the type of ownership does not substantially alter the result. It is common that lease payments be made on a monthly basis. Rented equipment is paid for on a monthly basis unless the equipment is rented for shorter periods, such as by the day or by the week.

How is cash flow impacted when the equipment is owned by the company? Obviously, a piece of equipment that has been purchased by the company, without any outstanding loans, does adversely impact the cash flow of a company. However, the impact on cash flow of a company may be different than the impact on cash flow on a project, because the company and the project are not viewed as being one and the same. For example, it is a common practice among construction firms, particularly those specializing in highway or heavy construction, to set up a "separate company," in an accounting sense, for the company-owned equipment. Under this arrangement, the equipment company is a separate "profit center" that "rents" the company-owned equipment to the company's projects. In some instances, the equipment company may even rent the equipment to other contractors. The principle behind this type of arrangement is that the rates charged for equipment must be realistic. By isolating the company-owned equipment in a separate company, the costs of owning and operating the equipment can be more accurately defined. For cash-flow purposes, payment for renting the company-owned equipment is generally an internal funds transfer that occurs at the end of each month. Thus, as far as the disbursement of funds is concerned, company-owned equipment is often similar to outside-owned equipment.

Materials: The costs of materials incorporated in a construction project, together with the labor and equipment costs, constitute the bulk of the costs of construction. How are payments made for materials? Materials are typically ordered from suppliers through purchase orders or purchase agreements. It is common for materials to be delivered to the construction site before the contractor has any obligation to pay for them. Shortly after the materials are delivered to the site, however, the contractor is generally expected to pay for them. The actual payment arrangements vary between suppliers, and these differences should be clearly understood. For example, the invoice for the materials received at the beginning of the month may state "2% 10, net 30." This means that a two percent discount is given if the invoice is paid by the 10th of the month and that the payment will be delinquent if the full amount is not paid by the 30th of the month. Some invoices give a one percent discount while others simply stipulate the date by which the payment must be made without any discount provision. In most cases, materials are expected to be paid for approximately a month from the time of delivery. The judicious contractor must be careful to take advantage of all discounts for early payment and to avoid any delinquent payments. Liens against a project for failure to pay for materials will seriously jeopardize a contractor's reputation with owners and suppliers.

Subcontractors: The costs of labor, equipment, and materials constitute the bulk of what goes into most construction projects. When a particular task requires specialized skills, certain work items might be subcontracted. There may be many other reasons for subcontracting portions of a project, including an advantage realized in cash flow. The full range of reasons for subcontracting work will not be discussed here, but suffice it to say that it is common to subcontract at least some of the work on virtually every project. Subcontractors typically request payment for services rendered on a monthly basis. It is also typical for general contractors to include a provision in the subcontract agreements that payments will be made to the subcontractors only after the general contractor has received payment for that work from the owner. This "pay when paid" provision essentially assures the general contractors that they will not suffer a negative cash flow on the subcontracted work. In addition, the general contractor will commonly withhold a portion of the payments due to the subcontractor as retainage. The retainage amount, stated as a percent, is typically the same as the retainage withheld from the general contractor by the owner.

Other: The general contractor should have a clear understanding of all expenditures to be incurred on a project. This includes the timing of the payments for these items. For example, several items—including insurance premiums, surety fees, various permits, and mobilization—occur early in a project. A careful analysis should be made of all overhead items, whether project or home-office overhead, to determine the timing of payments for these items.

Contract Provisions That Impact Cash Flow

The contractor must recognize that the cash flow to be realized on a particular project will be dictated to some extent by the contract. This is particularly true for the timing of the receipt of revenues, so it is important that the contract documents be examined closely to fully understand how the contractual provisions will impact the cash flow on a project. Some of the items to consider on virtually every project will be discussed.

Payment Schedule: The type of contract dictates the general nature of payment schedule. On a unit price contract, the various pay items are enumerated in the bid itself. The list of pay items gives a clear idea of the information needed to conduct a cash-flow analysis. Of particular interest are items that are specifically included as pay items or those that are specifically excluded from the pay items. For example, mobilization can be a large cost item for a contractor that occurs early in the project. An owner may permit mobilization as a pay item simply to prevent the contractor from incurring a large negative cash flow early in a project. Some owners pay for mobilization with the stipulation that an equal amount be assigned to demobilization. Some owners simply do not pay for mobilization, claiming that the value of the project itself is not enhanced by the mere presence on the project site of several pieces of equipment or a few temporary buildings. Other items that are not always pay items in a contract include formwork, scaffolding, and shoring. These items may be essential to deliver a project, but do not, in themselves, impart any value to the project. For example, if formwork is removed prior to concrete placement, no value is added to the project. The contractor will, of course, have to pay for these items. In order to be reimbursed for them, the contractor must simply allocate these costs to the other pay items in some fashion. The distribution may be "across the board" or the costs could be shifted to certain pay items that might yield a more favorable cash flow for the contractor. This allocation of costs must be done with care.

On lump-sum contracts, the actual payment schedule remains unresolved until after contract award. On these contracts it is common for the general contractor to submit a schedule of values or payment schedule to the owner for approval. Usually, no payments will be authorized until the contractor and the owner have agreed on an acceptable payment schedule. Thus, the contractor has an incentive to present a fairly realistic payment schedule to the owner. On these payment schedules, many owners permit some "front-end loading." Owners recognize that it is to their advantage that the contractor stay out of financial problems. Such owners do not object to fronting the contractor with some additional money early in the project. The risk to the owner is a real one in the event that the contractor defaults on the project; however, this can be minimized by prequalifying bidders or by requiring a performance bond.

Retainage: Whatever payment schedule is used on a project, it is common in the construction industry for the owner to withhold a stated percentage of the funds earned by the contractor as retainage. This retainage amount may be stated as being five, ten, 15, or even 20 percent, although ten percent seems to be the most common. Many public owners withhold ten percent but reduce the retainage withheld from further payments to zero if the contractor makes satisfactory progress by the time the project is 50 percent complete. Thus, such an owner is only withholding five percent by the completion of the project. Obviously, retention has an adverse impact on the contractor's cash flow. Assume that a contractor estimates construction costs to be $1,000,000 and desires a profit of eight percent. The contractor would then submit a bid for $1,080,000. If the retention policy for the project is ten percent for the project duration, the total amount withheld at the completion of the project will be $108,000. Thus, at project completion, the contractor will have spent $1,000,000 while receiving payments of only $972,000. Under such circumstances, it is clear that the contractor will be forced to finance a portion of the project.

Materials: For most materials it is clear when payment will be received from the owner. In most cases this is not a major concern for the contractor. The primary items that warrant further consideration are materials not immediately incorporated into the project. Of particular interest are items that must be purchased early in a project but cannot be installed until a later date. Long-lead-time items may be delivered early or scarce materials may be purchased early to ensure that they will be available when needed. The contract should be examined to determine if payments will be made for materials that are properly stored on site (or elsewhere) or if payment will be made only upon installation. Some owners allow payment for materials not installed if the materials are unique and not of a stock nature. Where such payments are permitted, the contractor has fewer worries about the consequences of the early procurement of certain materials. Of course, the storage of some materials may use up valuable space and require regular monitoring. These factors must be taken into consideration.

Mobilization: Mobilization can be a large cost item and the contract wording for the payment of mobilization should be closely examined. (See the section on payment schedule for a fuller discussion of mobilization.)

Monthly Payments: It is common for payments to be made to the contractor on a monthly basis. The contract may state that the contractor is to make the request for payment by the 30th of each month and that payment will be made to the contractor by the 10th of the following month. The impact on cash flow is more severe if the contractor is not paid until the 20th of the following month. These terms should be carefully examined and appropriately reflected in the bid.

Final Payment: The final payment is essentially the release of the retainage being withheld by the owner. The time in which these funds will be released should also be included in the contract. It is common for final payment to be made from one to three months after substantial completion. It should be recognized that the final payment is generally not made until after the punch list items have all been completed. In some instances, the owner may release some of the retainage amount and withhold only a sufficient amount to cover the items still included in the punch list.

Owner Policies and Practices That Impact Cash Flow

The contract provisions dictate the responsibilities of those parties entering into an agreement. Despite these provisions, the actual performance on a contract might be considerably different. The contract merely stipulates the limits within which the parties must act. Without prior experience with or "inside" information about a specific owner, the contractor may simply assume that the owner will act strictly within the limits of the contract. Such a contractor may bid a project differently than a contractor who knows about an owner's standard operating procedures. For example, the contract may give the owner tremendous latitude, to the contractor's detriment, but experience might show that the owner's on-site practices are not as severe as permitted in the contract.

Only experience or trade talk can help a contractor gain insights about an owner's on-site practices. For example, the contract may provide for the release of all retainage only after all punch list items have been completed. An owner may, however, elect to return most of the retainage at substantial completion and withhold only an amount to "cover" the anticipated costs of completing the punch list items.

The contract may permit the owner to withhold ten percent of all contractor payments for the entire project but permit withholding to drop to zero on those payments made after the project is 50 percent complete. This reduction in retainage is permitted only if satisfactory progress is being made at the time the project is half completed. If it is well known to the contractor that the owner rarely reduces the retainage at the 50 percent point in a project, a higher bid will be justified.

Whenever the discretion of the owner is part of the contract provisions, the contractor should obtain information on how the owner generally acts, possibly at a prebid conference. Obviously, the owner's actions either can be adverse to a contractor or they might be in the contractor's favor. In addition to the retainage practices mentioned above, owner discretion is involved in the payment for materials delivered but not installed in the project, the approval of monthly and final payments, and the timing of monthly and final payments.

The Cash-Flow Analysis

Cash-flow analysis consists of a detailed examination of the disbursement of funds and the receipt of revenues. There are two major reasons for conducting a cash-flow analysis. First, cash flow reveals if surplus funds are available during a project or if a negative cash position will occur during construction. The second purpose, related to the first, is to help establish the appropriate markup to apply on a bid.

The cash position of a contractor during a project, whether positive or negative, is of vital importance to the contractor. A negative cash position means that the revenues obtained on a project were insufficient to meet the financial obligations of the project. Thus, other funds within the company must be used or money must be obtained from an outside source. If the outside funds are to be obtained from a bank, it is important for the contractor to anticipate the need for such funds. A banker will not be favorably impressed with a contractor who did not anticipate the need for funds but suddenly needs them to satisfy an immediate indebtedness.

If the cash position is positive, the contractor may wish to invest the surplus funds for a short duration. These funds may be invested in a money market account, treasury notes, or certificates of deposit. The actual decision of where the funds will be deposited depends on the amount of funds available and the duration for which the surplus funds can be invested.

The flow of funds—whether revenues or disbursements—generally is plotted against project time. The abscissa is shown as the project duration and the ordinate is the cumulative funds flow (see Figure 7.1). The general shape of this information is in the form of a "lazy S." Although most curves take on a "lazy S" appearance, the relationship of the disbursements to the receipts dictates whether the cash position is positive or negative.

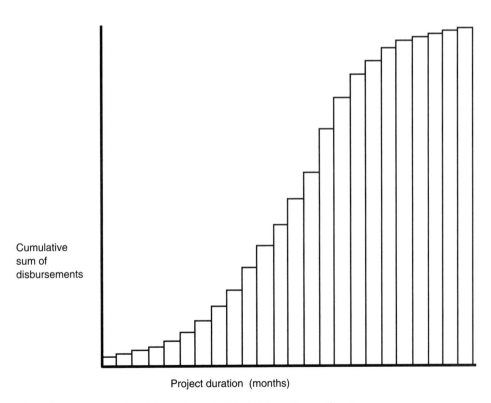

Figure 7.1 Example of the "Lazy S" Cash-Flow Curve for Expenses

The Sports Facility Project

The costs associated with the activities for the sports facility project described in Chapters 1, 3, and 6 are shown in Table 7.1. If the costs represent an accurate portrayal of the actual costs to be incurred, the information can be very valuable. For the contractor, the costs would be based on the estimated costs, while for the owner, the costs would be based on the schedule of values or the payment schedule.

The cash-flow analysis of the sports facility project is shown in Figure 7.2. The "lazy S" curve is somewhat apparent (cumulative cost line superimposed over the daily costs), but this would be clearer on a cash-flow analysis of a project with more activities that covered a longer duration. Also, if the activities were to be subdivided into smaller units of time, the cash flow would probably bear a closer appearance to the traditional "lazy S" curve. Note that the weekends are nonworking days in this schedule and do not impact the cash flow.

The Present Worth of Cash Flow

Although contractors tend to be most sensitive to their actual cash position throughout the life of a project, of equal importance is the consideration of the time value of money. Typically, the simplest analysis is to convert the value of all funds to an equivalent present worth amount at the time of bid submittal or contract award. This can be done by taking the present worth of the difference between the anticipated funds receipts and the anticipated disbursements. This analysis can be performed effectively if the flow of funds is accurately predicted and if an appropriate interest is utilized.

The example that follows presents the cash flow of a simple project. For this project, the total expenditures amount to $1,500,000 and the receipts are $1,700,000, for a total profit of $200,000. However, the present worth of this profit is not $200,000, and the

Table 7.1
COSTS OF THE ACTIVITIES FOR THE SPORTS FACILITY PROJECT

Activity	Dur	Predecessors	Cost
Mobilize	2	—	$5,000
Stock materials	4	Mobilize	$22,000
Clearing and grubbing	6	Mobilize	$8,000
Grading for road	7	Mobilize	$2,500
Finish grade	5	Stock materials, clearing and grubbing	$3,000
Prefab bleachers	16	Stock materials	$18,000
Landscape	12	Stock materials, clearing and grubbing	$35,000
Pave roadway	8	Grading for road	$30,000
Place tennis court	10	Finish grade	$20,000
Erect/paint bleachers	7	Prefab bleachers, clearing and grubbing	$16,000
Curbing	5	Landscape, pave roadway	$6,000
Final inspection/cleanup	3	Place tennis court, erect/paint bleachers, curbing	$4,000

Money and Network Schedules 173

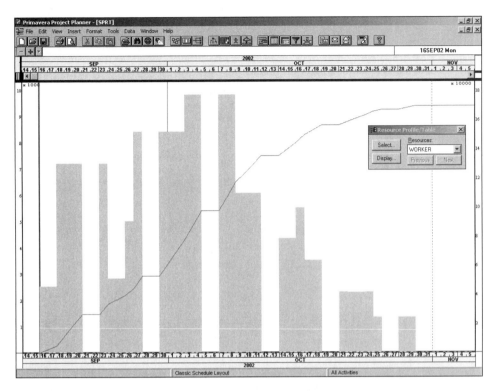

Figure 7.2 A Computer-Generated Cash Flow of the Sports Facility Project *(Primavera Project Planner®)* (Reprinted with permission from Primavera Systems, Inc.)

present worth will vary depending on specific conditions. In the example, all expenditures are simplified as being subcontractor payments. The retainage is ten percent, the final payment is made two months after substantial completion, the owner makes periodic payments on the 15th of the month, the subcontractors will be paid on the 25th of the month, and the project duration is 180 days. If the interest rate is zero percent, the present worth will be $200,000. Other present worth calculations are shown in Table 7.2.

The analysis could be carried on further to include the influence of varying the times when the monthly payments are received. Where major portions of the work are subcontracted, it is common for a general contractor's profit margin to be about two or three percent. With this in mind, it is obvious that the cash-flow analysis can have a significant impact on the decisions of a company.

The Value of Cash-Flow Analysis

The primary purposes for conducting cash-flow analyses have been discussed. Other parties will also have an interest in such analyses. These include the bankers, sureties, insurance carriers, equipment dealers, credit-reporting agencies, clients (owners),

Table 7.2
VARIOUS CASH-FLOW ANALYSES OF A PROJECT

Subs Are Paid on the 25th of the Month

Interest Rate	Retainage Percentage	Present Worth
0%	10%	$200,000
6%	10%	$192,520
6%	0%	$197,070
12%	10%	$185,420
12%	0%	$194,180
14%	10%	$183,150
14%	0%	$193,240

Subs Are Paid on the 20th of the Month

Interest Rate	Retainage Percentage	Present Worth
0%	10%	$200,000
6%	10%	$191,520
6%	0%	$196,070
12%	10%	$184,470
12%	0%	$192,230
14%	10%	$180,900
14%	0%	$190,990

If the Last Payment Is Received 15 Days After Substantial Completion

Subs Are Paid on the 20th of the Month

Interest Rate	Retainage Percentage	Present Worth
0%	10%	$200,000
6%	10%	$192,640
6%	0%	$196,070
12%	10%	$185,590
12%	0%	$192,230
14%	10%	$183,330
14%	0%	$190,990

If the Last Payment Is Received 75 Days After Substantial Completion

Subs Are Paid on the 20th of the Month

Interest Rate	Retainage Percentage	Present Worth
0%	10%	$200,000
6%	10%	$190,420
6%	0%	$196,070
12%	10%	$181,380
12%	0%	$192,230
14%	10%	$178,500
14%	0%	$190,990

stockholders, and business partners. The extent of detail required varies for different parties as different purposes are being served. Nonetheless, a carefully prepared cash-flow analysis will help the overall operations of a construction firm. Figure 7.3 shows the cost breakdown that one developer used to characterize the cost distribution through the life of one particular type of project, namely the construction of an apartment building. Figure 7.3 shows how expenditures might occur over the life of a construction project. Figure 7.4 is essentially the same cash-flow diagram with the revenues or reimbursements shown in addition to the disbursements. This type of analysis quickly shows when a "short fall" will occur in the cash flow (money needs to be borrowed) and when a surplus of cash might occur (investment opportunity).

Work Item Description	Percent of Total	Cumulative Amount
Slab (forms, rebar, concrete)	8	8
Plumbing (site and slab)	3	11
1st floor framing (through joists)	10	21
Deck 1st floor	2	23
2nd floor framing (through joists)	10	33
Deck 2nd floor	2	35
3rd floor framing (through joists)	11	46
Roof deck	2	48
Rough plumbing complete (through roof)	4	52
Roofing	3	55
HVAC rough in	4	59
Electrical rough in	4	63
Drywall, tape and float	5	68
Painting	4	72
Cabinets	4	76
Trim carpentry	3	79
Masonry complete	2	81
HVAC final	4	85
Plumbing final	3	88
Electrical final	4	92
Appliances	3	95
Carpet, vinyl flooring, ceramic tile	4	99
Final clean-up	1	100

Figure 7.3 Sample Cost Breakdown for an Apartment Building Project

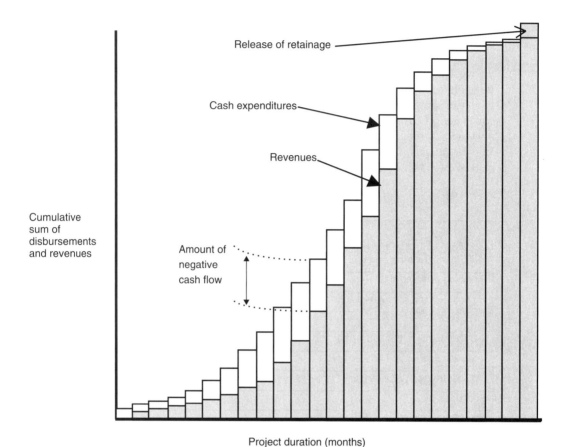

Figure 7.4 Example of the "Lazy S" Cash-Flow Curve for Revenues and Expenses

TIME–COST TRADE-OFFS

There is a saying that the whole is often greater than the sum of its parts. In some respects this is applicable to projects initially scheduled using a CPM-based approach, but in a manner that is quite different than the saying normally implies. The following discussion shows how it is possible that the originally estimated project duration is not necessarily the least time solution *nor* the least cost schedule for the project, in spite of the fact that each activity within the project was originally planned to be done in the most efficient (e.g., least cost) manner. The discussion will also provide the foundation for future contractor decisions related to determining whether it is economically advantageous to accelerate a project to make up for lost time. Fundamental to the discussion is the need to recognize that there are several different elements of project costs.

Direct Costs

These are the costs readily identified as being related to putting the facility components in place. They represent the costs of the resources used by activities, such as the materials installed, workers and equipment used, and subcontractors. The costs of materials and the costs of the subcontracted work tend to be relatively fixed or not subject to considerable variance. The biggest variable concerns the costs associated with labor and equipment. If the productivity estimates are overestimated, these costs will overrun. Overestimates on the productivity will result in a longer duration to complete the project. If the productivity is underestimated, the costs will be below budget for that work item and the time to complete the activity will also be reduced.

Indirect Job Costs (Job Overhead)

Indirect costs are those that are not specifically identified as being associated with a particular work item. These costs are generally incurred whether or not productive work is actually accomplished. Every project requires certain costs to be incurred as long as the project is underway. These are specifically related to one job, such as a site superintendent, job shack, or security fencing. They are generally not related to any particular activity in the project. Since these costs are incurred regardless of the amount of work put in place, longer project durations will result in higher indirect costs.

Overhead (Company Overhead)

This is the element of cost that covers the company-specific costs of running the business, such as those for corporate office personnel, office services and supplies, and administration. These costs continue even if only one job is being conducted by the company.

Profit

As with overhead, profit is company-specific and represents the excess monies earned by a firm over its total costs. As it provides investors with a return on their investment in the company, it is necessary to consider profit as a "cost" of a project. It is not specific to any project or activity, however, as investors are normally concerned only with overall company performance.

It is necessary to look at a combination of project costs to determine the overall cost of a project. By examining further the relationship between time (e.g., project and activity durations) and costs, it can be determined how decisions about one of these considerations relate to the other. Following is the development of the logic behind time–cost trade-offs and how an analysis can be performed using a CPM network.

Assumption: *Increasing or decreasing an activity's duration will lead to increased direct costs for that activity.* This impact is shown in Figure 7.5. Note that as the duration for the project is reduced, the incremental cost of further reductions becomes quite high. Ultimately a point is reached at which the project cannot be realistically reduced any further even with the infusion of additional resources. As the duration increases, the

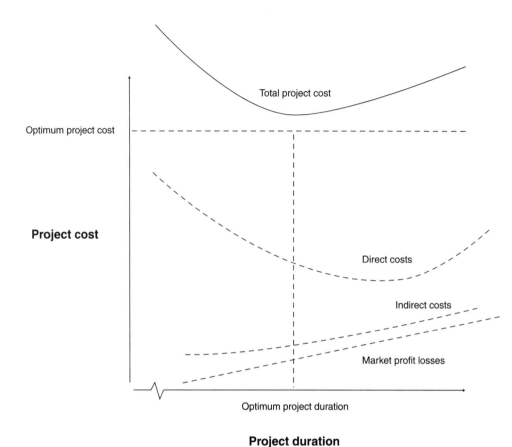

Figure 7.5 General Relationship of Project Cost to Project Duration

incremental costs increase but at a rate that is lower than the daily rate to reduce the duration below the optimum cost duration. The added costs associated with increasing the project duration consist of the direct costs (loss of worker morale, reduced productivity, material spoilage, etc.), indirect costs (managerial salaries, job site utilities, liquidated damages, etc.), and market losses (lost revenues when the facility is not put into use, other similar facilities completed ahead of the project in question, etc.) As the duration is extended further, a point will ultimately be reached where the owner will probably terminate the contract.

As discussed earlier, the duration of an activity is determined by selecting both a method for carrying it out and the amount of resources to apply to it. Presumably the selected resource application rate (for example, crew size) provides for carrying out the activity as efficiently as possible (meaning as inexpensively as possible). This can be considered to be the normal manner of accomplishing the task, as only unusual circumstances would cause the contractor to select other than the lowest-cost method of accomplishing the task.

Note that the duration of the activity is derived from decisions made based on the cost of the activity. Thus, it is possible that the activity could be done in less or more time. However, given the above assumptions, if the duration is either increased or decreased, the cost of the activity will always increase.

When discussing resources, an assumption is made that the relationship between daily resource utilization rates and an activity's duration is linear. Thus, an activity requiring a 20-day duration with a crew of four could be done in 16 days using a five-person crew. In both cases the total amount of resource required is 80 person-days. If true, then the cost for the activity will be the same regardless of duration, as long as no differential is paid to a smaller crew and no overtime is worked. This unrealistically assumes that productivity per worker is not impacted by changes in the crew size.

In discussing costs, the relationship between resource rates and durations is recognized as unlikely to be linear. Instead, it is realized that a variety of inefficiencies develop as greater or fewer resources than are "optimal" are applied. The inefficiencies result from congestion, unbalanced work-item-per-person relationships, added personnel idle time, loss of learning curve productivity benefits, or additional supervisory demands. Thus, some level of additional cost will be necessary to alter an activity's duration.

A major issue in dealing with the trade-offs between time and cost is determining exactly how much additional money is necessary to decrease an activity's duration. The specific problem is to determine the cost required to shorten an activity's duration by one day. This cost per day then represents the slope of a curve that plots the cost of an activity against the duration of the activity.

The shape of the time–cost relationship for activities varies depending on the nature of the means required to make the time reductions. The straight-line relationship between activity duration and cost will be assumed when project compression occurs. Many shapes may actually be applicable, including those shown in Figure 7.6.

In its simplest form a time–cost curve is a straight line. This indicates that each additional day that an activity is shortened costs the same additional amount as the preceding day. This can be seen as unrealistic when one recognizes that there must be some minimum duration for an activity when no further reductions can be made, regardless of the amount of resources (costs) that are applied. With this in mind, it is intuitive to expect that as this limit is reached, the incremental cost of reducing an activity by one unit of duration becomes greater than was previously the case. Consequently, activity time–cost curves tend to be of hyperbolic shape rather than straight lines between the normal duration–minimum cost point and the minimum duration–maximum cost point.

Assumption: *Decreasing a project's duration will lead to lower indirect costs.* Indirect costs are those incurred because a project is being undertaken. Some indirect costs are fixed, but most are variable. As long as the project is underway, these variable costs will be incurred; once the project ends, these costs should be eliminated. Thus, it can be assumed that the variable components of indirect costs for a project are the same for each day of the project. If a project's duration is plotted against its variable indirect cost, the curve will be a straight line with a positive slope. Consequently, if the project is finished one day sooner, the indirect costs will be reduced by one unit of the slope of the line.

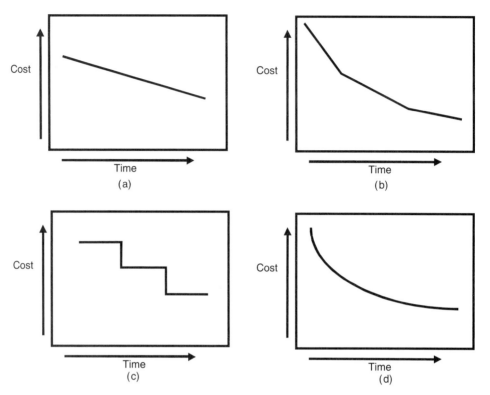

Figure 7.6 Sample of Time–Cost Relationships for Different Activities

Assumption: *A project's duration can be decreased by decreasing the duration of one or more activities on the critical path(s).* This observation is derived directly from the critical-path methodology and establishes the direct relationship between decreasing one or more critical activity durations in order to decrease the duration of the project.

Assumption: *Decreasing a project's duration may increase or decrease the total cost of a project depending on whether the additional direct costs required to decrease activity durations are greater than or less than the indirect cost savings of decreasing the project's duration.* A project's total cost combines both direct and indirect costs. Thus a graph of total cost versus duration involves adding the cost amounts contained in the direct and indirect cost curves. Remember that direct cost curves have a negative slope (costs increase as duration decreases) and indirect costs have a positive slope (costs decrease as duration decreases). The total cost-curve's slope at any point therefore depends on whether the slope (incremental cost) of the direct cost curve is less than that of the indirect cost curve.

This relationship suggests that in performing a time–cost trade-off analysis, it is necessary to determine the cost of decreasing the critical path by one day. Of course, the costs will vary depending on which activity's duration is decreased. Naturally, the critical activity with the least direct cost slope (lowest additional cost per day of shortening) will be selected first to minimize the cost of shortening the project.

As the critical path of a network is decreased, it must be remembered that all noncritical paths within the network lose a like amount of total float. Thus the extent to which an activity can be shortened and still obtain a shortening of the project is limited by the minimum amount of float that exists in any parallel chain of activities. As the duration of the project is reduced, the number of critical paths through the network increases. To reduce a project's duration, it is necessary to reduce the duration of all critical paths, and it can quickly become very expensive to shorten a project, as greater numbers of activities must be simultaneously shortened.

Four Different Solutions for Each Network

Whenever a construction project is undertaken, the client or owner will have certain objectives that are to be satisfied. Some will want to have the project completed in the shortest amount of time, some will want to have the project completed at the lowest cost, and other clients may have other project goals. Project scheduling can help meet the stated objectives. The schedule can be viewed in several different ways to determine the optimal way to satisfy the client. Some of these basic approaches will be described.

All Normal: The original network and activity durations result in what is termed the "all normal" solution because it is based on each activity being performed in its "normal," least cost manner. As noted, however, this is not necessarily the least cost or least time solution to scheduling a project.

Least Cost: Considering both direct and indirect costs, it may be possible to find a project duration that minimizes these total costs by paying more to decrease one or more critical activities, thereby recouping greater savings in indirect costs. This least cost solution will be shorter than the all normal solution.

Least Time: Normally a project can continue to be shortened beyond its least cost point. However, this additional shortening increases the total cost of the project. Eventually the least time solution is reached, which is the absolute minimum duration within which the project can be completed. This point is reached when no activities in one (or more) critical path(s) through the network can be physically shortened, no matter how many resources are applied.

All Crash: In this solution, every individual activity has been shortened as much as physically possible. Its duration is the same as the least time solution, but its cost is greater. Direct costs have continued to rise as remaining activities are taken to their minimum durations without realizing any additional indirect cost savings due to no further reductions in the project's duration. A fully crashed schedule or crash schedule is when all activities have been reduced to their shortest duration. This is not an efficient approach as some noncritical activities will be shortened without

Logically Reducing Project Duration

Figure 7.7 consists of a series of activities that can be reduced in duration (below those shown) in varying amounts and for differing costs. However, it would not be cost effective to randomly select any activity for shortening. This should be done in a rational manner. To begin the time–cost trade-off in a systematic fashion, it is important to first make some basic computations. First compute the early start and early finish times for each of the activities.

Once the early start and early finish times are computed, it is possible to compute the link lag values of each of the "links" between the activities. For example the link lag value for the link between Activities B and E is 0. This is computed by subtracting the early start value of the later activity (Activity E) from the early finish time of the prior activity (Activity B). The same procedure is used to compute a value of 4 as the link lag between Activities E and H (16–12). In Figure 7.8, the nonzero link lag values are shown on the respective links. The double connecting lines mean that the link lag value is 0.

From observation, it should be clear that there is at least one path between the first activity and the last activity in the network for which all the link lag values are 0. In this case, the activities so connected are A, B, F, H, and L. These activities form the critical path. If the project is to be shortened, it is imperative that the duration of one of the activities on the critical path be shortened. Without shortening the duration, the project can be completed in 27 days for a cost of $5,300. This is the normal duration

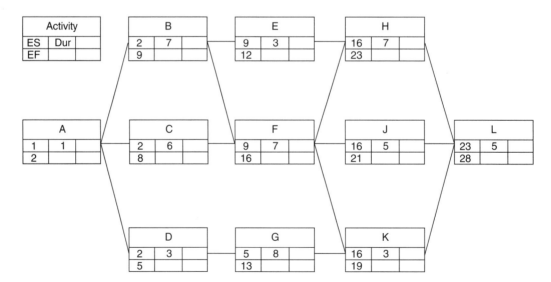

Figure 7.7 Example of Project for Time Compression

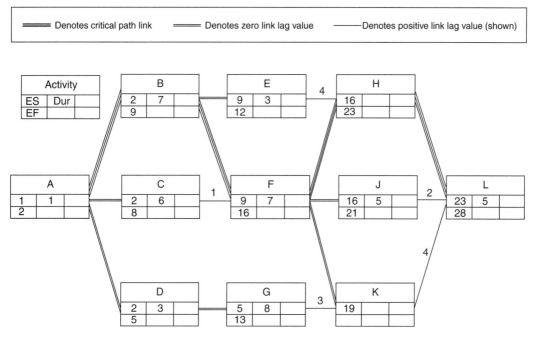

Figure 7.8   Sample Project Showing Link Lag Values

cost. It is assumed to be the normal cost solution for the project. Other considerations (such as overhead, liquidated damages provisions, changes, or needs of other projects) may make it desirable to complete the project in a shorter or longer duration. Any shortening of the project will result in an increase of the direct cost of the project. The specific information concerning activity costs is shown in Figure 7.9.

Assume that the following information exists about the various activities:

Of the critical activities (A, B, F, H, L), Activity A cannot be shortened. Of the remaining activities, Activity B can be shortened at a cost of $200 per day, Activity F for $150 per day, Activity H for $250 per day, and Activity L for $350 per day. The logical choice for the first activity to shorten is Activity F, which has the lowest incremental cost for making duration reductions. Note that the cost of the cycle will be $300 when Activity F is shortened by two days, but the selection of the activity to shorten must be based solely on the minimum cost per day and not the cost of the cycle (see Figure 7.10). It is evident that Activity F can be shortened two days (see Figure 7.9), but how many days should the activity be shortened? The answer lies in computing the network interaction limit (NIL).

The network can be considered as something similar to a rigid frame. If one activity is shortened, it is possible that other activities might be impacted. This can be determined graphically by drawing a line vertically down through the network. To determine the network interaction limit, a vertical line will pass through the activity or

184   Chapter 7

Activity #	Normal Duration	Crash Duration	Normal Cost	Crash Cost	Days to Shorten	Cost per Day
A	1	1	$ 800	$ 800	0	—
B	7	4	1,000	1,600	3	$200
C	6	4	300	500	2	100
D	3	2	400	800	1	400
E	3	1	100	200	2	50
F	7	5	500	800	2	150
G	8	4	200	1,400	4	300
H	7	6	350	600	1	250
J	5	3	700	850	2	75
K	3	2	500	1,000	1	500
L	5	4	450	800	1	350
			5,300			

Figure 7.9   Duration–Cost Data for Project Activities

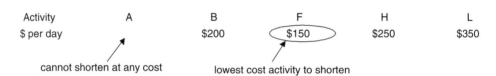

Figure 7.10   Identifying Activities to Select for First Compression Cycle

activities being shortened and through any positive link lags that may be in the path of the line. A zero-value link line can be crossed if its value will be increased by the activity duration being reduced. Examine Figure 7.11. Note that if Activity F is shortened, the link lag values of E–H and G–K will be reduced by the same number of days.

Since the link lag value of E–H is 4, this is the number of days that Activity F could be shortened before this link lag value becomes 0. Similarly, the link lag value of G–K will become 0 after Activity F has been shortened three days. The point at which one of the link lag values becomes 0 is known as the *network interaction limit*. When Activity F is being shortened, the network interaction limit is 3, the smaller of the link lag values of E–H and G–K. By examining Activity F, it is observed that it can only be shortened two days. Thus, Activity F should be shortened by two days, which is the smallest of the link lag values or the NIL and the maximum number of days that the activity can be shortened. This information is reflected in Figure 7.12.

Figure 7.13 has been updated to reflect the link lag values after Activity F has been shortened by two days. The project duration is now 25 days and the cost per day to shorten it by two days was $150 per day. The project cost at 25 days' duration is $5,600.

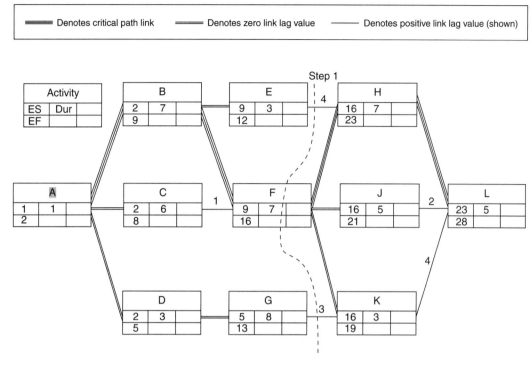

Figure 7.11 Location of the First Compression Cycle

Cycle #	Activity to Shorten	Can Be Shortened	Nil	Days Shortened	Cost per Day	Cost for Cycle	Total Cost	Project Duration
0								27
1	F	2	3	2	$150	$300	$5,600	25

Figure 7.12 Summary of the First Compression Cycle

A rational approach to reducing the project duration in a logical fashion requires some bookkeeping of what has been done and what can be done concerning the reduction of activity durations. For example, the link lags must be updated as noted, and it must be noted that Activity F cannot be shortened further (note shading of Activities A and F in Figure 7.13). We can now consider making a further reduction in the project duration. As no additional critical paths were created when Activity F was shortened by two days, the same activities are still under consideration. As is readily observable (Figure 7.14), Activity B should now be shortened.

Activity B can be shortened a total of three days at a cost of $200 per day. This is the lowest cost activity to shorten. How many days should Activity B be shortened (see Figure 7.15)? The line drawn through the Activity B and the nonzero link lag values

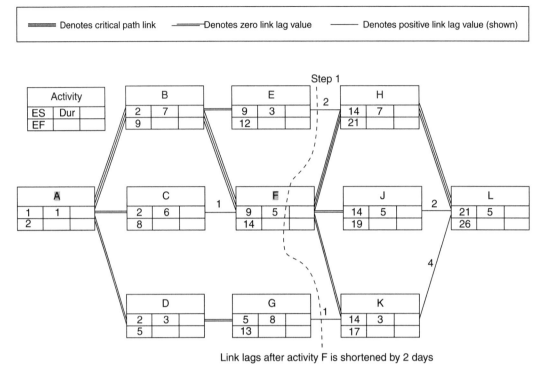

Figure 7.13 Updated Network After First Compression Cycle

Figure 7.14 Identifying Activities to Select for Second Compression Cycle

shows that the network interaction limit (NIL) is one day. The NIL is the number of days an activity can be shortened before some other link lag value becomes 0. In this second step, the NIL is one day. Note that if Activity B is shortened one day, link lag values for C–F and G–K become 0. Thus the NIL is one day. Always examine the number of days that an activity can be shortened and the number of days in the NIL and select the smallest of these two values.

The new project duration is 24 days and the cost of completion is $5,800. This last day of compression was achieved at a cost of $200 per day (see Figure 7.16).

The new updated link lag values are shown in Figure 7.17. Note that Activity C is now an additional critical activity and the link lag value of G–K is 0.

To further reduce the project duration, consideration must be given to shortening Activities B and C (jointly), Activity H, or Activity L. In this case, the decision is

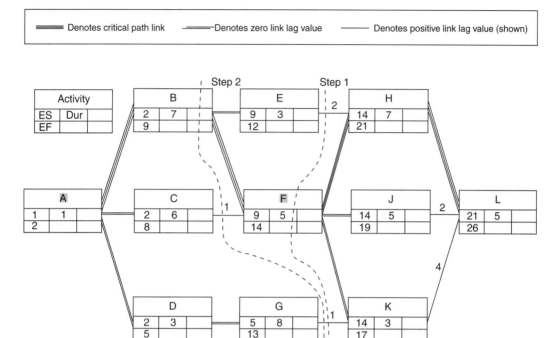

Figure 7.15 Location of the Second Compression Cycle

Cycle #	Activity to Shorten	Can Be Shortened	Nil	Days Shortened	Cost per Day	Cost for Cycle	Total Cost	Project Duration
0							$5,300	27
1	F	2	3	2	$150	$300	$5,600	25
2	B	3	1	1	$200	$200	$5,800	24

Figure 7.16 Summary After Second Compression Cycle

based solely on costs. Activity H is selected for shortening. This is more easily seen when the costs are written out in Figure 7.18.

The NIL for shortening Activity H is two days (link lag value of J–L), but Activity H can be shortened only one day, so Activity H will be shortened by one day (see Figure 7.19). The project duration is now 23 days and the cost is $6,050 (see Figure 7.20).

After Activity H is shortened by one day, the network link lag values are now as shown in Figure 7.21. Note that no link lag values became 0 after step 3.

The next iteration of time compression in the network considers Activities B and C (together) and Activity L (see Figure 7.22). The least costly option is to shorten Activities B and C. Note that if only B or only C were to be shortened, the project duration would remain the same.

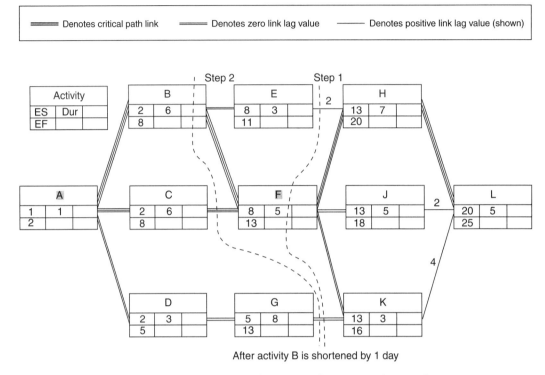

Figure 7.17 Updated Network After Second Compression Cycle

Figure 7.18 Identifying Activities to Select for Third Compression Cycle

As shown in Figure 7.23, when Activities B and C are jointly shortened, the section cut in the network passes through link line K–L and link line F–K. The NIL can now be determined. To shorten Activities B and C, the NIL is 3, the link lag value of link line K–L. Careful study of the network will reveal that in this compression step the link lag value of F–K will increase as Activities B and C are shortened. This compression cycle will result in the link line F–K becoming positive. By shortening the durations of Activities B and C, the start and finish dates of Activity F will also be shortened, but this will not shorten the start date of Activity K. The start date of Activity K will be dictated by the finish of Activity G. As Activity G is not altered by this compression step, the start of Activity K is similarly unaltered. A review of the activity data shows that Activity B and Activity C can each be shortened by only two days. Thus, the compression in this step will be two days. The project information is now as in Figure 7.24.

Money and Network Schedules 189

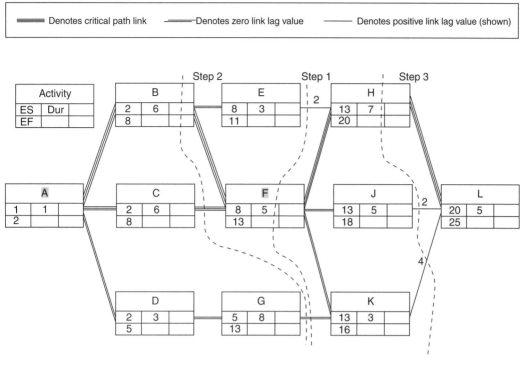

Figure 7.19 Location of the Third Compression Cycle

Cycle #	Activity to Shorten	Can Be Shortened	Nil	Days Shortened	Cost per Day	Cost for Cycle	Total Cost	Project Duration
0							$5,300	27
1	F	2	3	2	$150	$300	$5,600	25
2	B	3	1	1	$200	$200	$5,800	24
3	H	1	2	1	$250	$250	$6,050	23

Figure 7.20 Summary After Third Compression Cycle

The network has been updated. Note the changes in the link lag values in Figure 7.25. The link lag value for link line K–L was reduced by two days, as determined in prior steps. Note that the link lag value of link line F–K was increased by two days. While this may appear odd, it simply means that Activity F will be completed earlier than originally scheduled but that the completion date of Activity G has not changed. Thus, the start date of Activity K is not changed in this compression cycle.

At this stage of the network compression, it is obvious that Activity L is the only activity remaining that can be shortened (see Figure 7.26). By observation, the NIL for Activity L is infinity. A section cut through the network will not cut through any link

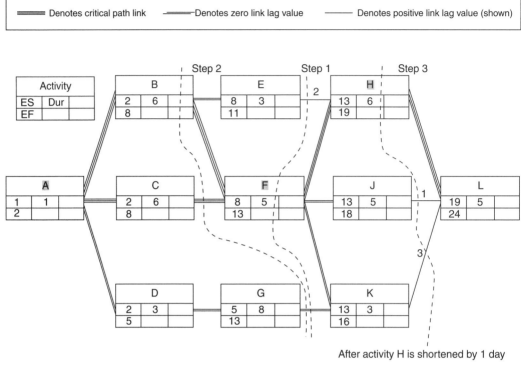

Figure 7.21 Updated Network After the Third Compression Cycle

Figure 7.22 Identifying Activities to Select for Fourth Compression Cycle

lines, so the NIL is limitless. The compression is determined by the number of days that Activity L can be shortened, namely one day. Following this step the project summary is now as in Figure 7.27.

At this point, the network cannot be compressed further. This point is reached whenever all the activities on a critical path can no longer be shortened. While there are other activities that can be shortened, the project duration would not be altered by doing so. Figure 7.27 summarizes all the compression information. Note especially the values in the column entitled "Cost per Day." Note that the values increase consistently. This is a quick check to determine if an obvious error might have been made. The values in this column must always increase in each successive compression step. In some rare cases the values could be equal, in which case the scheduler actually may have more than one possible activity or activities to shorten at a given cost per day. When this occurs, the scheduler must decide which activities to shorten first. Some

Money and Network Schedules

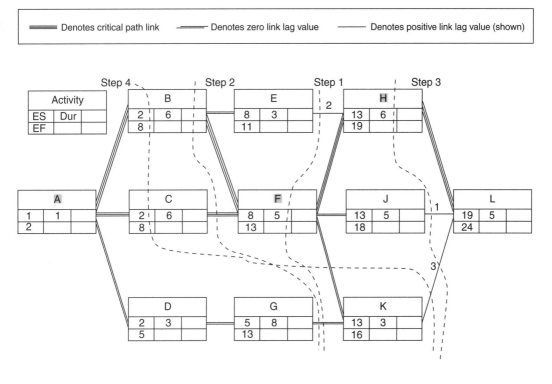

Figure 7.23 The Location of Fourth Compression Cycle

Cycle #	Activity to Shorten	Can Be Shortened	Nil	Days Shortened	Cost per Day	Cost for Cycle	Total Cost	Project Duration
0							$5,300	27
1	F	2	3	2	$150	$300	$5,600	25
2	B	3	1	1	$200	$200	$5,800	24
3	H	1	2	1	$250	$250	$6,050	23
4	B, C	2	3	2	$300	$600	$6,650	21

Figure 7.24 Summary After Fourth Compression Cycle

sort of prioritization must occur. As a general rule, under such circumstances higher priority should be given to the activity that occurs earlier in the network. Other factors may also be present that might establish a different priority. When this project has been fully compressed, the link lag values would be as shown in Figure 7.28.

The steps that have been outlined for the compression of a network can lead to a very rational approach to reducing the duration of a project. While this approach may be applicable in many situations, the model cannot account for all the unique circumstances that may exist on a project. For example, if a project duration were to be compressed for

192   Chapter 7

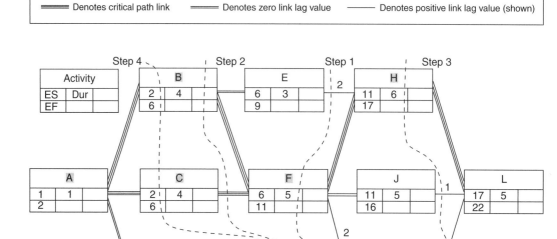

Figure 7.25   Updated Network After Fourth Compression Cycle

Activity	A	B, C	F	H	L
$ per day					$350

Figure 7.26   Identifying Activities to Select for Fifth Compression Cycle

Cycle #	Activity to Shorten	Can Be Shortened	Nil	Days Shortened	Cost per Day	Cost for Cycle	Total Cost	Project Duration
0							$5,300	27
1	F	2	3	2	$150	$300	$5,600	25
2	B	3	1	1	$200	$200	$5,800	24
3	H	1	2	1	$250	$250	$6,050	23
4	B, C	2	3	2	$300	$600	$6,650	21
5	L	1		1	$350	$350	$7,000	20

Figure 7.27   Summary After Fifth Compression Cycle

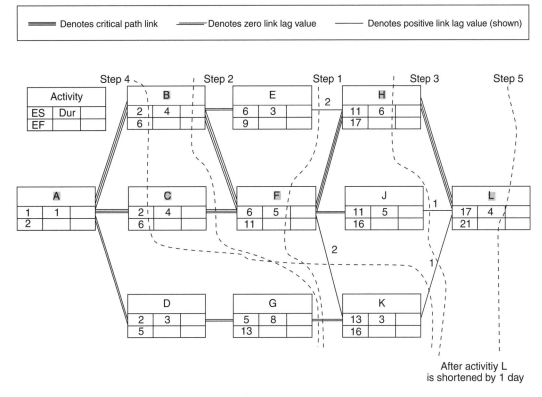

Figure 7.28 Updated Network After Fifth and Final Compression Cycle

a duration of five days only, the project scheduler and others with responsibility for the project will probably be reluctant to accept a solution that calls for the reduction of activities that occur at the end of the project. There is greater confidence if the reduction can be made earlier in the project as it affords the opportunity to shorten other activities if subsequent reductions become necessary.

FINAL COMMENTS

The time consumed by a project directly influences the costs incurred on a project. The adage of "time is money" is not only true but applicable on construction projects. Many aspects of the costs on a project are linked to issues also related to time. These are established in the contract, by law, and through industry practice. Ideally, the duration established for a project places the costs at the lowest possible level for that project. When this occurs, the project costs will increase whether the duration is reduced or increased. On some projects it is necessary to reduce the project duration. When this occurs, the reduction should be made systematically so that the cost increases are minimal at each progressive stage of project compression.

REVIEW PROBLEMS

1. Determine the logical steps to reduce the following project to its shortest duration.

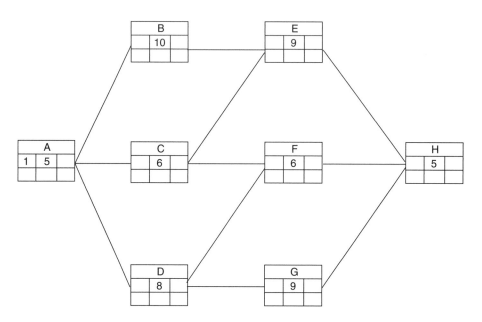

Activity #	Time Normal	Time Crash	Cost Normal	Cost Crash	Time Difference	Cost Difference	Slope $/Day
A	5	5	$100	$100			
B	10	7	$500	$560			
C	6	4	$300	$400			
D	8	6	$400	$600			
E	9	5	$200	$320			
F	6	5	$100	$140			
G	9	7	$320	$400			
H	5	4	$410	$500			

Cycle #	Activity to Shorten	Can Be Shortened	Nil	Days Shortened	Cost per Day	Cost for Cycle	Total Cost	Project Duration
0	—	—	—	—	—	—		
1								
2								
3								
4								
5								
6								
7								
8								
9								
10								

2. Determine the logical steps to reduce the following project to its shortest duration.

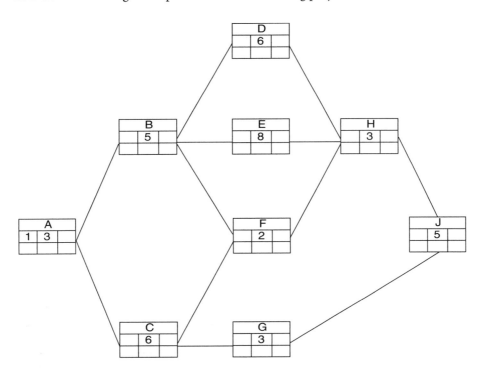

Activity #	Time Normal	Time Crash	Cost Normal	Cost Crash	Time Difference	Cost Difference	Slope $/Day
A	3	3	$600	$600	0	0	∞
B	5	4	$250	$340			
C	6	3	$180	$840			
D	6	3	$300	$450			
E	8	4	$200	$340			
F	2	1	$500	$560			
G	3	2	$700	$740			
H	3	2	$900	$1000			
J	5	5	$100	$100	0	0	∞

Cycle #	Activity to Shorten	Can Be Shortened	Nil	Days Shortened	Cost per Day	Cost for Cycle	Total Cost	Project Duration
0	—	—	—	—	—	—		
1								
2								
3								
4								
5								
6								
7								
8								

196 Chapter 7

3. Determine the logical steps to reduce the following project to its shortest duration.

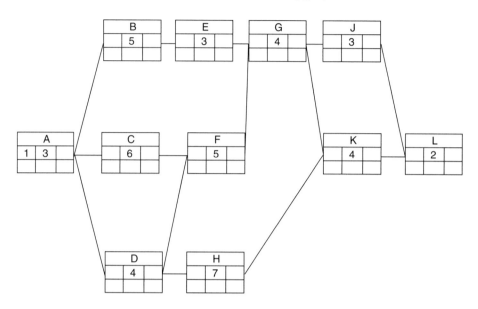

Activity #	Time Normal	Time Crash	Cost Normal	Cost Crash	Time Difference	Cost Difference	Slope $/Day
A	3	3	$100	$100	0	0	∞
B	5	3	$200	$500			
C	6	4	$400	$560			
D	4	3	$200	$240			
E	3	2	$200	$260			
F	5	2	$250	$520			
G	4	2	$200	$270			
H	7	5	$400	$440			
J	3	3	$200	$200	0	0	∞
K	4	1	$300	$600			
L	2	2	$500	$500	0	0	∞

Cycle #	Activity to Shorten	Can Be Shortened	Nil	Days Shortened	Cost per Day	Cost for Cycle	Total Cost	Project Duration
0	—	—	—	—	—	—		
1								
2								
3								
4								
5								
6								
7								
8								

4. Determine the logical steps to reduce the following project to its shortest duration.

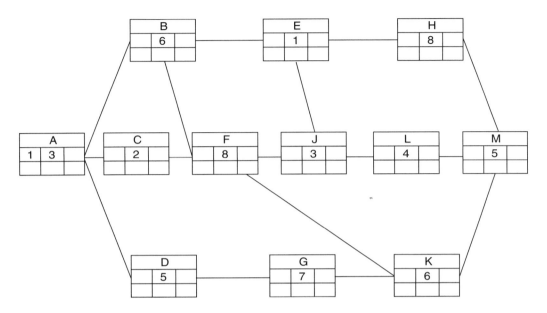

Activity #	Time Normal	Time Crash	Cost Normal	Cost Crash	Time Difference	Cost Difference	Slope $/Day
A	3	1	$500	$900			
B	6	3	$400	$700			
C	2	1	$800	$970			
D	5	3	$1000	$1240			
E	1	1	$100	$100	0	0	∞
F	8	4	$300	$540			
G	7	3	$1200	$1920			
H	8	5	$400	$1000			
J	3	2	$600	$690			
K	6	2	$700	$1340			
L	4	1	$600	$960			
M	5	5	$600	$600	0	0	∞

Cycle #	Activity to Shorten	Can Be Shortened	Nil	Days Shortened	Cost per Day	Cost for Cycle	Total Cost	Project Duration
0	—	—	—	—	—	—		
1								
2								
3								
4								
5								
6								
7								
8								
9								

5. Determine the logical steps to reduce the following project to its shortest duration.

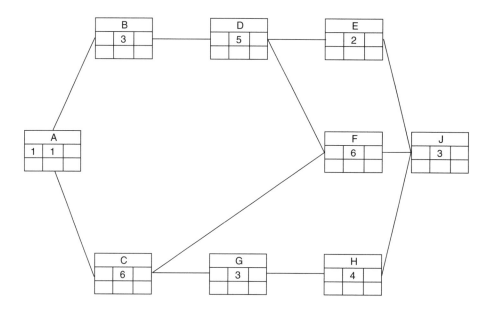

Activity #	Time Normal	Time Crash	Cost Normal	Time Difference	Cost Difference	Slope $/Day
A	1	1	$1,000	0	0	
B	3	2	$2,000	1	$100	$100
C	6	1	$1,000	5	$450	$90
D	5	3	$1,000	2	$60	$30
E	2	1	$1,000	1	$40	$40
F	6	2	$1,000	4	$320	$80
G	3	1	$1,000	2	$150	$75
H	4	3	$1,000	1	$60	$60
J	3	2	$1,000	1	$150	$150
			$10,000			

Cycle #	Activity to Shorten	Can Be Shortened	Nil	Days Shortened	Cost per Day	Cost for Cycle	Total Cost	Project Duration
0							$10,000	18
1								
2								
3								
4								
5								
6								
7								
8								
9								

8
Project Monitoring and Control

It's important to have a schedule, but it's more important to use it.

Project control can be reduced to two basic components. The first, monitoring, consists of a means of understanding what is happening on a project, obtaining information about the project by some means. The second component, control, consists of action taken in response to the information. Thus, it is not sufficient to simply know what is happening on a project; knowledge carries with it the responsibility to respond to the available information.

CONSTRUCTION TIME

Effective information retrieval requires that communications be clear. Where scheduling is concerned, it is important for all parties to have an understanding of the terminology used in relation to the schedule. One very important aspect of this includes the definition of *construction time*.

Construction progress is generally measured in terms of days. It might appear to the casual observer that a day is a discrete event and that there should be no confusion about its definition. However, two general uses of the term are commonly encountered, namely, working days and calendar days.

When preparing the estimate, the estimator generally thinks about the project duration in terms of working days. Working days consist of weekdays, excluding holidays. Since most activities occur on "working days," it is common for schedules that are prepared prior to contract award to be based on working days. Such schedules are easier to prepare than are schedules based on calendar days. A preliminary calendar-day schedule for which the start date has not been established cannot accurately predict when weekends and holidays will occur, so calendar-day durations are difficult to estimate.

Many owners prefer to use calendar-day schedules. A calendar day is any day of the year. The definition of a calendar day is universal, so little confusion exists when calendar days are used. Some activities are better measured in calendar days, and, in fact, sometimes working days may not be meaningful to an activity. For example, the procurement of materials is best defined in terms of calendar days. Likewise, the curing time for concrete will extend over weekends and holidays. It may also be simpler to establish regular updating on the basis of calendar dates.

As can be seen, both working-day schedules and calendar-day schedules have shortcomings. Some shortcomings are common to both. For example, prior to establishing the start date, the estimator cannot fix exactly the timing of specific activities on the calendar. For example, the timing of activities may be defined in general terms, but it may not be known whether an activity will occur in October or November. The difference of one month in the scheduling of some activities can have a significant impact on the efficiency with which they are performed. This is especially true in areas that have particularly rainy months.

Some projects must be completed by specified dates. These milestone dates are used on various types of projects. For example, the construction contract for a school typically requires the project to be completed prior to the first day of school in the late summer. A contract for the construction of a retail store may stipulate that the project be completed several weeks prior to major shopping seasons, such as before Christmas.

Regardless of the way time is measured on a project, it is important to be able to convert a working-day schedule to a calendar-day schedule and vice versa. Thus, the duration of a project can be stated in terms of both calendar days and working days. To properly convert between the two types of durations, one must have a clear impression of the definition of a working day. If a working day is defined as being any day, then calendar-day and working-day schedules will be the same. However, most working days exclude weekends and holidays. Some may exclude only Sundays and holidays. Some may even exclude a block of months during the winter. Other variations include each weekday and any other day that is worked being defined as a working day.

When manual calculations are made on a network, it is simplest to use either a beginning-of-day convention or an end-of-day convention. The preference typically depends on whether one focuses on the start time or the finish time of a particular activity. In most cases, it would appear prudent to use the beginning-of-day convention because the starting dates serve as a stronger coordination goal than the ending times. Ideally, the start times should be stated as being based on the beginning-of-day convention while the finish times use the end-of-day convention. While one must decide on the appropriate convention to use on manually solved networks, computer programs typically do not present this dilemma. Schedulers working manually can choose either beginning of day or end of day. Computer software will generally assign a beginning-of-day start date and an end-of-day finish date to activities. Thus, a computer program might show a one-day-duration activity as starting on day 29 and ending on day 29.

One issue not often clearly resolved concerns the definition of *float*. Are the total float and free float measured in terms of working days or calendar days? It is essential that it be defined in terms that match the time convention used in the definition of activity durations. For example, if durations are expressed in working days, then the float should be defined in the same manner.

EFFECTIVE SCHEDULING

The effective use of scheduling techniques requires control. To simply plan a project is not enough; the plan or schedule must serve as a guidance document. In order for a plan to be an effective guidance document, the plan must incorporate any limitations that may arise during the construction operations. This includes constraints imposed by limited resources. To be an effective project control tool, it is important that the schedule have the following attributes:

- It must be an accurate reflection of the way the job is actually constructed.
- It must be sufficiently flexible to accommodate changes and to also predict their impact.
- It must allow for corrections as necessary to keep the schedule current or up to date.

Control consists essentially of monitoring progress and updating the schedule as needed. Instrumental to the success of the monitoring effort is the need for clear communications. Perhaps two aspects of communication are worth pointing out at this time. This first consists of the need to communicate relevant information about the schedule to the field personnel. This must be directed in a form that will be clearly understood. For example, it would be inappropriate to send a craft foreman a copy of a two-thousand activity network as the sole means of relaying information about the project schedule. Perhaps the foreman is able to comprehend an entire network, but a bar chart might be more meaningful as a way to communicate general information about an entire project schedule. For more detailed information, again, it would not be prudent to expect a foreman to organize the crew members by studying a two-thousand activity network. A short interval schedule will be more effective and less time will be consumed in developing a plan of activities for the crew.

Second, the information obtained from the field personnel is important. An effort should be made to educate the field personnel about how information from the field will be used. If the field personnel recognize that the information they generate is important and will be used, the chances are greatly increased that the information from the project will be reasonably accurate.

It must not be forgotten that on long-duration projects communications should constantly be two-way in nature. This is perhaps the major shortcoming of communications on many projects. It is not sufficient to install a mechanism of data retrieval and have information channeled in only one direction. If the field personnel diligently provide information regarding project status, but never receive any indication that the information is used or is of value, the quality of data generated in the field will quickly deteriorate.

MONITORING PROJECT STATUS

In order for a schedule to be an effective management tool, it must be used. Whether or not a schedule is used, the act of developing it will pay dividends because thought processes have already focused on how a project is to be put together. However, this is not making optimal use of a schedule and it certainly does not constitute a schedule becoming a management tool.

The first step to using a schedule is to monitor progress in relation to the schedule. In order for the monitoring of construction activities to be a manageable task, the scheduled items must be organized in a rational manner. One common means by which projects are organized is by work breakdown structures (WBS). This is a systematic approach of defining the project so that each work item can be readily identified and controlled. In a WBS, each work item has a unique identification number that is linked directly to the cost codes. Thus, the WBS helps to monitor progress in terms of both time and money.

Many firms elect to develop their own unique numbering system for their projects. The objective is the same regardless of the actual system being used. The WBS may be set up so that the project is divided into the different physical work areas on the project. The system may also identify the work activities attributed to a particular subcontractor. The work may be described by the divisions in the technical specifications. In addition, the detail may identify the work being performed by the crew of a particular foreman. The general use of the WBS is to be able to highlight the work to be accomplished or attributed to a particular party. Thus, the masonry subcontractor's work can be reported in a simple report that has been filtered to identify only the masonry work. The work of the roofing subcontractor can be similarly reported. The work to be performed by a particular concrete foreman's crew can also be isolated. The work of the area superintendent might include the work of several crews and several subcontractors, but this could also be reported. It is also possible to report those work items that are associated with a particular phase of construction, such as the foundation work.

A carefully crafted numbering system that describes the work activities on a project can be useful for isolating desired work items. This numbering system might be quite complex. In some cases the numbering system might consist of up to a dozen or more characters. The numbering system may be similar to the example in Figure 8.1.

Ten character spaces are required for this numbering system. In the past this was essentially the only way for a WBS to be devised. With current scheduling software, the system need not be as complex. The WBS can be described in simpler terms with many of the new software programs. For example, the subcontractor or the supervisor responsible for a work item can be directly identified, eliminating the need to decipher complex numbers. While the traditional cost codes may still be employed by some firms to identify tasks to be scheduled, this is no longer a requirement for effective project monitoring.

Monitoring consists of receiving or maintaining accurate reports of the events that have taken place on the project and the specific work items that have been accomplished. Monitoring is essentially a means by which feedback is obtained on the progress on the project.

Project Monitoring and Control 203

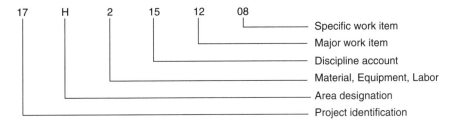

Figure 8.1   Example of a Cost Code System

Monitoring can be accomplished through direct observation. This consists of field visits made to the site to determine the actual status of a project. In very small companies, it is common for the owner of the firm to be resident on the project. Such an owner is not in need of any formal mechanism of obtaining feedback; the owner will process ongoing observations intuitively. Direct observations could also be made by a project manager who may visit the site on a daily basis. As with an owner's visits, little formality is needed to collect the needed information. If a scheduling network has been developed, it is a simple matter to determine the project's current status. Direct observation is appropriate as the sole monitoring mechanism where there is a short span of control or a shallow hierarchy.

As firms get larger and as projects get more complex, it becomes necessary to formalize the method of capturing information about construction progress. One common method uses photography to record progress and to provide documentation of work done. Sequential pictures taken at regular intervals (every week, every two weeks, or every month), provide clear evidence of the potential for meeting a specified completion date. On some projects, progress photos are required by the owner as part of the contract. Of course, photographs also serve a number of other useful purposes, including training of workers and documentation for claims. However, photographs do not capture or record everything. For example, they do not accurately disclose the length of time required to accomplish a specific activity. Time-lapse photography is useful to overcome this shortcoming. The time-lapse camera is set up to record one frame within every stated time interval, such as one second, five minutes, or one hour. The length of time required to perform each task can then be determined with reasonable accuracy.

In many instances the information obtained from photography is inadequate. Some cameras will not take pictures of or capture information on all work tasks. Check-off lists may then prove to be more effective. Where check-off lists are used, field personnel are asked to simply indicate the status of various work items. Specifically, the activities can be noted as "started," "finished," or "in progress." It might be overly cumbersome to field personnel to review a list of several hundred activities. This can be simplified by having the field personnel review a list of activities scheduled to take place during a specified time interval, namely, two weeks or one month. Many computer software packages can print out only the activities that have been scheduled during a specified block of time. Obviously, a judgment must be made as to the appropriate block of time. On large networks it might be best to use a small time interval in order to limit the number of activities being listed. An extension of a check-off list is the use of slightly more detailed schedule status forms. In addition to being useful in capturing basic activity status data, they can be used to record actual start/end dates, revised activity durations, and "percent complete" figures (see Figure 8.2). For these forms to be useful, it is important that they be simple, easy to understand and short.

Similar to the use of check-off lists, feedback about job progress may also be obtained directly on bar charts sent out to the field. The bar chart would have information similar to that on a check-off list. The "marked up" bar chart is then sent back to the main office, where the information is transferred to the home office chart or the information could even be incorporated in the computerized schedule information. The use of bar charts can be an effective way of getting the desired information; however, it is best when the project is not complex. Bar charts may have limited value on large projects as the activities do not entail sufficient detail to serve as a viable management tool. The level of detail must be commensurate with the needs of the intended user.

If the project is fairly simple and the schedule has few activities, the feedback information can also be obtained directly on the network. Notes on the network will indicate which activities have started, which ones are in progress, and which ones have been completed. Special notes can also be added on the network. On very complex projects with thousands of activities, it would be unwieldy to ask field personnel to

Activity		Scheduled		Actual/Expected		Status	Notes
ID	Description	Start	Finish	Start	Finish	%	

Figure 8.2   Activity Status Report to Monitor Project Schedule Status

review the status of the entire project. To simplify the task, it is best to print out a network that is broken down by subgroups, such as designated areas on the project or specified crafts. Networks have some distinct advantages. If the network is used to organize work tasks on the project, field personnel will already be familiar with the network. Thus, no orientation will be required. An additional advantage of using the network is that field personnel are forced to focus on how the current project status relates to the "big picture." Of course, networks have little value for field personnel who are not acquainted with them and who are simply confused by them. A management task includes the determination of the appropriateness of using networks for obtaining feedback on job progress.

Some aspects of the construction project warrant particular attention in terms of monitoring, namely long lead-time items and safety. Although these are very important to the success of a construction project, they might be ignored or dismissed if one becomes overly engrossed in the busy day-to-day operations of a construction project.

It is common to include procurement activities on project schedules, especially where major purchases are involved. With their inclusion, the schedule will help alert the construction personnel of the need to issue purchase orders or purchase agreements on these items. While this is of valuable assistance to ensure materials, supplies, and equipment are on order, the issue of additional monitoring should not be ignored. If the item is readily available with no additional unique requirements, there may be no cause for further monitoring until shortly before the item is actually needed on site. When a purchase involves unique items, more intense follow-up may be warranted.

Long-lead items warrant particular attention. Examples of these include structural bar joists, pressure pumps for sprinkler systems, electrical panels, electrical controllers, chillers, air handling units, metal door frames, specialty glazed tiles, elevators, escalators, specialty casework, and other items that are specially made for the project. While all of these items require some degree of monitoring prior to their actual delivery on the job site, the duration of their estimated fabrication times will dictate the extent to which periodic monitoring is warranted.

Because of the importance of some of the long-lead items to the success of the project, monitoring them should be a regular function of project management personnel. This monitoring is done though a variety of mechanisms. Perhaps the most basic technique is to simply place a telephone call to a supplier or fabricator to inquire about the status of a particular order. The reliability of the information obtained from a production manager on prior purchases will dictate the extent to which one can rely on the current status report on a particular order. If there is a concern, based on experience with previous orders, or if this is a new supplier for the firm, the monitoring information obtained in a telephone call may not be sufficient. The information may be more reliable if the supplier provides the detailed status information in a letter. When the delivery of a particular item is critical to the success of a project or if the supplier's information is placed in doubt, a visit to the supplier's premises may be justified. Such factory visits can be costly to a firm, but they do send the message that the delivery of the item in question is important. It can also give the necessary assurances to project personnel that all items are on track.

Just as long-lead items are tracked by the general contractor, so too might subcontractors track the requisition of materials and equipment. To assist in the process, it is important for the general contractor to first establish the right to access the necessary information from each subcontractor's suppliers. Is this a justified approach? If the general contractor's own purchases are tracked with care, is there any reason that less care should be exercised with the purchases made by the subcontractors? Clearly, the success of the project is the objective, and effective monitoring must include all aspects of the project. Thus, the general contractor is well advised to include, in the subcontract agreement, the right to be involved in monitoring purchases made by the subcontractor. At project start the general contractor could ask each subcontractor to provide a list of all major pieces of equipment and materials to be provided. Schedules can then be established to include those items to be monitored. While each subcontractor should be responsible for tracking these materials and equipment, the general contractor should have the right to verify the status of manufacture and prospective project delivery dates, especially of the more important materials and equipment.

When preparing the schedule, safety considerations must also be addressed. Every activity should be considered in terms of any related or specific safety needs. First, consideration must be given to the need to acquire any safety equipment that is required to perform the work safely. All work activities must be planned so they can be performed safely. Thus, each work activity must be examined in terms of any unique requirements that exist in safety plans and training. Examples of tasks with particular safety concerns include trenching, setting structural steel, and conducting any elevated work involving the need for scaffolding. Just as the general contractor's work must be performed safely, so too must the work of subcontractors be performed safely. With overall control of the project operations, the general contractor must assume an authoritative role to ensure that all tasks are performed with safety in mind.

Safety is a topic that must underlie every activity that is included in a schedule, and it is clearly not a subject to be included only as an afterthought. If a serious injury occurs on a project due to poor project planning, the project undertaking is a failure.

DIFFICULTIES IN ASSESSING PROGRESS

The need for accurate information cannot be overstated. In addition, there must be an understanding of the need for a certain level of detail. For example, the field reports about project status should be provided with the same level of detail as was used in developing the original schedule. This means that the scheduler must use sufficient detail when developing the schedule. If this is not done, errors or inaccuracies will be more common. This point can be clarified by an example.

When resources are allocated to an activity, it is assumed that there will be a linear or uniform consumption of the resources over the project duration. This may result in inaccurate information, particularly if an activity has a long duration or if the activity is very segmented, consisting of a composite of several activities that use different resources to differing degrees.

Suppose an activity is entitled "Erect retaining wall." This activity consists of excavation work, formwork erection, steel reinforcement installation, and concrete placement. For such an activity, what would "50% complete" imply for resource utilization? It could mean that half of the formwork is complete and that half of the concrete has been placed. However, it could also mean that all formwork is complete, but that no concrete has been placed.

It should be obvious that the user of the information has to exercise some judgment. Unfortunately, when several hundred activities exist in a network, it may not be easy to analyze each activity with this level of detail. It might be simpler, and certainly more accurate, to break the activity into two or three basic activities. Another example of an activity that might require judgment is one entitled "Removal of existing bridge abutment." In reality, such an activity might consist of drilling, blasting, loading, and hauling. In fact, drilling may consist of 15 days, blasting one day, with loading and hauling requiring five days. The percent complete could relate to the percent of time used or the percent of the bridge abutment actually removed. Again, in such a case it might be advisable to break the work down into several activities.

The primary issue is to attempt to define activities that have durations with approximately the same order of magnitude as the monitoring period. For example, if monitoring will be weekly, then it is helpful to have activities defined with durations that are typically between five and ten days. This reduces the number of activities for which "percent complete" figures would have to be determined as part of the data-collection phase, compared to activities with durations of weeks or months. When "percent complete" figures need to be estimated, one could express them relatively "grossly," such as in terms of 20% or 25% ranges, without serious problems.

UPDATING THE SCHEDULE

Careful monitoring of a project is an effective means of giving insights to management about the need for action. However, the schedule is an effective management tool only as long as it bears a reasonable resemblance to the actual project. When discrepancies are noted, adjustments must be made to the network. This process is known as *updating*.

Project progress must be accurately described through field personnel. Thus, feedback is important, and it is important that the feedback come from competent personnel.

Updating may be necessitated for a number of reasons. The person who originally developed the network may have envisioned one way of sequencing the project activities, while the field superintendent now thinks another approach is more expedient. In some cases the logic of the network must be reorganized in order to accommodate new constraints imposed on the project, such as limited resources. In some cases the logic must be reorganized simply because some of the activity durations exceeded expectations. Overall, updating is required whenever the schedule no longer reflects the actual operations taking place on the project. Updating is not needed if all duration/cost estimates are accurate and if the sequencing of the activities is exactly as originally scheduled.

It is important that an informed decision be made about the need to update a schedule. Such a decision should be made on a periodic basis, such as biweekly or monthly. Many construction contracts, which require schedules to be prepared, include a provision requiring that the schedule be updated when the owner or the owner's representative requests it. As a rule, updating should be done when the benefits derived from updating the schedule exceed the cost of doing the updating.

Updating can become a normal routine using computer project management software. Data collected in the monitoring process should be used to update the computerized network information. A copy of the original schedule should be retained for comparison purposes. As time passes, the "updated" schedule turns into an "as-built" schedule as the actual set of activities, their sequence, durations, and start and finish dates are entered into the computer. As each update of the network is performed, a copy of the file can be made and retained as a "snapshot" of the project at that point in time.

CONTROLLING THE PROJECT

As was previously mentioned, project control consists of two components: one, accurate information about the schedule status of a project, and the second, actions taken in response to the status reports. The driving force for keeping a project "on schedule" is money. Typically contracts include liquidated damages for a project not being completed on time or bonus/penalty clauses that provide an incentive to the contractor to complete the project as soon as possible. These considerations are in addition to the normal cash-flow implications discussed earlier and the possible opportunity costs associated with not completing the project in a timely manner.

Schedule status information allows one to make an assessment about whether the project is on schedule relative to the original plan and to individual activities. Remember that the status of individual activities does not necessarily indicate whether the entire project is likely to be finished on schedule. With a CPM approach the project manager is able to determine the impact of any activity schedule deviations by evaluating the deviations relative to available float on the activity based on its scheduled start.

As long as an activity started on or before its late start date (and/or finished by its late finish date), the project itself will not be delayed. If delay of the activity from its scheduled start is less than the available free float, then the project manager need not be concerned as absolutely no other activity start dates will have been affected. Subsequent activities will need rescheduling, however, if an activity falls behind its scheduled time by more than its free float. The implications of such "delays" are primarily related to changes in when resources (subs, materials, equipment, and labor) need to be available (and used) for these "delayed" activities. Project duration will not be impacted and thus no penalties, damages, or additional costs will be levied. Particular attention should be paid to any delays that were equal to the available total float, as all activities subsequent to such an activity will be forced to their late dates and thereby must be considered critical in all future analyses.

Activities started after their late start (or finished after late finish) will cause a delay in the project. In these cases all available float will have been used *and* additional delay encountered. The start date of all subsequent activities must be adjusted and they will become critical. All subsequent analysis must recognize that a different set of activities may have become critical than was previously the case. Only the shortening of the duration of one or more subsequent activities on the critical path will permit the contractor to recover from the project's delay.

Contractors must determine whether it is to their benefit to recover from delays. There are two primary considerations to take into account:

1. Who is responsible for the delay?
2. Is it economically advantageous to recover lost days?

The issue of delay responsibility is related to whether the contractor is awarded, or is liable for, costs and additional time to complete the project. The categories of responsibilities are

- Owner (or Agent) Responsible: Contractor will be granted time extension and additional costs (indirect), where warranted.
- Contractor (or Subcontractor) Responsible: Contractor will not be granted time or costs and may have to pay damages/penalties.
- Neither Party (e.g., Act of God) Responsible: Contractor will receive additional time to complete the project but no costs will be granted and no damages/penalties assessed.
- Both Parties Responsible: Contractor will receive additional time to complete the project but no costs will be granted and no damages/penalties assessed.

Unless the contractor is responsible for the delay, he or she will be granted additional time to complete the project as long as the delay has impacted an activity on the critical path or delayed a noncritical activity by an amount greater than its total float. This means that "delays" caused by weather, stop-work orders, and change orders will lead to a longer project without penalty or damage to the contractor. In such cases the only real incentives for a contractor to take action to recover these days are if the contract includes bonuses for completing ahead of schedule, or if opportunity costs will be incurred because the project will take longer than originally expected.

In cases where an owner is responsible for a delay, contractors receive not only more time, but also additional money. Owners generally compensate contractors for direct costs incurred as a result of any additional work required and also indirect costs associated with any extension in the project duration caused by the delay. Conversely, contractors may be assessed damages or penalties for delays that they cause. Damages may be "liquidated damages," standard per-day charges established as estimated damages incurred by the owner for not completing the project on time, or they may be "actual damages" that are specifically calculated for this individual project. Actual damages are based on an assessment of the real cost to the owner, such as lost rent or other income, or, in the case of a public agency, costs incurred by the public, such as additional travel time and operating expenses if a highway is not open as expected.

The CPM-based procedure used to assess the cost/benefit of attempting to recover from a delay is essentially the same as that explored in the discussion of time–cost trade-offs earlier. Using a current network schedule, it would be necessary to identify the critical activities and determine the additional costs of shortening the durations of one or more of them. These additional direct costs need to be considered relative to the cost savings of bringing the project back "on schedule." The cost savings must take into account indirect costs, damages/penalties, bonuses, and opportunity costs. Where damages and other costs are not substantial, it may not be economically advantageous for a contractor to recover from a delay even if the contractor is responsible for it. On the other hand, a contractor may determine that it is to his or her advantage to recover from delays, even if the contractor is not responsible for them!

Special note should be made of how the impacts of change orders (both time and cost) are evaluated using a CPM network. First, the specific point at which the change order will impact the existing network must be identified. It is here that the set of activities associated with the change order will be tied into the network. Next, the contractor has the right to identify all of the activities that are required to implement the change (e.g., change proposal, additional mobilization, materials procurement) as well as the activities associated directly with the changed work itself. All of these activities represent a subnetwork, or "fragnet" (network fragment), that must then be added onto the existing network. The dates of the initial change order request and the Notice to Proceed are used with this subnetwork in determining the overall impact of the change order on the project. Assessing project completion dates before and after, including the change order subnetwork, establishes the number of additional project days to grant because of the change. This also gives a basis for requesting additional indirect costs as well as any other direct costs associated with preparing and implementing the change order.

The Sports Facility Project

The sports facility project described in prior chapters will be re-examined. The bar chart for the sports facility is shown in Chapter 1 and the precedence diagram for this project is shown in Chapter 3. These are representations of the "as planned" schedule for the project. On working day 15 (at the end of three weeks) of the project, it was determined that significant changes had occurred in the schedule and that an update of the schedule was warranted. The following information was provided for the updating:

Mobilization activity went as scheduled.
Stock materials activity went as scheduled.
Clearing and grubbing finished two days late, i.e., duration increased by two days.
Grading the road finished two days late, i.e., duration increased by two days.
Finish grade started on day 14, with the duration expected to remain the same.

Project Monitoring and Control   211

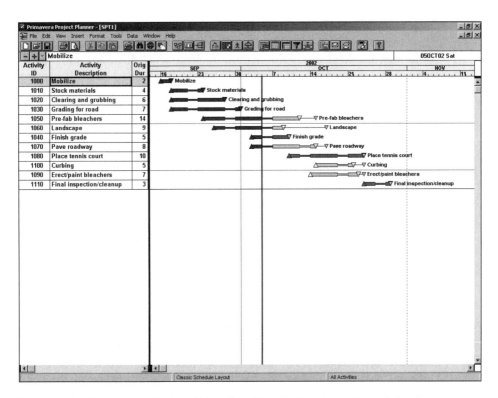

Figure 8.3   Computer Updated Bar Chart for the Construction of the Sports Facility Project *(Primavera Project Planner®)* (Reprinted with permission from Primavera Systems, Inc.)

Prefab bleachers started on time, duration changed to 14 days (reduced by two days). Landscape started on time, duration changed to nine days (reduced by three days). Pave roadway started on day 14, with the duration expected to remain the same.

With the schedule updated, the project duration is now shown to be 31 days (see Figure 8.3). Note also that the critical path has changed. At this point the project manager should evaluate the schedule and decide whether any changes are warranted. Perhaps the logic could be altered to finish even earlier than planned. Of course, the resource utilization should also be taken into consideration, which might justify schedule changes that might actually increase the project duration.

AS-BUILT SCHEDULES

As-built schedules are similar to as-built drawings. Whereas as-built drawings are corrections made on the original set of construction drawings to reflect the actual locations of all construction features, as-built schedules reflect the actual occurrence times for all major project activities. As-built schedules are not commonly prepared on most projects. Note that an updated schedule merely shows the correct status of the project activities beginning at a designated point in time. In the updated schedules, the actual occurrence time of past activities is not of primary interest. It is only when the true representation of the entire project schedule is needed that an as-built schedule is warranted. Since these take time to prepare, a decision must be made if the effort is justified.

What is the purpose of as-built schedules? The most widely recognized use is in the area of claims, especially delay claims. From a contractor's point of view, an as-built schedule might be prepared to demonstrate how a particular owner-caused delay or unforeseen condition resulted in a significant project delay. This will often be difficult to show without an as-built schedule. To make the impact of a delay clear, comparisons must be made against other schedules of the project, namely those that were prepared without the anticipation of a delay.

Of course, to show the impact of a specific delay, the as-built schedule may be prepared for only that portion of the schedule that was altered by a particular event. It is often easier to make a more compelling case if an as-built schedule is prepared for an entire project. For example, the as-built schedule and the original schedule may be very similar in the beginning of the project. If the variation between the as-built and original schedules becomes most noticeable after a specific event or delay took place, it is much easier to demonstrate the magnitude of the impact of the event on the entire schedule.

From the owner's point of view, the use of an as-built schedule covering the entire duration of a project may show that variations from the original schedule were significant, even when no intervening events took place. If this is the case, the as-built schedule might be successful in showing that the original schedule exhibited limited credibility.

If a claim is being made that one party significantly impacted a schedule, the deciding individual or individuals might want to examine an as-built schedule. The as-built schedule might be requested by an arbitrator, an arbitration panel, a disputes review board, or any third party that is asked to render a decision on the merits of a claim related to impacts on the schedule.

As-builts might be prepared to develop a clear historical record of a project. This is not to be confused with the historical record that shows the number of worker hours required to perform different tasks. Such cost records are presumably being maintained for all projects. It should be noted that cost records are regularly examined to determine the amount of time to allocate for a particular activity. With a good historical record, it is generally possible to determine the number of hours of work associated with all activities.

Since the cost records already capture the time required to perform a given task, can as-built schedules really yield that much additional information of value? The answer is a qualified "yes." While the cost records do provide information on the time required to perform each activity (depending on the level of detail of the cost records), they may fail to present information on the time that lapses between activities.

A simple example will clarify this point. The network in Figure 8.4 has been prepared to demonstrate the concept of as-built schedules. Assume that several structures or facilities have essentially the same features. Assume that the network is a model of a

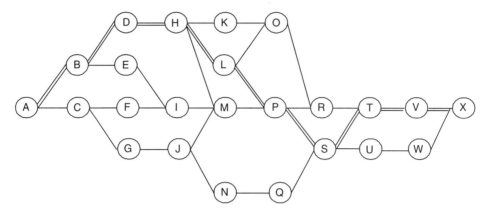

Figure 8.4 Sample Network for an As-Built Schedule

building project. There are many such projects, so this network schedule will receive wide usage in the firm. The activity durations may change between projects, but the logic will remain the same. Assume that the network is adequate to describe the sequence of activities of most company project activities. Note that the critical path is indicated by the double link lines. It should be evident that these projects are "cookie cutter"-type projects in which little changes from project to project, with size being the most significant variable that may change.

Assume further that the critical path tends to invariably include the same activities. We will assume also that the company has an effective historical cost record system from which to make accurate future cost estimates. To simplify the project in terms of monitoring, the company has redrawn the network to reflect only the bare essentials, namely the critical path activities (see Figure 8.5). These are reduced to a bar chart that can be readily understood.

Schedule status reporting is closely linked to the abbreviated schedule. One project has 20 similar buildings for which 20 of the above network schedules have been prepared. Some of the durations for the activities of different projects will vary to reflect the major differences. When the company superintendent reports on the status of the project, a simple note is made of the status of each of the buildings. For example, the foundation work of the project may be reflected by Activity B. Over time, very valuable information can be obtained on the durations of the different buildings. Note that the duration of the project is not simply the sum of the durations of activities; that information could be obtained from the weekly cost reports. What this type of schedule reporting will document is the time to complete the project, which includes

A	B	D	H	L	P	S	T	V	X

Figure 8.5 Sample Bar Chart for an As-Built Schedule

the lag period that exists between the finish of one critical path activity and the start of the next or succeeding critical path activity. This lag is largely an indication of the efficiency of the project coordination effort. If the activities of subcontractors are not coordinated well, the inefficiency will be reflected in the lag between activities. Note that this information will not be captured in conventional cost reports.

As-built schedules are particularly valuable on development projects. On apartment complexes consisting of 30 to 50 buildings, one can document project status by monitoring the construction status of each building as described above. A development project consisting of many single-family residences or several office buildings could also use this approach. For each building, on each project, the same precedence diagram schedule can be used with changes being made only in the activity durations. Since the buildings are essentially the same, differing primarily in their square footage, project monitoring focuses on the standard precedence diagrams that are reduced to the single bar of a bar chart. This method is sufficiently simple that superintendents are not reluctant to provide daily reports on the construction operations. This information proves to be valuable in scheduling company projects.

As noted earlier, as-built schedules are not prepared on most projects. The circumstances warranting their use will vary considerably. The use to be made of the information will dictate the type of approach used to capture the necessary information. One method of capturing the information was described above. The scheduler's own particular needs and creativity will determine the nature of the as-builts to be developed on specific projects.

FINAL COMMENTS

Control of a project involves evaluating the impacts of schedule deviations on the scheduling of future activities and the project as a whole. Often the response consists of simply notifying subcontractors and suppliers of changes in the dates that their services or products will need to be available. Obviously, it may not always be possible to accommodate these changes and an evaluation must be made of how to get the project back on schedule and what it will cost. Alternatives must be identified (e.g., line up new sub, accelerate project, allow project to slip), and these must be evaluated according to the economic costs and benefits as determined by performing time–cost trade-offs with critical path method networks and principles.

REVIEW QUESTIONS

1. In general terms, what are the two basic components of project control?
2. What are some of the inherent advantages and disadvantages in measuring project duration in calendar days? Working days?
3. What are some of the different ways that progress on construction projects can be monitored?

4. What are the benefits of using a work breakdown structure on a construction project?
5. In general, when should construction schedules be updated?
6. How can the owner of a construction firm increase the probability of obtaining accurate scheduling information from the field personnel?
7. Describe the type of information that an as-built schedule can provide to a firm that might not otherwise be obtained.

9

Computer Scheduling

Projects should be scheduled one byte at a time.

For many construction projects, a multitude of activities are required for completion. As a project increases in size or complexity, so does the scheduling effort. Large and complex projects contain thousands of unique or repeated activities, each with various interdependencies to other related activities. For these types of projects, the tasks of manually creating, monitoring, updating, and reporting a schedule using the various techniques outlined is very time consuming and often difficult. For example, performing a forward and backward pass by hand to determine the total and free float values of a network of activities for the construction of a new underground light rail system is not recommended and, some might say, impossible. On the other hand, while manually scheduling a simple, smaller project, such as the remodeling of a single-family home or the repavement of a county road, is quite feasible, the effort may mean time spent unproductively for the particular scale of work and short project timeline. As a result, schedules for most complex construction projects and many simple ones are now created using a computer.

The capabilities of computers make them ideal tools for creating and manipulating project schedules. Their large data storage capacities allow computers to calculate schedule information for very large projects with thousands of activities or for many different projects that are to be kept on file for future reference. The schedule from a past project can be saved and used in the future to develop a schedule for a similar project. This capability eliminates the time required to create a schedule from scratch. Schedules from similar previous projects are merely copied and then altered accordingly to reflect the current project's time frame, unique design elements, and current availability of construction resources. If needed, portions of or entire schedules from multiple past projects can be taken from the computer files and merged to develop a new schedule. Construction firms that consistently perform a specific type of work can especially benefit from this capability.

The computational speed, power, and accuracy of computers match the extensive computational requirements of project scheduling. CPM requires numerous computations to determine activity start and finish dates, float values, and the overall project duration. Additional calculations related to resource allocation, resource leveling, and cash-flow restraints greatly increase the computational effort and complexity. Repeatedly performing these calculations by hand for projects with numerous activities is very time consuming, with added effort required to verify accuracy. Computers are ideal tools for repeatedly performing mathematical computations on a set of data. The mathematical coprocessors within current computers can perform thousands of computations with relative ease and tremendous accuracy in a matter of seconds. This capability is one of the main reasons that computers are used extensively not only for project scheduling, but also throughout the construction and engineering fields for other highly computational tasks.

Computers enable and facilitate mobility of the project schedule. Schedule information can be transferred quickly between computers, within an office or to remote locations, for access by any individual. This capability is especially helpful for delivering schedule information between the home office and job site. Short interval, look-ahead schedules, or information relating to schedule changes, can quickly be created and sent to the job site for immediate review and incorporation in the construction effort. Work performance and task completion updates from the field can be sent to the home office via the computer and other electronic technologies. This allows immediate updating of the planned project schedule to create an as-built schedule. On many projects the contractor is required to submit an initial project schedule along with periodically updated schedules to the owner or architect for review. This task is greatly facilitated when the schedule can be sent over a network or delivered on a CD rather than printed on paper.

Developments in computer software have supplemented and enhanced the use of computers. Current Windows®-based software programs provide a user-friendly working environment. Programs are laid out and designed to provide an intuitive and comfortable interface with the computer. Colors and graphics are frequently employed to help the user remember the software features and quickly orient him- or herself to the program's screens. Often-used functions are quickly accessed by a simple click of a mouse button. Pull-down menus provide an organized view of and access to all of the software's features. Most programs come with on-line help instructions for reference. Scheduling programs have greatly benefited from these software developments. Scheduling information can be quickly and easily input and viewed on the screen in a variety of formats. Colored activity bars and graphical icons highlight project milestones or key schedule features. These and other capabilities of current scheduling software facilitate the creation, review, and verification of activity networks during development and management of a project schedule. Throughout this text comments have been made about computer capabilities as they pertain to various scheduling functions.

The capabilities of computer hardware and software have been successfully captured in current personal computers (PCs). As a result, personal computers have become standard work tools in virtually every construction office. PCs can quickly be set up and brought online for immediate use. Laptop computers provide a means of transporting the computer's capabilities anywhere. A network of PCs provides multiple users access to centralized information from many different computers at remote locations. One major reason for their popularity is that PCs provide a powerful tool at a cost-effective price: only minimal monetary investment is required to obtain and operate a PC. In addition, advances in the computer industry have continually led to increased computational power at a lower price. This trend has opened the door for many construction companies to purchase and utilize computers for scheduling efforts as well as other construction-related tasks.

COMPUTER SCHEDULING TERMS

Effective and productive use of the CPM techniques for project scheduling requires knowledge of various terms related to the scheduling methods. The development of scheduling software has led to a collection of commonly used computer scheduling terms in addition to those employed in the basic CPM techniques. For the most part, the computer scheduling terms are used in the same fashion in most scheduling programs. As a result, it is easy to learn and use different scheduling programs. Following are common terms used within many computer scheduling programs.

Activity Code: A value assigned to each activity to help organize the activities into manageable groups. Similar to cost codes, activity codes are typically used to facilitate sorting and filtering activities according to specific criteria. For example, a common set of activity codes may be given to all activities performed by the electrical subcontractor or all activities that require use of a tower crane. The schedule can then be filtered to view only those activities related to the electrical subcontractor or to review the utilization of the tower crane.

Activity Form: A window that displays detailed information about a specific activity. While typically only one of several tools that the scheduling program makes available for inputing, viewing, and modifying activity information, the activity form provides convenient access to the most commonly referenced activity data.

Base Calendar: A calendar that applies to all of the activities in a project. The base calendar describes the days on which work can be performed. It also indicates such days as holidays and weekends on which work cannot be performed. A modified base calendar can be created for a task that cannot be performed or is prohibited from being performed on specific working days. These calendars are typically based on either a working-day or calendar-day schedule and are developed when the project schedule is initially created. Most scheduling software programs provide a default base calendar that reflects the normal Monday-through-Friday working-day schedule and all major recognized holidays. Additional holidays can be added with ease.

Baseline Schedule: The original schedule created at the beginning of the project against which the project's progress is measured. This schedule is typically saved as a separate, read-only file so that it cannot be overwritten or modified and can be referenced later in the project.

Collapsing the Schedule: Consolidating the subtasks within their respective summary tasks so that only the summary tasks are shown. This action helps one view the main project activities without cluttering the screen with the minor activities. It provides a way to broadly view the overall project schedule.

Constraint: A restriction imposed on the start or finish of an activity. Constraints are used to prescribe limitations on the schedule based on external conditions such as contractual restrictions or imposed float requirements.

Data Date: The date used as the starting point for schedule calculations. During development of the schedule prior to the construction phase, this date is the anticipated day on which construction is to begin. During the construction phase, it is the date on which schedule data are input or modified.

Data Date Line: A vertical line on a Gantt chart representing the current date. The number of days to completion is calculated from this date. This line provides a helpful graphical view of the current date on the project timescale.

Expanding the Schedule: Showing the subtasks within their respective summary tasks. This action shows the detailed activities of the project schedule rather than just the general activities.

Filtering: Searching through the project activities and showing only the activities that match specific criteria. Filtering does not delete activities from the schedule, but merely hides activities from view. Filtering allows for focusing on specific activities, such as those performed only by the mechanical subcontractor during the upcoming month. Filtering also allows the user to view those activities starting after a given date, activities completed in a given time period, or those meeting a variety of other criteria.

Global Change: A change that affects all activities or a selected group of activities at the same time, even those filtered from view.

Layout: The appearance of the schedule on the computer screen. The layout is essentially how the visual elements of the schedule, such as the format and color of the bars, organization of the activities, and visible columns, appear when viewed on the computer screen. Different layouts can be saved for convenient viewing each time the schedule is accessed.

Link Line: The line that connects the bars of linked tasks on a Gantt chart to show logic relationships.

Linking: The process by which relationships between activities are created. Activities are linked *finish-to-start, start-to-start, finish-to-finish,* or *start-to-finish* and can include lag or lead times.

Milestone Activity: A zero-duration activity that signifies the start or finish of an activity or group of activities.

Network Loop: Circular logic within a network of activities that prevents progression across or through the network. When illogical scheduling relationships are created, calculations cannot be performed; that is, all loops must be eliminated for computations to be made.

Progress Bar: A bar on a Gantt chart that represents the progress of a particular task. Progress bars are used to provide a graphical representation of the percent completion of an activity at a specific date. The bars are typically shown adjacent to or within, and a different color than, the activity bars.

Recurring Task: A task that occurs at regular intervals in a project, such as a weekly project meeting or a safety walk-through of the job site. Recurring tasks can be input only once and the program will automatically place them at the appropriate dates.

Resource Calendar: A calendar that applies to a specific resource to define when the resource is available. Resource calendars are utilized for limited resources that may control the scheduling of multiple activities at the same time. The scheduling of an activity that utilizes a limited resource is controlled by both the base calendar (or modified base calendar) and the resource calendar.

Sorting: Organizing the project activities according to a specific format. Sorting allows for grouping activities to control the order in which the activities are shown on the screen and presented in the printed schedule. Sorting does not eliminate tasks from view (as is done when filtering), but merely rearranges them according to a specified format. Colors are often used to highlight grouped activities.

Subtask: A minor task typically representing detailed effort. This task is subordinate to a summary task.

Summary Task: A task representing a general activity of construction. Summary tasks contain multiple subtasks. Summary tasks provide an outline structure to the schedule that identifies the project's major phases. Typically, no duration is associated with summary tasks.

SCHEDULING SOFTWARE

There are many computer scheduling programs on the market. While all of the programs essentially perform the same principal task—namely, to assist in the development and management of schedules—the scheduling capabilities and mix of functions and features differ with each program. Some programs are designed to handle very large projects and can accommodate many thousands of activities within one network. These programs, which also tend to contain many different functions and features with fancy graphics and reporting capabilities, are typically quite high in price. Other programs that have a lower limit on the number of activities and do not contain as many colors and graphics are offered at a much lower cost. The lower-cost programs are typically marketed for use on smaller projects constructed by smaller construction firms.

Several popular computer scheduling programs are described below. The programs described are certainly not the only scheduling programs available, but represent computer scheduling programs commonly used in the construction industry. No endorsement is to be inferred. The program's features and cost, along with the needs, resources, and type of work performed by the contractor, should be taken into consideration when purchasing a program. When choosing a program to use on a specific project, selection should consider the characteristics of the project, including the construction contract, which may require the contractor to submit schedules using a particular program in order to easily interface with the owner's computer capabilities.

Primavera (P3®)

Primavera Project Planner® (*P3*®) is one of the most popular computer scheduling software programs used in the construction industry. Its popularity has led to continued development of additional features and functions to make it a full-service program. With a maximum capacity of one hundred thousand activities on a single project, plus the ability to group an unlimited number of projects together, *P3* can be used on virtually any project. Creation of a schedule is facilitated by templates made from similar projects and the use of various assistants, called Wizards, which lead the user through the input of specific data or the use of program features. Schedule development can be speeded up by writing subroutines to automatically input or modify activity information. To expedite reviewing and revising the schedule, the activity predecessors and successors can be traced throughout the network to verify schedule logic. Global changes are allowed. Project performance can be measured according to earned value, BCWS, and costs. Cash-flow diagrams can be displayed showing the costs, revenues, and the net amount as a function of time. *P3* offers over 150 predefined reports and related graphic representations of the schedule. Module attachments to *P3* provide additional project management tools for reports, cost control, resource management, and estimating.

While the many functions and features of *P3* comprise a very powerful project management tool, the program's price tag of several thousand dollars makes it a costly investment. This is often this program's main drawback for many construction firms. As a result, it is generally used by large firms that are able to absorb the initial investment and work on large projects requiring the greater capabilities and additional features provided. (Primavera Systems, Inc., Three Bala Plaza West, Bala Cynwyd, Pennsylvania 19004. Telephone: 610-667-8600 or 800-423-0245. Fax: 610-667-7894. Web site: www.primavera.com)

SureTrak Project Manager

SureTrak Project Manager® by Primavera is basically *P3*'s little brother. *SureTrak* provides most of the same functions as *P3* at a much lower price. While *SureTrak* allows a maximum of only ten thousand activities on a single project, it does allow projects to be grouped. This feature essentially allows much larger projects to be represented as a group of smaller projects. Resource leveling in *SureTrak* is performed only forward through the network, whereas it is performed both backward and forward in *P3*. Global changes are allowed in *SureTrak*, but are limited to specific default options. *SureTrak* provides 40 preprogrammed reports and related graphic representations. Accessory modules to *P3* are not available for *SureTrak*. Except for other minor differences, the functions and features of *SureTrak* are similar to *P3*. *SureTrak* and *P3* read and write in the same format as well, which allows for interchanging project schedules between programs. (Primavera Systems, Inc., Three Bala Plaza West, Bala Cynwyd, Pennsylvania 19004. Telephone: 610-667-8600 or 800-423-0245. Fax: 610-667-7894. Web site: www.primavera.com)

Microsoft Project

Microsoft Project® is a scheduling program preferred by many construction firms. Its popularity is greatly attributed to its ease of use and similarity to other Microsoft programs. *Project* provides most of the functions and features as other, more expensive programs and allows a maximum of 10,000 activities. It includes an interactive assistant, pop-up screen tips, and context-sensitive help information. The screen tips, or Smart Tags, appear when changes are made to the schedule to alert the user of alternative scheduling options or verify that the information is input correctly. Twenty predefined reports and graphic representations are included. The program layout and operations are designed to be user friendly. A major advantage of *Project* is that it is designed to interface with other Microsoft programs. This feature allows easy transferring of information between programs and creation of scheduling reports containing both text and graphics along with the schedule. Additionally, tasks entered in *Microsoft Outlook* can be displayed in *Project* to maintain assigned activities in one convenient location. *Project* is comparable in cost to *SureTrak*. (Microsoft, One Microsoft Way, Redmond, Washington 98052–6399. Telephone: 800-426-9400. Web site: www.microsoft.com)

Web-Based Programs

Several software companies offer computer scheduling capabilities via web-based programs. *Primavera Engineering and Construction (P3e/c)*, for example, is an on-line version of *P3*. *P3e/c* provides integrated team communication and collaboration; coordinated, schedule-based procurement; and project planning and control to ensure successful design, construction, and facility management projects. *Primavera* additionally offers *Contractor*, a scaled-down on-line version of *P3e/c* that provides scheduling capabilities for projects with up to 2,000 activities. These and other web-based programs provide the convenience of access to schedules and project management information from any computer via the Internet. The need for multiple user access to project information from different locations, and the efficiencies gained through mobile, wireless technologies, make web-based scheduling programs highly desirable.

CREATING A SCHEDULE

The modeling of network schedules on almost all programs is performed by the CPM. While the more sophisticated programs also allow scheduling using PERT (described in Chapter 15), CPM is by far used most frequently. Most programs only offer Gantt chart and precedence-diagram views of the activity network. The Gantt chart is typically the default view shown on the screen because its graphic representation of the

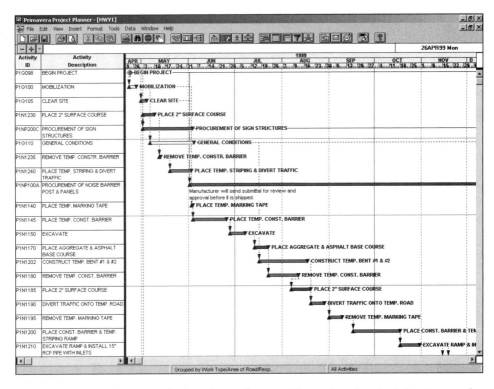

Figure 9.1 Activity Description List and Gantt Chart Showing Activity Bars and Relationships (Primavera Project Planner®) (*Reprinted with permission from Primavera Systems, Inc.*)

network can be grasped quickly and easily (Figure 9.1). Activity information can also be viewed in spreadsheet format (Figure 9.2). Spreadsheets provide a tabular listing of all the activities and related information.

The process of creating a schedule begins with inputting activities. Programs typically allow input of activity data in two ways. Data can be input directly into the spreadsheet like any set of related data (Figure 9.2). Each row of the spreadsheet represents one activity, and the columns list such information as the activity name, duration, predecessors, successors, cost, date constraints, and resources. After filling out the spreadsheet, the Gantt chart or precedence diagram can be viewed for a graphic representation of the network. Most programs also provide an activity form, or dialog box, for inputting activity information (Figure 9.3). This preformatted window helps guide the user in entering the necessary activity data. One advantage of using the activity form to input data is that the Gantt chart can be viewed simultaneously and provides an immediate view of the activity network during development of the schedule.

The basic information required for each activity is the activity name, duration, predecessors, and successors. Each activity is given a unique ID number, if not already chosen by the program. If relevant to the schedule, the type and amount of resources utilized by

Primavera Project Planner - [HWY1]

File Edit View Insert Format Tools Data Window Help

21APR99 Wed

Activity ID	Activity Description	Closed Lane	%	Original Duration	Remaining Duration	Early Start	Early Finish	Late Start	Late Finish	Total Float	Free Float	Budgeted Cost
P1G098	BEGIN PROJECT		100	0	0	23APR99A		23APR99A				0.00
P1G100	MOBILIZATION		100	4	0	23APR99A	27APR99A	23APR99A	27APR99A			100.00
P1G105	CLEAR SITE		100	4	0	01MAY99A	04MAY99A	01MAY99A	04MAY99A			3,583.68
P1N1230	PLACE 2" SURFACE COURSE	NO7	100	3	0	02MAY99A	09MAY99A	02MAY99A	09MAY99A			24,147.36
P1NP200C	PROCUREMENT OF SIGN STRUCTURES		100	10	0	02MAY99A	03JUN99A	02MAY99A	03JUN99A			20,000.00
P1G110	GENERAL CONDITIONS		100	13	0	07MAY99A	04JUN99A	07MAY99A	04JUN99A			19,500.00
P1N1235	REMOVE TEMP. CONSTR. BARRIER	NO7	100	1	0	13MAY99A	13MAY99A	13MAY99A	13MAY99A			1,666.52
P1N1240	PLACE TEMP. STRIPING & DIVERT TRAFFIC	NO7	100	1	0	20MAY99A	02JUN99A	20MAY99A	02JUN99A			248.00
P1NP100A	PROCUREMENT OF NOISE BARRIER POST & PANELS		0	10	19	02JUN99A	01FEB00	02JUN99A	02AUG00	130	128	12,000.00
P1N1140	PLACE TEMP. MARKING TAPE	NO7	100	1	0	03JUN99A	03JUN99A	03JUN99A	03JUN99A			288.32
P1N1145	PLACE TEMP. CONST. BARRIER	NO7	100	1	0	04JUN99A	25JUN99A	04JUN99A	25JUN99A			1,790.52
P1N1150	EXCAVATE	NO7	100	5	0	28JUN99A	07JUL99A	28JUN99A	07JUL99A			2,846.60
P1N1170	PLACE AGGREGATE & ASPHALT BASE COURSE	NO7	100	6	0	14JUL99A	22JUL99A	14JUL99A	22JUL99A			107,477.52
P1N1202	CONSTRUCT TEMP. BENT #1 & #2	NO8	100	20	0	17JUL99A	17AUG99A	17JUL99A	17AUG99A			22,004.40
P1N1180	REMOVE TEMP. CONST. BARRIER	NO7	100	1	0	23JUL99A	11AUG99A	23JUL99A	11AUG99A			1,790.52
P1N1185	PLACE 2" SURFACE COURSE	NO7	100	3	0	07AUG99A	19AUG99A	07AUG99A	19AUG99A			23,865.36
P1N1190	DIVERT TRAFFIC ONTO TEMP. ROAD	NO7	100	1	0	20AUG99A	28AUG99A	20AUG99A	28AUG99A			248.00
P1N1195	REMOVE TEMP. MARKING TAPE	NO7	100	1	0	01SEP99A	09SEP99A	01SEP99A	09SEP99A			288.32
P1N1200	PLACE CONST. BARRIER & TEMP. STRIPING RAMP	NO7	100	4	0	17SEP99A	17OCT99A	17SEP99A	17OCT99A			2,588.12
P1N1210	EXCAVATE RAMP & INSTALL 15" RCP PIPE WITH INLETS	NO7	100	10	0	18OCT99A	30OCT99A	18OCT99A	30OCT99A			11,149.20
P1N1220	PLACE AGGREGATE & ASPHALT	NO7	100	4	0	15NOV99A	29NOV99A	15NOV99A	29NOV99A			75,774.08

Grouped by Work Type/Area of Road/Resp. All Activities

Figure 9.2 Spreadsheet View of a Project Schedule (Primavera Project Planner®) (*Reprinted with permission from Primavera Systems, Inc.*)

each activity, plus any associated costs, are input at this time. To facilitate sorting, filtering, and reporting the schedule, activity codes and a Work Breakdown Structure identification code are defined and assigned to each activity. Activity codes should be defined and organized to effectively represent different facets of the project. Common activity codes include areas of the job site, worker responsibilities, and type of work performed. A Work Breakdown Structure (WBS) generally represents a hierarchy of tasks required to complete the job. The WBS is typically arranged to define the work involved in major features of the project. These work tasks are typically divided into specific identifiable activities.

A base calendar must also be specified for the project. Programs typically provide a default base calendar, which reflects the normal Monday-through-Friday workweek and includes all the major holidays. The base calendar can be modified to match the workweek planned for the project and those holidays offered by the company. Some project managers add artificial holidays to the calendar to reflect days on which little work is typically performed, such as the Friday after Thanksgiving or the days between Christmas and New Year's Day. If any tasks cannot be performed on a specific workday, a modified calendar can be created for that particular activity. If, for example, the city's electrical inspector is not available to inspect the work on Mondays, a calendar tied to all electrical inspection activities would show Mondays as nonworking days.

Computer Scheduling 225

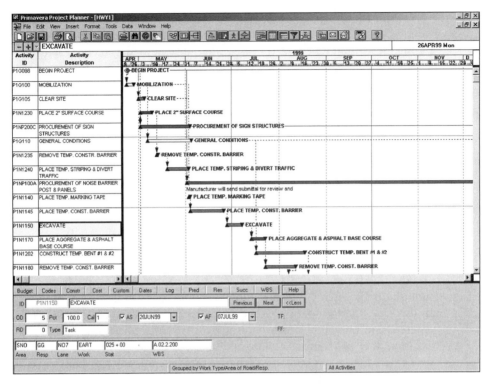

Figure 9.3 Gantt Chart and Activity Form Used to Input Activity Information (Primavera Project Planner®) (*Reprinted with permission from Primavera Systems, Inc.*)

In addition to the activity calendars, resource calendars are created at this time. Resource calendars reflect the availability of project resources. They describe the days, or periods during the day, when resources are available. For example, on some projects pile driving is limited to specific periods of time during the day due to the noise and vibrations created. For these projects, resource calendars would be developed for the pile driving equipment that reflect the time periods when pile driving is allowed. If no resource calendars are created, the program assumes that resources are available during all working hours on every working day.

Once the activity information and any desired resource and cost data are input, schedule calculations can be performed. In most programs, CPM calculations are initiated simply by clicking on a button on the screen. The program determines the activity dates, project duration, total and free float values, and the critical path. If network loops are encountered, the program terminates the calculations. Elimination of the network loops is required before the calculations can be completed. The results of the calculations can be seen graphically on the Gantt chart according to the timescale. Activity dates and total and free float columns can also be added to the spreadsheet to view the results in a tabular format.

Prior to the start of construction, the planned construction schedule is typically saved as a baseline schedule. The baseline schedule provides a measuring stick for comparing the as-built schedule. It is used not only as a management tool for determining the accuracy of planning efforts, but also as a basis for any construction delay claims.

UPDATING A SCHEDULE

As the project progresses, tasks may not start and finish exactly as calculated in the planned schedule. Tasks may be delayed or extended for such reasons as weather delays, late material deliveries, or inaccurate estimates of their expected duration. Consideration of these events is required in order to properly adjust the schedule to meet the planned completion date.

Periodic updating of the schedule can be performed as soon as as-built information is acquired. This information might include actual activity start dates or the percent completion of various tasks. Suppose, for example, on the construction of a new house, completion of the foundation was delayed two days due to bad weather and the wall framing is 25% complete. The activity form on which activity information is initially entered typically allows for input of as-built information (Figure 9.3). In this case, the foundation would be listed as 100% complete, an actual start date entered, and an actual finish date of two days later than planned also entered. In addition, the wall framing task would be listed as 25% complete and an actual start date entered. A typical way in which programs graphically represent this information is by changing the color of the bar on the Gantt chart. In this case, the entire bar for the foundation task would be changed to a different color and its length increased by two days. For the wall framing, 25% of the bar would be shown in a different color.

Once all of the as-built information is entered, CPM calculations of the network are again performed. The calculations are based on the current data date and include the changes made to the schedule. Revised activity start dates and float values are generated. If activities have been delayed, the project completion date may be affected. Most programs provide the option of holding or fixing the completion date when performing the CPM calculations. In order to accomplish this without creating any negative float, activity relationships or anticipated durations may need to be revised.

Comparison of the as-built and baseline schedules is generally shown using the Gantt chart. Most programs will show both the as-built and as-planned activity bars side-by-side on one Gantt chart. The bars will also be different colors to help understand and digest the information quickly.

PRESENTING A SCHEDULE

Presentation of the schedule is an important task. Effective communication of activity information to subcontractors and workers is essential for coordinating the work. Presenting the schedule in a clear and concise format is facilitated by the tremendous graphic capabilities of current computer scheduling software. A multitude of graphics, fonts, line styles, and icons are available to help display the schedule. These capabilities are enhanced by the clear quality of laser printers and further highlighted by the use of color printers.

Most scheduling programs offer a wide variety of printed views of the schedule. Gantt charts are the most commonly printed views because they are easy to read and understand. Programs typically allow the timescale to be reduced or increased so that the schedule can be printed to a specified size or on a specific number of pages. For short-interval schedules, a specific time frame along the Gantt chart can be targeted and then printed. Most programs also allow precedence diagrams to be plotted to assist in viewing the network logic.

The reporting capabilities of scheduling programs play an important role for project managers and are one reason why the programs are powerful and important management tools. When resource and cost data are entered along with each activity, diagrams and charts allow for tracking the resources and costs as the project progresses (Figure 9.4). Each program typically includes a number of formatted reports that present the information in different ways. Reports of material utilization or monetary expenditures during different periods of the project are typically printed for evaluation. Most programs offer the ability to create a report on any specific resource and for any time frame during the project.

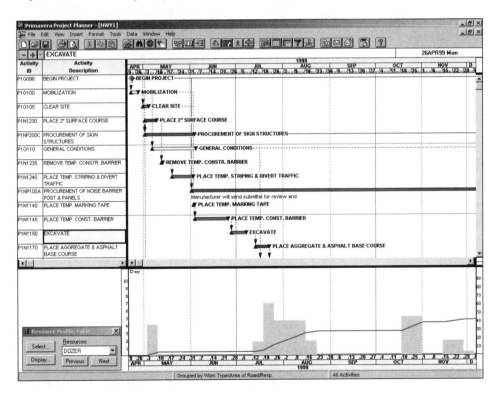

Figure 9.4 Resource Profile Showing Utilization of Bulldozers (Primavera Project Planner®) (*Reprinted with permission from Primavera Systems, Inc.*)

USEFUL SOFTWARE FEATURES

Current computer scheduling programs offer a wide variety of features. While all of the features are designed to assist in the development and management of schedules, some are more useful than others. Features that are easy to use and allow precise control and management of the schedule greatly enhance program effectiveness. Several useful features are described below.

Sorting and Filtering

Activity networks for many projects can get quite large. When a network includes many hundreds of activities, it is easy to get lost in the schedule. Sorting, which groups the activities according to some common criteria, helps the user to focus on the schedule. Programs that include the sorting feature typically allow grouping the activities according to one or more criteria (Figure 9.5). Color coding the grouped activities enhances the sorting feature. Filtering provides for a greater level of focus on the schedule. Activities

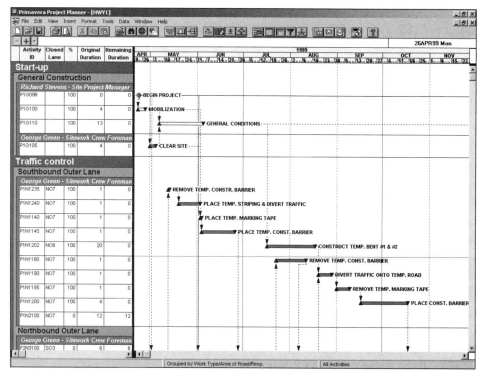

Figure 9.5 Activities Grouped According to Type of Work, Area of Construction, and Responsibility (Primavera Project Planner®) (*Reprinted with permission from Primavera Systems, Inc.*)

can be filtered from view so that only activities that meet certain criteria are visible. Filtering is quite useful for creating "mini-schedules" covering specific portions of the work, such as all of the tasks performed by the electrical subcontractor. Both sorting and filtering are almost essential for managing, updating, and reporting large project schedules containing many activities performed by numerous subcontractors.

Global Editing

Development of a schedule requires a great amount of data input. For large activity networks grouped together, changes to the schedules can be very difficult and time consuming. Global editing provides a quick and easy way to modify the information of many related activities. For example, two different project schedules may be linked together because they share common limited resources. If additional resources become available, global editing allows all activities utilizing the resources to be updated when the resource limit is increased. Global editing can also be applied to changes in activity durations, relationships, and as-built activity start and finish dates.

Cash-Flow Analysis

If cost information about resources utilized is entered (labor, material, and equipment), cash-flow diagrams can typically be generated. These diagrams provide a useful means of monitoring costs over the life of the project. When comparing the as-built and as-planned schedules, reports of the actual cost of work performed (ACWP) and budgeted cost of work performed (BCWP) can be combined to show the cost variance (CV) for the project. Reports are also typically available to determine the schedule variance (SV) by comparing the cost of work performed to the cost of work scheduled. More sophisticated programs allow for income revenue streams to be modeled across the activity network as well. Revenues can then be combined with costs to obtain a report of the net dollar value of the project over time.

Resource Leveling

Resource leveling techniques are utilized when the project duration is fixed. Maximizing the effects of resource leveling requires performing both a backward and forward pass through the network. Some scheduling programs carry out only a backward pass when performing these calculations. As a result, back float made available during the backward pass does not get used to further level the resources. More sophisticated scheduling programs perform both the backward and forward passes.

LINKING TO OTHER PROJECT MANAGEMENT SOFTWARE

Computer scheduling software is designed to assist with the scheduling and control of construction projects. Most other project management tasks associated with the planning, design, construction, and operation of constructed facilities are also performed using computers. A wide variety of computer programs are available for such activities as estimating, document control, safety planning, and project communications. The individual capabilities of these programs along with scheduling software allow for the complete management of the construction phase using a computer. Integration of these programs, however, provides an even greater benefit. The linking of estimating and scheduling software, for example, allows for the efficient management of both time and money during the course of a project. Cost information developed for the estimate can be electronically added to a schedule for cost and resource control during construction.

The integration of schedule information with the design documents provides additional benefits. Four-dimensional (three coordinate axes plus time) computer-aided design programs (commonly referred to as 4D-CAD) are designed to permit dynamic presentation of the construction process. These programs allow for the design information to be created in layers that depict the timing and duration of the activities to be undertaken in the construction process. Using this capability, the project team can review the construction sequence prior to the start of construction to eliminate design conflicts and constructability problems. Individual layers can also be associated with specific work crews, materials, or equipment to view the resource usage during the project. Such capabilities help to eliminate the amount of reworking and design changes that might be required, plus the overlapping of trades.

4D-CAD is useful for project monitoring as well. As construction progresses, the initial CAD files can be updated to create progress (as-built) drawings that depict the actual construction work accomplished up to a specific point in time. From a common vantage point, the progress drawings can then be compared with the as-planned drawings at the same point in time. This comparison allows for a graphic determination of whether the project is ahead of or behind schedule and the progress of specific phases or features of the project.

The benefits provided by 4D-CAD are amplified further when estimating capabilities are added. By integrating the design, schedule, and cost information, the types of materials and equipment along with associated quantities are automatically available for estimating and scheduling functions. This capability eliminates the need to perform quantity take-offs, helping to minimize errors associated with estimating quantities and recognizing the required materials of construction. Purchasing can be performed directly from the CAD files using layers that list the materials of construction. Including the time component provides procurement and delivery information as well so as to allow for just-in-time delivery.

The management and communication of project design, schedule, and cost information can be quite time-consuming and complex. Completion of construction projects requires communication of project information among many different project team members. Owners, designers, constructors, and construction managers, together with many other parties involved in the construction process, exchange all sorts of information during the course of a project. This communication of information is facilitated through Web-based project management services. These services provide project management software capabilities in one central location that all project team members can access. Project information such as design drawings, construction schedules, and cost data can be viewed, manipulated, and printed without the need for specialized computer programs. Most Web-based services allow for access rights that can be set to manage access to project information. These capabilities permit efficient access to and communication of any project information between project team members.

FINAL COMMENTS

Personal computers are the standard means by which construction scheduling is performed today. With the widespread availability and acceptance of computer software to perform scheduling tasks, manual computations to obtain scheduling information are rarely performed on construction projects. Several viable user-friendly scheduling programs are available. Since the scheduling computations can be made more quickly, more contractors are beginning to develop formal schedules on their projects. When contemplating the acquisition of a scheduling program, consideration should be given to the various features that different programs offer. While some schedulers opt for the most expensive programs, others may decide to obtain programs that are affordable and that perform specific tasks. Regardless of the software chosen, the personal computer has had a significant impact on the way scheduling is performed on construction projects.

REVIEW QUESTIONS

1. Since virtually all major construction scheduling tasks are now performed with the use of scheduling software on personal computers, why are manual computations still taught in scheduling classes?
2. What is the essential difference between sorting and filtering as it applies to construction scheduling programs?
3. What is an example of a construction activity that is properly scheduled in terms of working days? Calendar days?
4. What are the benefits, and possible problems, of having the construction schedule on a network?
5. What is an example of a network loop?
6. What is a possible drawback of performing all scheduling data input directly on the computer?
7. How has the use of scheduling programs on personal computers changed or impacted the construction industry?

10

Earned Value: A Means for Integrating Costs and Schedule

Time is of the essence.

The critical path method system is largely time oriented even though preceding discussions have often involved the analysis of project costs. Something that has not yet been discussed in detail is the extent to which CPM can be used as a cost-monitoring system. The ability to assess adherence to a proposed time schedule is clearly available. But is it possible to also determine how well the project is adhering to its "cost schedule," or estimate? Both issues are important and, as has already been discussed, they can be expected to be interrelated.

The integration of cost and schedule control systems is of natural interest to construction professionals, because the true "status" of a project can only be assessed if both cost and schedule data are examined in conjunction with one another. For example, a project may appear to be well under budget based on the amount of money spent to date compared to what was projected. However, this figure alone could be very misleading; costs may be very low because the project is well behind its schedule! Thus it is necessary to know what actual project costs are relative to expected costs while also knowing where the project is on a time basis.

A common question asked in the construction industry is, "What is the stage of completion of the project?" A common reply might be 30% or 85%. But what is the basis of that assessment? It could be based on the percent of time that has been utilized, the percentage of the value of construction put in place, or some other measure. There are rules of thumb in the industry that are often applied to given types of structures. For example, on a given type of building it might be typical to state that a project is 10% complete once the foundation is completed or that the building is 50% complete once it is "dried in." Of course, these percentages may vary considerably with changes in the features specified for the structure. To a large extent, this percentage completion assessment is based on a visual inspection of the project. A project involving the placement of underground piping from one point to another, for example, may be considered 75% complete if the piping has been placed for 75% of the required length. Determination of the percent complete on this type of project is quite simple. It certainly becomes more difficult to make an accurate assessment from a visual perspective when the roofing on a building is almost completed but a significant amount of work has already begun with the interior finishes. This difficulty arises because it is not easy to assess the amount of work accomplished when one building component consists of concrete, another consists of lumber, another consists of drywall, another consists of piping, and so on. A visual inspection of the different building systems, in varying stages of completion, will result in a guess at best. If an accurate assessment is needed, it will be necessary to utilize a more objective means of determination for the percent completion.

In the early to mid-1960s, the U.S. defense agencies promulgated guides and requirements related to the integration of cost and schedule data into a single system. A key concept that evolved was termed *earned value*. *Earned value* refers to the determination of how much work has been performed on the basis of what was budgeted for the work that has actually been completed. The idea is that a contractor has "earned" whatever amount was budgeted for the work that has in fact been completed. This value then becomes a measure against which other cost and schedule data can be compared in order to determine actual cost and schedule status.

Two systems using this technique emerged within the government in the mid-1970s. The Department of Defense developed a Cost and Schedule Control System Criteria (C/SCSC) technique and the Department of Energy a system called Performance Measurement System (PMS). While these systems were oriented toward owners, the earned value concept found its way into contractor systems as well.

THE EARNED VALUE CONCEPT

The earned value concept (sometimes also called *achieved value*) compares several measures to obtain an overall picture of project status. The following are the three primary data requirements:

BCWS: Budgeted Cost of Work Scheduled: This measure, also called the project plan, is developed at the outset of the project as it involves assigning to activities the amount budgeted (estimated) for every activity. Knowing the project schedule, it is possible to determine the amount of money budgeted relative to time. This is essentially the same as developing the cumulative cost curve discussed earlier.

BCWP: Budgeted Cost of Work Performed: This is the earned value as it indicates what the budgeted costs are for the work that has actually been performed to date. Note that this represents the periodic worth of the work that is actually performed, based on the initial estimate, and has nothing to do with the actual costs incurred in performing the work. It requires making an assessment of the amount of work completed (time schedule) to date and then applying the appropriate budgeted amounts for this work.

BAC: Budgeted Cost at Completion: This is the original cost estimate of the total cost of construction. It is ideally the total estimated cost to complete a project.

EAC: Estimated Cost at Completion: This is the forecast of the total actual costs required to complete a project based on performance to date and estimates of future conditions.

ACWP: Actual Cost of Work Performed: This is the measure that brings together the monitoring of both time schedule (work performed) and cost records (actual cost). It gives the actual costs to date. (Note that ACWS, Actual Cost of Work Scheduled, is not a component of the system, because if the work scheduled has not yet been performed, it is impossible to determine its actual cost!)

Comparing the first (BCWS) and last (ACWP) values to the earned value (BCWP) gives us an indication of "variances" from the expected:

SV: Schedule Variance; BCWP–BCWS: Subtracting the budgeted cost of work scheduled from the budgeted cost for work performed provides an indication of schedule deviance in terms of dollars of work. The cost side of the system remains constant (using budgeted cost figures) so the difference between the two amounts must be due to schedule deviations alone. A negative value indicates that the project is behind schedule as the "value" of work performed is less than that scheduled. This is shown graphically in Figure 10.1. Note that the schedule variance is presented in terms of dollars, while in reality this is an indication of project status in terms of scheduled time.

$$SV = BCWP - BCWS \quad (SV > 0: \text{ahead of schedule}, SV < 0: \text{behind schedule})$$

CV: Cost Variance: BCWP–ACWP: The difference between the budgeted and actual costs of work performed clearly indicates the status of the project in terms of cost. The work performed provides the common unit of comparison. A negative CV indicates the project is over budget as the actual costs exceed the budgeted costs. This is shown graphically in Figure 10.2.

$$CV = BCWP - ACWP \quad (CV > 0: \text{under budget}, CV < 0: \text{over budget})$$

These variances can be further expressed as ratios in percentage figures:

% SV: % Schedule Variance: 100 × SV/BCWS: This ratio indicates the percent deviation from the schedule based on budgeted cost values.

% CV: % Cost Variance: 100 × CV/BCWP: This measure gives the percentage deviation of the actual costs from what was budgeted for the work performed to date.

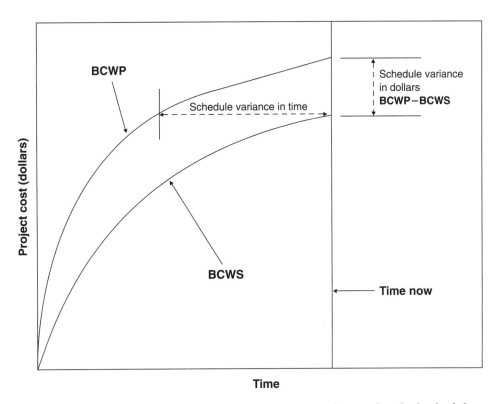

Figure 10.1 Earned Value Analysis Showing a Project that Is Ahead of Schedule

Similarly, cost and schedule indicators can be measured as:

SPI: Schedule Performance Index: BCWP/BCWS: This ratio provides a direct relationship between work performed and work scheduled based on budgeted costs. A value greater than 1.0 indicates that more work has been performed than was scheduled, and thus the project is ahead of schedule.

SPI = BCWP/BCWS   (SPI > 1: ahead of schedule, SPI < 1: behind schedule)

CPI: Cost Performance Index: BCWP/ACWP: This ratio directly compares budgeted costs to actual costs based on work performed. A value greater than 1.0 indicates that the budgeted amount of the work is higher than what it actually cost, and therefore the project is under budget. Since dollars are not represented in the CPI, the results are sometimes easier to interpret.

CPI = BCWP/ACWP   (CPI > 1: under budget, CPI < 1: over budget)

It can be expected that these performance indices will show variations throughout the life of the project. Generally, it can be expected that the indices will be less than 1.0 early

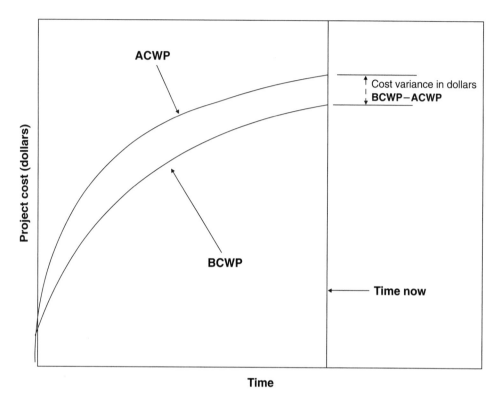

Figure 10.2 Earned Value Analysis Showing a Project that Is Over Budget

in the project and rise as the project progresses. This is due to start-up and "learning curve" factors that normally inhibit efficient operations early in a project. If calculated separately for each time period, the values may also show a ragged series of peaks and values. Calculating a cumulative index from project start to date shows a smoother transition and over time should approach 1.0 for a project that is close to schedule and budget.

Determining the status of percent complete of a project can also be accomplished with the variables already defined. The percent complete (PC) is simply the ratio BCWP/BAC, expressed as a percent. Note that this assessment merely defines the percent completion based on original estimates and is not influenced by actual cost figures. Another way of viewing earned value (BCWP) is given in the following expression:

$$\text{Earned Value} = (PC) \times (BAC)$$

The percent completion can be compared with the expected or anticipated percent completion at a given point in time. The planned percent complete (PPC) is derived at a point in time with the ratio BCWS/BAC, expressed as a percent. The PPC is easy to grasp as a measure of the project status when compared to the PC. For example, if the PPC is determined to be 60% and the PC is determined to be 65%, it is clear that the project is running ahead of schedule.

Information on the cost index and how it has changed over time can be used to forecast what the costs will be at the end of the project, referred to as Estimated at Completion (EAC). The forecast of the EAC can also be obtained at a given point in time with the variables that have already been defined. A simple computation of the EAC at a given point in time is computed as follows:

EAC = ACWP + (BAC − BCWP) where BAC − BCWP represents unearned hours

The EAC is the latest revised estimate of the total cost of the project and can be compared to the BAC (original estimated project cost) to determine an estimated At Completion Variance (ACV). Since the EAC is based on costs incurred to date, this estimate will be accepted as being more accurate than the BAC. The ACV is a simple computation and provides an estimate of the amount that the total cost of the project is expected to exceed or to be less than the budgeted amount (ACV = BAC − EAC). If ACV > 1, the project is expected to be completed under budget, and if ACV < 1, the project is expected to overrun or exceed the budget. Because of the importance of the value of ACV, it is important that the determination of EAC be as accurate as possible. The projection of EAC may use a cumulative average CPI, the CPI for the most recent period, or some other selected figure. Applying the selected CPI ratio to the BAC value will yield an estimated EAC (EAC = BAC/CPI).

Example

A simple problem will help to illustrate the application of the principles of earned value. A project has been defined that consists of 12 activities (each requires one month to complete) for which the estimated costs and durations have been defined (see Figure 10.3).

The project costs are estimated to be \$257,000 (BAC = sum of the costs of all activities). The project is scheduled to be completed in six months. After three and a half months, the sitework, excavation, foundation, fencing, and rough electrical work are completed. The framing is one-half complete, the rough plumbing is three-fourths complete, and the paving is half complete. The incurred costs to date are \$152,000. What is the status of this project in terms of the schedule and the budget?

ACWP = \$152,000 (given)

BCWS = (\$22,000 + \$30,000 + \$50,000 + \$10,000 + \$18,000)
  $+ \frac{1}{2} \times$ (\$40,000 + \$6,000 + \$16,000) = \$161,000

BCWP = (\$22,000 + \$30,000 + \$50,000 + \$10,000 + \$6,000)
  $+ \frac{1}{2} \times$ (\$40,000) $+ \frac{3}{4} \times$ (\$16,000) $+ \frac{1}{2} \times$ (\$18,000) = \$159,000

Project Status:

Schedule Variance = BCWP − BCWS = \$159,000 − \$161,000 = −\$2,000
SPI = BCWP/BCWS = \$159,000 / \$161,000 = .99
% Schedule Variance: 100 × SV/BCWS = 100 × (\$2,000 / \$161,000) = −1.2%

<u>Slightly behind schedule</u>

Activity	Cost	Month 1	Month 2	Month 3	Month 4	Month 5	Month 6
Sitework	$22,000	■					
Fencing	$10,000		■				
Paving	$18,000			■			
Excavation	$30,000		■				
Foundation	$50,000			■			
Framing	$40,000				■		
Rough electric	$ 6,000				■		
Rough plumbing	$16,000				■		
Drywall	$13,000					■	
Suspend ceiling	$ 4,000					■	
Interior finish	$34,000						■
Carpeting	$14,000						■

Figure 10.3   Schedule of Activities and Their Associated Costs

$$\text{Cost Variance} = \text{BCWP} - \text{ACWP} = \$159{,}000 - \$152{,}000 = \$7{,}000$$
$$\text{CPI} = \text{BCWP/ACWP} = \$159{,}000 / \$152{,}000 = 1.05$$
$$\%\ \text{Cost Variance}: 100 \times \text{CV/BCWP} = 100 \times (\$7{,}000 / \$159{,}000) = +4.4\%$$
$$\underline{\text{Below budget}}$$
$$\text{Percent Complete} = \text{BCWP/BAC} \times 100\% = \$159{,}000 / \$257{,}000 \times 100\%$$
$$= \underline{62\%\ \text{complete}}$$
$$\text{EAC} = \text{ACWP} + (\text{BAC} - \text{BCWP}) = \$152{,}000 + (\$257{,}000 - \$159{,}000)$$
$$= \underline{\$250{,}000}$$

Although the project is slightly behind schedule, it is performing under budget. The project is currently at the 62% completion stage and is estimated to be completed for a revised estimated cost of $250,000, a decrease from the original estimate.

DIFFICULTIES IN INTEGRATING COST AND SCHEDULE SYSTEMS

A fundamental requirement of a system that can support the integration of cost and schedule data is a set of common, or at least compatible, data collection as well as reporting units. It is necessary to develop and collect both cost and schedule data in a manner that allows the two to be associated with one another. Historically, a major

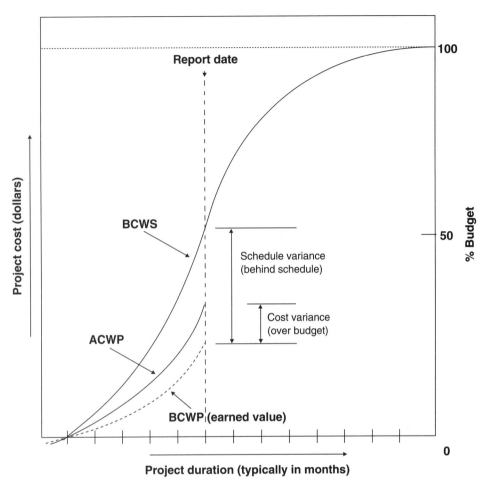

Figure 10.4   General Concepts of Earned Value Analysis

difficulty with such systems has been that the common unit between the two has been at such a detailed level that it has been burdensome to maintain. On the other hand, the reporting level of the data cannot be so generalized that operational detail is masked, making it difficult, if not impossible, to ascertain which specific areas of the project or estimating process need attention.

The dimensions of the problem are suggested in Figure 10.4. The view of a project will vary depending on whether the focus is the project itself or the functional organization structure, that is, the set of departments (scheduling, estimating, quality control, engineering) and personnel responsible for work on the project. The intersection of a project breakdown (Work Breakdown Structure [WBS]) and the functional organization structure represents what can be called a "cost account." This must represent a unit at which data can be collected on both cost and schedule. It is assumed in this paradigm that the work included within a cost account can be defined as a set of one or more "tasks" on which progress can be determined and that has a measurable finish point. Likewise, the unit must relate to the organizational structure of the company such that reporting and management of both cost and progress reporting can be expected from company personnel. The potential complexity and magnitude of such cost accounts can be seen as potentially overwhelming.

As discussed earlier under project monitoring, it is imperative to measure the progress of work within each cost account. Only then is it possible to assess what work has been performed. This may involve establishing identifying milestones, quantities in place, observation, resources expended, or some other measure. While inherent in measuring any schedule progress using CPM, the problem is complicated in this situation because of the simultaneous need to relate progress to cost data.

Of the various measures for assessing work performed, perhaps a strong argument can be made for utilizing work hours. Expressing all work in terms of a single unit, namely work hours, is quite desirable. Work hours do not carry some of the common problems that are associated with budgeted costs. Costs can be distorted with lump sum payments being disbursed at varying times. Since costs are distorted if the pay schedule is front-loaded, it is obvious that the cost estimate should not be confused with the schedule of values agreed upon between the owner and the contractor. Costs will be further distorted with the manner in which overhead is distributed to the various cost items, as overhead is rarely equitably distributed across all cost items. Since work hours are a common element across many work items, some of the disadvantages associated with costs can be avoided. Of course, work hours are more difficult to monitor when much of the work is subcontracted.

Some contractors have determined that progress can be monitored either by worker hours expended or by construction expenditures. An example breakdown of these costs is shown in Table 10.1. From the information provided, it is apparent that worker hours expended are not always the same as the actual costs. This is because

Table 10.1
PROJECT BREAKDOWN BY WORKER HOURS AND BY COSTS

Building Component	Worker Hours	Cumulative Hours	Cost	Cumulative Cost
Mobilization	3.0%	3.0%	0.5%	0.5%
Foundation excavation	6.0%	9.0%	3.0%	3.5%
Foundation concrete	6.0%	15.0%	8.9%	12.4%
Structural framing	11.6%	26.6%	17.8%	30.2%
Roof framing	6.0%	32.6%	6.0%	36.2%
Exterior wall cladding	8.8%	41.4%	12.1%	48.3%
Roofing	6.0%	47.4%	9.6%	57.9%
Electrical system	11.6%	59.0%	6.0%	63.9%
Mechanical system	8.8%	67.8%	2.1%	66.0%
Interior finishes	29.2%	97.0%	31.0%	97.0%
Demobilization	3.0%	100.0%	3.0%	100.0%

some building components are labor intensive and others are not. For example, once the roofing work is done, the project is considered 47.4% completed on the basis of worker hours expended but is considered 57.9% completed on the basis of costs. This points out that the definition of percentage completion might not be consistently applied in the industry. In some cases it might be desirable to use worker hours expended, and in other situations the costs might seem more appropriate. The decision as to the appropriate unit of measure should be made early in the life of a project. Depending on the type of project and contracting arrangement, different approaches might be selected.

FINAL COMMENTS

It is vital that construction contractors and owners of construction projects have viable measures of project performance. With front-end loading on billings, it is clear that the cash position on a project can be very misleading. That is, a project could have a strong cash position in terms of costs and revenues, yet be headed for a net loss. Similarly, a negative cash-flow position may not be an automatic indication of an impending loss on a project. The measure that is needed to establish the proper status of a project must not be biased. Earned value measurements provide the means of assessing the status of a project without the bias introduced by many other measures commonly used in the construction industry.

For the owner wishing to monitor the progress of a contractor, the pitfalls of using the schedule of values must be recognized. While the owner may have little recourse other than using the budgeted costs as a measure, the schedule of values and the project schedule should be carefully evaluated before using these as a means of evaluating performance. Only then can earned value be a reliable method of monitoring progress.

REVIEW QUESTIONS AND PROBLEMS

1. When using earned value, under what circumstances can the Actual Cost of Work Scheduled (ACWS) be utilized?
2. The following bar chart shows the activities to complete a project in six months (each column represents one month). The total project cost is $1,480.

Activity	Cost	Month 1	Month 2	Month 3	Month 4	Month 5	Month 6
A	$100	■					
B	$200		■				
C	$ 60		■				
D	$ 80			■			
E	$ 40			■			
F	$150			■			
G	$300				■		
H	$100				■		
I	$ 50					■	
J	$400						■

After three months, Activities A, B, C, E, and half of F are completed, for a cost of $490. What is the status of the project in terms of the budget (over or under) and the schedule (ahead or behind)? (Give each answer in terms of a percent variance.)

3. The following bar chart shows the activities to complete a project in six months (each cell represents one month). The total project cost is $700.

Activity	Cost	Month 1	Month 2	Month 3	Month 4	Month 5	Month 6
A	$100	■					
B	$ 30		■				
C	$ 40		■				
D	$ 80			■			
E	$ 20			■			
F	$ 50			■			
G	$ 70				■		
H	$ 30				■		
J	$200					■	
K	$ 60					■	
L	$ 20						■

After two months, Activities A, B, half of C, and half of D are completed, for a cost of $200. For this point in time (two months after the start date), what are the values of the budgeted cost of work performed, the budgeted cost of work scheduled, and the actual cost of work performed?

4. Cost breakdown structures (CBS) were compared with work breakdown structures (WBS). What is the primary problem (as related to CBS and WBS) in trying to integrate budgets and schedules?

5. Consider a simple project in which four activities describe the basic tasks. This project is presented here as a bar chart drawn as a time-scaled schedule. The sum of the amounts for the activities in the schedule of values approximates the originally anticipated cost of the project. Project duration is 5 weeks.

Activity	Cost	Week 1	Week 2	Week 3	Week 4	Week 5
A. Brush clearing	$1000	■				
B. Tree cutting and removal	$3000			■		
C. Fencing	$1500					■
D. Site grading	$2500					■

After 3 weeks, the following has been accomplished:

Activity A 100% complete, total cost $900
Activity B 40% complete, cost to date $1,400
Activity D 50% complete, cost to date $1,500

For this project, using earned value concepts, determine the schedule variance and the cost variance. State if the project is over or under budget and if it is ahead of or behind schedule.

6. Which of the following earned value terms has the least meaning for analyzing project status?

——— ACWP ——— BCWP ——— ACWS ——— BCWS

7. The following depicts a six-month schedule for a project.

Activity	Cost	April	May	June	July	August	Sept
A	$ 50	■					
B	$120		■				
C	$ 80		■	■			
D	$200			■			
E	$120				■		
F	$300				■	■	
G	$ 60				■		
H	$ 70					■	
J	$ 50					■	
K	$ 60						■

After three months on the job (July 1), the contractor concluded that 20% of Activity F (Activities A, B, and D are completed) was completed, 80% of Activity C was completed, and 40% of Activity E was completed. The project has cost the contractor a total of $535. For this project, using earned value concepts, determine the schedule variance and the cost variance. State if the project is over or under budget and if it is ahead of or behind schedule.

8. The following eight activities constitute an overall bar chart portrayal of a four-month project.

Activity	Cost	Month 1	Month 2	Month 3	Month 4
A	$ 2,000	■			
B	$ 1,000	■			
C	$ 6,000		■		
D	$ 2,000		■		
E	$ 7,000		■	■	
F	$ 9,000			■	
G	$ 3,000			■	
H	$10,000				■

After one and one-half months, Activities A, B, D, and half of E were completed, for a cost of $8,000. For this project, using earned value concepts, determine the schedule variance and the cost variance. State if the project is over or under budget and if it is ahead of or behind schedule.

9. The following six activities constitute an overall bar chart portrayal of a four-month project.

Activity	Cost	Month 1	Month 2	Month 3	Month 4
A	$ 2,000	■			
B	$ 8,000		■	■	
C	$ 2,000		■		
D	$ 5,000			■	
E	$ 1,000			■	
F	$15,000				■

Two months after the start of construction, Activities A, C, D, half of E, and one-fourth of B were completed, for a cost of $10,500. Evaluate the status of the project. Using earned value concepts, determine the schedule variance and the cost variance. State if the project is over or under budget and if it is ahead of or behind schedule.

10. The following six activities constitute an overall bar chart portrayal of a four-month project.

Activity	Cost	Month 1	Month 2	Month 3	Month 4
A	$ 7,000	■			
B	$12,000		■	■	
C	$ 6,000		■		
D	$ 8,000			■	
E	$10,000			■	
F	$25,000				■

One and one-half months after the start of construction, Activities A, B, half of C, and half of E were completed, for a cost of $29,500. Evaluate the status of the project. For this project, using earned value concepts, determine the schedule variance, the cost variance, and the percentage completion. State if the project is over or under budget and if it is ahead of or behind schedule.

11

The Impact of Scheduling Decisions on Productivity

Gifts delayed are not prayers denied.

As a contractor undertakes a construction project, he or she begins to make decisions that are intended to contribute to the efficient delivery of the final constructed facility to the owner. These decisions ideally take into account the impact that given strategies will have on the project costs, schedule, and quality. This requires the decision maker to be cognizant of the impact that his or her directives will have on the overall project. While very little is published on the relationship among project costs, schedules, and quality, it should be clear that there are often very close associations between them.

Many decisions made regarding the schedule are those intended to reduce the duration of selected activities. While the rationale for these decisions may appear logical at first glance, they may have adverse impacts on other aspects of the project, especially in terms of productivity.

WORKING OVERTIME

A common response to the need to accomplish more work in a smaller time frame is to have workers work overtime. The workers are already on site and they simply need to be informed of the need to work additional hours each day or to work on the weekends. What will the impact be on productivity if workers are asked to work overtime? Perhaps, if they are asked to work overtime for two or three days, the adverse impact may hardly be noticeable or may be nonexistent. However, it is generally accepted that working overtime for an extended period of time adversely impacts productivity. While little has been published on this topic, some have used the following formula to predict the productivity impact of working overtime for an extended period:

$$Eff (\%) = 100\% - 5\,[(\text{days} - 5) + (\text{hours} - 8)]\%$$

Where:

Eff = Worker efficiency based on 100% for a regular 40-hour week
days = Number of days worked per week
hours = Number of hours worked per day

While generic in its format, the equation does give some insight into the impact that overtime can have on worker productivity. Other considerations have to be taken into account before placing heavy reliance on the equation. For example, since setting up an operation consumes a considerable amount of time, it would seem logical that the overtime should first be applied to the existing workdays rather than adding another workday to the schedule. If workers carpool, the impact of having only some workers working overtime may have an adverse impact on morale. If workers commute considerable distances to get to the work site, there will be additional considerations.

By examining the above formula, one can get a quick glimpse of the impact of overtime on labor productivity. Based on the formula, the associated efficiencies have been determined for given overtime schedules (see Table 11.1). Note that the normal 40-hour workweek has been defined as the benchmark for 100% efficiency. For example, if 12 hours are worked each day for seven days per week (84 hours per week), the efficiency drops to 70%. For this scenario, it can be said the productivity of this

Table 11.1
IMPACT OF SCHEDULED OVERTIME ON LABOR PRODUCTIVITY

Hours/Day	Days/Week	Hours/Week	Efficiency (based on 40 hr.)	Effective Hr./Week
8	5	40	100%	40
9	5	45	95	42.75
10	5	50	90	45
11	5	55	85	46.75
12	5	60	80	48
8	6	48	95	45.6
9	6	54	90	48.6
10	6	60	85	51
11	6	66	80	52.8
12	6	72	75	54
8	7	56	90	50.4
9	7	63	85	53.55
10	7	70	80	56
11	7	77	75	57.75
12	7	84	70	58.8

84-hour workweek is equivalent to working 58.8 hours at 100% efficiency. In addition to the drop in productivity, the decision maker must also evaluate the impact of this scheduling decision on the costs of wages. Obviously, in an 84-hour workweek, more than half of the time will be worked at premium wages.

Example:

A contractor employs a workforce of 20 workers on a construction site for a private owner. The workers are paid an average wage of $12.00 per hour. Because of slippage in the schedule, it now appears that the project will not be completed on time, and there is a liquidated damages provision of $1,000 per day for every day that the project is extended beyond the contractual deadline. The contractor is now contemplating working 12 hours per day for five days each week. If the contractor expects to make up or shorten the project duration by ten days, is this a viable option?

Considerations:

By using the formula on overtime or by referring to the information in Table 11.1, it is easy to determine that the 60 hours worked each week per worker will be at a level of productivity achieved with 48 hours of work (using a 40-hour workweek as a baseline). Thus, if the 20 workers are assigned to this same work schedule, the project duration will be reduced by eight hours or one day for each week of such work. Therefore, it will take ten weeks to reduce the schedule by ten days.

What Is the Cost of Overtime?

Since the average wage of the workers is $12.00 per hour and since any work hours above 40 hours are paid at an overtime rate of $18.00 (assuming time and a half), it can be determined that the premium pay, the overtime portion of the base wage, is $6.00 per hour for 20 hours each week, or $120.00 per worker. The cost of the overtime pay for the entire workforce is $2,400 per week. From this, the contractor might very well conclude that each day that the duration is shortened will cost $2,400 (the cost of the overtime portion of the wage).

What Is the Cost of Lost Productivity?

It must be realized that working 60 hours each week nets only 48 hours of regular performance. The contractor's estimate assumed one hour of production for every hour worked. This means that there are 12 hours of regular work hours lost each week per worker, which equates to $144 per week per worker, or $2,880 for the 20 workers.
Another way of examining the production and overtime cost issue is to simply look at the additional labor costs incurred each week. Since the contractor will be paying straight time wages for 40 hours and time and a half for 20 hours, this is essentially the same (in terms of costs) as incurring a straight time labor wage cost for 70 hours. Since only 48 hours of production are realized, it can be said that the contractor loses 22 hours of wages each week that are attributable to the overtime cost and the lost productivity. The 22 hours at straight time would cost $264 ($12.00 × 22) per worker each week. This equates to a weekly cost of $5,280 for the crew of 20 workers. Note that

this is the same as the sum of the overtime cost of $2,400 and the lost production of $2,880. Thus, at a cost of $5,280, the contractor can avoid a liquidated damages charge of $1000.

What Are the Savings?

When compared to the liquidated damages of $1,000 per day, it appears that the added cost of overtime and the lost production far exceeds the cost of damages. If the numbers were a little closer, as they might be if there was a five-person crew only, some additional considerations should be addressed. For example, every day that the project duration is shortened will also reduce the number of days that the equipment and staff will be required on the project. The superintendent might be paid $1,000 per week, resulting in an additional savings of $200 per day. Depending on the type of equipment being used on the site, there will be additional savings, especially if the equipment is rented or could be utilized on another site. There are certainly other considerations that are difficult to quantify. For example, the company's reputation might be bolstered if it can deliver the project on time, and this might be beneficial for being awarded future contracts.

INCREASING THE WORKFORCE (CROWDING)

Working overtime may be desirable in situations where a quick response is needed or where training costs for certain skills are high. However, a contractor may also decide to add extra workers to the workforce. In some cases, this added workforce may be asked to work on a second shift. On new construction projects, this may not be explored due to the requirements for additional lighting and the lack of supervisory or managerial personnel for the second shift. If additional workers are employed on the regular shift, it stands to reason that some complications can arise, especially if the workforce is increased by a significant amount.

When additional workers are hired on a construction site, management must consider the impact that these workers will have on the existing resources. For example, is there an adequate supply of hand tools and various pieces of equipment? It would be foolish to double the crew size for concrete placement without also increasing the supply of concrete vibrators and screeds. If the resources can be provided, the next consideration is the impact that the additional workers on the site might have on labor productivity. This possible reduction in productivity can be attributed to the crowding that can occur in a work area, shown in the following formula:

Eff (%) = 115% − 15 (size of expanded workforce/size of normal work force)%

Where:

Eff = Worker efficiency based on 100% for a normal workforce

From this formula, the productivity can be computed when workers are asked to work under conditions of crowding (see Table 11.2).

Based on the computed impacts of crowding on productivity, it can be seen that a 10% increase in the crew size will have only a minor impact on worker productivity.

Table 11.2
THE IMPACT OF CROWDING ON LABOR PRODUCTIVITY

Relative Size of Workforce	Relative Productivity
Normal	100%
10% above Normal	98.5%
2 Times Normal	85%
3 Times Normal	70%
4 Times Normal	55%
5 Times Normal	40%
6 Times Normal	25%

Often, the evaluation should not be based on the total workforce but rather on the actual work being done. For example, if a project has 100 workers and 10 additional pipefitters are hired, the impact on productivity, as assessed with the use of the crowding formula, will be small. However, the impact might be quite adverse if the pipefitters are assigned to a work area where 2 pipefitters were already working. This would result in 12 workers in an area previously occupied by 2 workers.

The impact will be compounded if crew stacking occurs. This will occur when concurrent activities are scheduled for the same location as a result of a change or where some type of project compression is required. Crew stacking is not specifically addressed by the above formula because it is focused primarily on an increase in crew size. If other crews are also forced to work in the same area, productivity will be compromised further. This loss in productivity will arise from crews competing for access to the same doorways and passageways, and they may also compete for the use of the same tools or power sources. In some cases, their productivity will be compromised simply through the inability to store materials or tools conveniently and to move freely. Such crew stacking conditions might require crew sizes to actually be reduced.

Example:

A specialty contractor employs a workforce of 20 carpenters on a construction site who are paid an average wage of $12.00 per hour. The carpentry work is running behind schedule and the contractor is being faced with a liquidated damages provision of $1,000 per day for every day that the carpentry work is extended beyond the contractual deadline. The contractor would like to explore the possibility of increasing the workforce to 30 carpenters. If the contractor expects to make up or shorten the project duration by five days, is this a viable option?

Considerations:

By using the formula on crowding, it is computed that the 50% increase in the number of carpenters will result in a level of productivity that is 92.5% of that to be expected in a 40-hour workweek. Normally, the 20 workers would each accomplish 40 hours of work each week for a total of 800 productive hours per week for the entire crew. With the additional workers, the productivitiy level per worker will be reduced to 37 hours each week per worker, or a total of 1,110 productive hours per week for the crew. From this, it is determined that the 30 workers will reduce the schedule by 0.38 week ((1110–800)/800), or 1.94 days for each week that is worked. If the additional ten carpenters are employed for three weeks, the carpentry work will be reduced by more than five days.

What Is the Cost of Lost Productivity?

Working with the additional carpenters, the workforce will have 37 productive hours each week. Thus, the contractor will lose three productive hours for each carpenter each week, which equates to 90 hours lost each week for the entire crew, or 270 hours over a three-week period. This will cost the contractor $3,240 in lost production over the three-week period.

What Are the Savings?

By working a crew of 30 workers, the schedule will be shortened by five days. When considering the liquidated damages of $1,000 per day, it is apparent that $5,000 will be saved in liquidated damages by employing the additional ten workers. Of course, there are other considerations. Are the additional carpenters as skilled as the 20 already on the project or are they relatively unskilled and unfamiliar with the company? This will dictate to a large extent whether the formula to compute productivity is even realistic. The assumption in the formula is that the additional workers are of the same skill level as the existing crew. If the added workers are less skilled than the existing carpenters, the productivity will probably be well below 92.5%.

The specific job site conditions will also impact the productivity of the workers. Judgment will need to be exercised when considering the addition of workers to a crew. For example, suppose an office building is being constructed. Production losses might be minimal if an additional crew is brought in to perform drywall installation. The drywall crews would probably be assigned to install drywall in different rooms. Thus, there would probably be a minimal adverse impact on productivity. If the additional workers are assigned to perform work in the same location as the existing workers, the impact on productivity could be considerable. For example, suppose two plumbers are installing plumbing fixtures in a bathroom. If two additional plumbers are assigned to assist with the installation of these plumbing fixtures, the impact on productivity could be considerable. There may be instances in which the cramped conditions could result in even less work being performed with the additional workers. The unique conditions at the job site and the specific tasks being performed will dictate the reliability of the formula for computing productivity with the addition of workers.

INCREASING THE NUMBER OF STARTING POINTS

To avoid crowding, it might be possible to assign workers to different areas. For example, instead of doubling the number of workers at the face of a tunnel, it might be more effective to use the additional workers at a second tunnel face. The same might apply to having additional sheet-metal workers assigned to work on different floors. Naturally, this will spread out the job and some losses in productivity can be expected. With the additional starting places, making material deliveries and satisfying equipment needs on each floor will become more complex. Use the following equation to compute the amount of schedule reduction that can be expected by increasing the number of starting points:

$$T_{new} = \frac{T_{old}}{(Points_{new}/Points_{old})^{2/3}}$$

Where:

T_{new} = Time required to complete a new project/task
T_{old} = Time to complete a past project/task
$Points_{new}$ = Number of starting points on the new project/task
$Points_{old}$ = Number of starting points on a past project/task

This equation determines the amount of time required to complete a given project or task when the number of starting points and the duration of a similar completed project/task are known. The equation determines the duration of a similar existing project/task. Table 11.3 summarizes some computations made with this equation.

Table 11.3 presents information on the relative productivity of work crews when additional starting points are utilized. Note that productivity actually decreases at a smaller rate as more starting points are used. This may or may not flag an error in the formulation; however, this is the only known formula to address this issue.

Example:

A contractor has a contract to erect a 45-story high-rise building. To make up for bad weather that occurred early in the project, the contractor is considering the possibility of employing additional crews on the building. Since the building is dried in, the contractor is thinking about tripling the workforce from what was originally anticipated. The original plan was to do the finish work inside the building by progressing from the first floor to the top floor. The contractor is now contemplating having work take place at

Table 11.3
IMPACT OF NUMBER OF STARTING POINTS ON LABOR PRODUCTIVITY

Relative to Known Project Task	Relative Duration	Relative Productivity
Same	100%	100%
2 times the number	63%	79.4%
3 times the number	48%	69.3%
4 times the number	40%	63%
5 times the number	34%	58.5%
6 times the number	30%	55%
8 times the number	25%	50%

three locations. These would be the first, sixteenth, and thirty-first floors. From their respective starting points, the work would proceed upward to finish the interior work. Initially, 30 workers would have been employed and they were expected to finish the work in 80 work days. The workers are paid an average wage of $15.00 per hour, and the liquidated damages provision in the contract is for $3,000 per day for each day of late completion. Should the venture with three starting points be pursued?

Considerations:

By increasing the number of starting points from one to three, the duration is expected to drop to 48% of the originally estimated duration of 80 days to about 38.4 (say 39) days (refer to the formula or to Table 11.3). The decision to be made then is whether it is more economically feasible to have 30 workers working for 80 days or 90 workers working for 39 days.

What Is the Cost of Lost Productivity?

The worker wages are an average of $15.00 per hour per worker, so 30 workers will accumulate 240 hours each day, for which the pay will be $3,600. Thus, the wages for working 80 days will be $288,000. With the workforce tripled, there will be 90 workers, who will accumulate 720 hours each day, for which the pay will be $10,800. The total wages over a period of 39 days will be $421,200.

What Are the Savings?

By tripling the number of starting points in the building, the number of days that the schedule is shortened is 41 days, for an additional cost of $133,200. This equates to a cost of $3,249 for each day of schedule reduction. This is quite comparable to the liquidated damages amount and warrants further consideration of other variables. Some considerations will be purely monetary, including savings in overhead. Goodwill established with the owner by completing the project earlier than currently scheduled may also pay dividends. Of course, the feasibility of actually tripling the workforce must also be assessed. Factors to address will include the availability of workers, ability to supply equipment and materials for all workers, and the logistics of actually having all floors ready for the crews.

IDENTIFYING THE CAUSES OF DELAYS

It is an ideal situation when a single cause of productivity losses can be isolated and the impact quantified. In reality, there are many causes that can contribute to losses on a construction project. Once identified, steps can be implemented to address those causes. One general source of productivity losses is delays. When workers wait idly at their work stations with no ability to proceed with their work, costs of production can suffer severely. Unfortunately, the causes of delays are numerous. They include delays resulting from the lack of supervision, interference with another crew, waiting for equipment and/or materials, and so on. To address construction delays, it might be prudent to implement a program involving foreman delay surveys.

Since the late 1970s, programs utilizing foreman delay surveys have been used effectively to bolster productivity. These delay surveys generate more detailed information than do most other means commonly employed to control labor costs. Cost reports can identify where problems exist, but they do not identify the root cause of those problems. Delay surveys have been developed which not only quantify the amount of delay time incurred, but also isolate the source of the delays.

The most popular form of delay survey is known as the "foreman delay survey" or crew delay survey. It is a survey that is completed by the foremen or first-line supervisors on the project. Each foreman estimates the total amount of time lost by the crew during each day because of specifically noted sources. When multiplied by the number of workers in a crew and the number of hours worked, an idea of the magnitude of the problem becomes apparent.

A program involving foreman delay surveys must be carefully implemented in order to receive valuable information. First, the foremen must be educated about the purpose of the foreman delay surveys and why it is important for them to provide their honest input. Next the foremen are asked to complete the foreman delay survey, preferably at the end of each day. The foreman delay survey will ask each foreman to estimate the total number of hours that their workers were delayed for various reasons on a specific day (see Figure 11.1). The foreman delay survey could be filled out on a weekly basis, but the results will be less accurate since there will be less recall of all the causes of delays after a few days have passed. When possible, the best results can be obtained when the foremen complete the survey every day for one to three weeks.

Once the delay surveys have been collected, the data can be analyzed to identify particularly problematic causes of delays. The results are generally shared with the foremen and their input is sought in identifying solutions. After various suggestions have been received, steps are taken to address the problem of the delays.

After approximately six months, the foreman delay surveys are again provided to the foremen for their input on the causes of delays in their crews. If conditions have improved adequately, no further action may be required. Naturally, project management must be responsive to the information that is garnered from these delay surveys.

A note of caution should be offered concerning the delay surveys. It is human nature to blame others for your own problems. This might cause the initial results of the delay survey to point the guilt at other parties than the foremen. On the other hand, when the second series of delay surveys are conducted, the foremen might feel as if they are a part of the solution, causing them to be biased and subconsciously reporting fewer delays. The data must be viewed objectively to see where changes are possible and most likely warranted. Look for consistency of reporting from the foremen. Outliers may have to be excluded from consideration due to the subjectivity of the survey responses.

INTERRUPTION OF WORK ON MULTIPLE UNITS (IMPACT OF LOST LEARNING)

Repetition of the same operation generally results in a reduction in the time and effort required to perform subsequent operations; that is, repetitive tasks become easier to perform with experience. This reduction tends to be regular and predictable. The

Foreman Delay Survey

Date of Survey: _____   Craft: _____
Foreman's Name: _____   Work Area: _____
Number of Workers in Crew: _____   Type of Work Done: _____

Manhours Lost Per Day

Cause of Delay	Number of Hours Delayed	x	Number of Workers Delayed	=	Worker Hours
1. Waiting for materials	_____	x	_____	=	_____
2. Waiting for tools	_____	x	_____	=	_____
3. Rework from design errors	_____	x	_____	=	_____
4. Rework from field errors	_____	x	_____	=	_____
5. Waiting for instructions	_____	x	_____	=	_____
6. Waiting for inspection	_____	x	_____	=	_____
7. Work area not ready	_____	x	_____	=	_____
8. Work area overcrowded	_____	x	_____	=	_____
9. Waiting for equipment	_____	x	_____	=	_____
10. Equipment breakdown	_____	x	_____	=	_____
11. Delayed by owner change	_____	x	_____	=	_____
12. Interference with another crew	_____	x	_____	=	_____
13. Other: _____	_____	x	_____	=	_____

Comments: _____
_____
_____

Figure 11.1 Foreman Delay Survey Form

observed characteristic of the improved performance is known as *learning*. This learning can be shown graphically or mathematically on a learning curve. A learning curve is also known as a *manufacturing process function*, an *experience curve*, and a *dynamic curve*. In simple terms, it means that the costs of production can be lowered by increasing quantity of production or increasing learning.

The principle of learning curves can be used effectively in procurement and production. If the learning rate is known, it may be possible to estimate the cost of producing additional units based on the learning information. In cost estimating, the bidding, pricing, and capital requirements are based partially on the learning curve concept.

Learning is realized at two levels. The first, and most obvious, is with the learning or experience acquired by direct labor. As a worker continues to produce a particular unit, he or she naturally improves. The second is in the learning acquired in the management process. This is accomplished through engineering programs that improve production, encourage high-quality production, reduce design complexity, create technological progress, or foster product improvement.

The subject of learning curves frequently comes up on construction sites. While contractors may utilize learning-curve principles when they assemble a cost estimate, learning does not normally become an issue until there is a delay in the work sequence. It is especially when the owner delays the work of a contractor that the contractor may make a claim for compensation, partially to regain the losses resulting from the loss of learning caused by the owner-caused delay. Learning-curve concepts are frequently brought up, so it behooves the scheduler to be aware of them.

Learning-curve principles can be applied to such diverse operations as ship building, aircraft industries, computers, machine tools, building construction, and refinery construction. The concept of learning curves applies best to a very specific type of operation with these criteria:

- High cost, especially in terms of labor
- Low volume
- Discrete item production

Thus, the concept does not apply to operations involving a high volume of units or those with a low unit cost (labor cost). Some general points that apply to learning curves are as follows:

- The amount of time and cost required to produce each unit tends to decrease for successive units.
- The amount of time to produce each unit decreases at a decreasing rate.
- The reduction in time required to produce each unit follows a specific estimating model; that is, the rate of improvement (learning) can be predicted by mathematical models (see Figure 11.2).

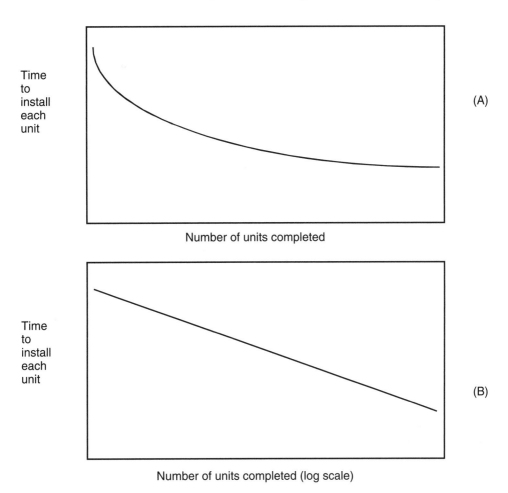

Figure 11.2 General Shapes of the Learning Curve

LEARNING APPLIED TO INDIVIDUAL UNITS

There are two types of mathematical formulations for learning-curve applications. The first relates to the time (generally in worker hours) required to perform each unit of work and the other relates to the average time per unit for a given number of units. First, the model for predicting the time to perform each unit of work:

$$T_N = K_T \times N^s$$

Where:

T_N = Effort required to complete the nth unit
N = Unit number
K_T = Constant (theoretically $K_T = T_1$)

s = Slope parameter or slope factor (this is a negative value)
s = Log ϕ / log 2
ϕ = Rate of improvement (generally based on doubled units—log 2 implies doubled units). If ϕ = .80, then the second unit is done with 80% of the effort of the first unit. The 4th unit would require 64% of the effort of the first unit.

Example:

The first unit of construction is completed in 10,000 hours. A learning rate of 80% is expected on doubled units (always assume doubled units unless specifically stated otherwise). How much time will be required to complete the eighth unit?

$$s = \log \phi / \log 2 = \log .8 / \log 2 = -.3219$$
$$T_N = K_T \times N^s$$
$$T_8 = 10{,}000 \times (8)^{-.3219}$$
$$T_8 = 5{,}120 \text{ hours}$$

Example:

If the learning rate is 95%, how much time will be required to complete the eighth unit?

$$s = \log .95 / \log 2 = -.074$$
$$T_8 = 10{,}000 \times (8)^{-.074}$$
$$T_8 = 8{,}574 \text{ hours}$$

What happens if the learning rate is not known? It must then be computed from the available information. The learning rate can be determined if information on two units is known. Suppose that the time of effort is known for two units.

$$T_i = K_T \times N_i^s \qquad T_j = K_T \times N_j^s$$

By dividing:

$$T_i / T_j = (K_T \times N_i^s) / (K_T \times N_j^s)$$

or

$$T_i / T_j = N_i^s / N_j^s$$

Take the log:

$$\log(T_i / T_j) = s \times \log(N_i / N_j)$$
$$s = (\log T_i - \log T_j) / (\log N_i - \log N_j)$$

Then solve for K:

$$T_j = K_T \times N_j^s$$
$$\log T_j = \log K_T + (s \times \log N_j)$$

Note the form:
$$Y = A + BX$$
$$K_T = \text{Intercept}$$
$$s = \text{Slope}$$

Example:

Suppose the fifth unit was completed in 200 hours and the tenth unit was completed in 150 hours. Find the time required to complete the twentieth and the thirtieth units.

$$s = (\log T_i - \log T_j) / (\log N_i - \log N_j)$$
$$s = (\log 200 - \log 150) / (\log 5 - \log 10)$$
$$s = .1249 / (-.301) = -.415$$
$$\log T_j = \log K_T + (s \times \log N_j)$$
$$\log 150 = \log K_T + ((-.415) \times \log 10)$$
$$\log K_T = 2.5911$$
$$K_T = 390$$

Learning rate:

$$s = \log \phi / \log 2$$
$$-.415 = \log \phi / \log 2$$
$$\log \phi = -.415 \times \log 2 = -.1249$$
$$\phi = .75 \text{ or } 75\%$$

(Note: This is observable by inspection because 150 is 75% of 200.)

For unit 20:

$$T_{20} = K_T \times N^s$$
$$T_{20} = 390 \times (20)^{-.415}$$
$$T_{20} = 112.5 \text{ hours}$$

(Note: This is observable by inspection because 75% of 150 is 112.5.)

For unit 30:

$$T_{30} = K_T \times N^s$$
$$T_{30} = 390 \times (30)^{-.415}$$
$$T_{30} = 95.08 \text{ hours}$$

If several data points (more that two) are known, a more accurate and, perhaps, more realistic learning curve can be developed as a predictive tool. There are essentially two methods of obtaining learning information from the data. The first is simply to plot the data points on log-log paper. A "best fit" straight line should then be drawn "through" those points. The slope of the line will be the learning

Table 11.4
USE OF THE LEAST SQUARES FIT METHOD

N	T	log N	log T	(log N)²	(log N) × (log T)
10	510	1.0000	2.7076	1.0000	2.7076
30	210	1,4771	2.3222	2.1818	3.4301
100	190	2.0000	2.2788	4.0000	4.5576
150	125	2.1761	2.0969	4.7354	4.5631
300	71	2.4771	1.8513	6.1360	4.5859
		9.1303	11.2568	18.0532	19.8443

rate (it will be negative). The second is to use the least squares fit method (see also Table 11.4).

$$s = \frac{M\Sigma(\log N \times \log T) - \Sigma \log N \times \Sigma \log T}{M\Sigma(\log N)^2 - (\Sigma \log N)^2}$$

$M = $ number of data points

$$\log K_T = \frac{\Sigma \log T \times \Sigma(\log N^2 - \Sigma \log N \times \Sigma(\log N \times \log T)}{M\Sigma(\log N)^2 - (\Sigma \log N)^2}$$

$$s = \frac{5(19.8443) - 9.1303(11.2568)}{5(18.0532) - (9.1303)^2} = -.515$$

$$\log K_T = \frac{11.2568(18.0538) - 9.1303(19.8773)}{6.9036} = 3.19208$$

$$K_T = 1556$$

Once s and K_T are determined, other units can be easily estimated.

LEARNING APPLIED TO CUMULATIVE AVERAGE UNITS

Although the preceding discussion of predicting the amount of effort required to produce a given unit is helpful, it is not as useful as a predictor of the average unit cost of producing a given number of items. Since estimating is generally based on the overall average cost per unit (rather than the discrete cost associated with each individual unit), a predictor based on unit averages is more meaningful.

The cumulative average time for completing a given number of units can be related directly to the individual times, as determined earlier. This cumulative average time (hereinafter called A) can be shown algebraically as follows:

$$T_{1-N} = (T_1 + T_2 + T_3 + T_4 + \ldots\ldots T_N)$$
$$A_N = (T_1 + T_2 + T_3 + T_4 + \ldots\ldots T_N)/N$$
$$T_{1-N} = N \times A_N$$

Another method (approximate and not to be used if accuracy is desired):
$$A_N = K_T \times (N^{s+1} - 1) / ((s + 1) \times (N-1))$$

Example:

Given

$$K_T = 1{,}000 \quad \phi = 80\% \text{ or } s = -.3219$$
Find the A value for 100 units or find A_{100}
$$A_{100} = 1{,}000 \times (100^{-.3219+1} - 1) / ([-.3219 + 1] \times [100 - 1])$$
$$= 21{,}706 / 67.129 = 323$$

Another method (more approximate than the above method):

Reasonably close if $N > 20$, but not recommended if accuracy is desired
$$A_N = K_T \times N^s / (1 + s)$$

Example:

Using the method applied to the same problem as above:

$$A_{100} = 1{,}000 \times 100^{-.3219} / (1-.3219)$$
$$= 335 \text{ (versus 323 obtained with previous method)}$$

Although these approximations may have their place, one never has a clear idea of how "approximate" the answer really is. Thus, a more accurate and reliable procedure is more desirable. The format of the formula that will be used is identical to that already discussed:

$$A_N = K_a \times N^s$$

Regardless of the apparent similarity between this equation and the one used earlier, there is no apparent direct relationship between the values of K_a and K_T or between the values of s that are used in the formulas. The mechanics of solving the problems are, of course, the same. That is, if the learning rate for individual units is known, it is not easy to determine what the learning rate is for cumulative units. Using A values does, however, give one greater flexibility. The following formulas should make that clear:

$$T_{1-N} = N \times A_N$$
$$T_N = (N \times A_N) - ([N - 1] \, A_{N-1})$$

So it is easier to determine individual unit times to complete a unit from A values than it is to determine A values from information about individual units (T values). At any rate, the use of A values gives more accurate information. As before, the formula to remember:

$$A_N = K_a \times N^s$$

The mechanics of solving problems with this equation are identical to the method used to solve for T values.

WHAT HAPPENS WHEN WORK IS INTERRUPTED?

If the flow of work is interrupted for some reason (inclement weather, labor strike, an extended holiday season, a plant shutdown, or job reassignment), the learning that has occurred will also be affected. Eventually, the learning will regress to the level that existed earlier when the first unit was produced. This forgetting function has been formulated by Gates and Scarpa (1972) as follows:

$$F = 1 - (1 / \log [D + 10])$$
F = Forgetting function
D = Number of time units of delay, in calendar days

This function is then used in the following equation:

$$A_{AD} = A_{BD} + (F \times (1 - A_{BD}))$$

Where:

A_{AD} = A after the delay (fraction of A_1 or K_a)
A_{BD} = A before the delay (fraction of A_1 or K_a)

Example:

Given

$$K_a = 1{,}000 \qquad \phi = .90 \qquad s = -.152$$

interruption of five days after 100 units

Find the A for units 101 to 105 or find $A_{101-105}$

$$F = 1-(1 / \log (D + 10)) = 1-(1 / \log 15) = .1497$$
$$A_{BD} = 100^{-.152} = .4966$$
$$A_{AD} = A_{BD} + F(1-A_{BD})$$
$$= .4966 + (.1497 (1 - .4966)) = .57196$$
$$.57196 = N^s$$

(solve by taking the log of each side of the equation)

$$\log .57196 = s \log N = -.152 \log N$$
$$.24263 = -.152 \log N$$
$$1.5963 = \log N$$
$$N = 39.47$$

As a result of the delay, learning has regressed to a point as if only 39.47 units had been produced. After the delay, the rate of production will be the same as it was for units 39.47 through 44.47. That is, after the work resumes, the learning rate will resume at the same pace as it had before. The delay simply shifted the production rate "back up" the learning curve. So the production for units 101 through 105 will be calculated as follows:

$$A_{44.47} = 1{,}000 \times (44.47)^{-.152} = 561.68$$
$$A_{39.47} = 1{,}000 \times (39.47)^{-.152} = 571.96$$
$$A_{39.47-44.47} = [(44.47 \times 561.68) - (39.47 \times 571.96)] / 5$$
$$= 480.53$$

Example:

Suppose a contractor has a labor contract with a facility owner to construct four buildings that are essentially identical. The contractor has estimated the actual labor cost (with benefits) of each building to be $26,671. Allowing for an approximate 10% margin to cover overhead (7%) and profit (3%), the amount to be paid to the contractor for each building is $29,400. After the third building is completed, the owner determines that the fourth building will not be needed, so the owner elects to cancel the contract. The contractor asks the owner for additional money. The contractor's records of labor costs per building have been accurately maintained, as shown in Figure 11.3. If the learning-curve principle is applicable, what will the profit margin be if the contractor is paid $29,400 for the three buildings that were completed? What sum would be appropriate to provide the contractor with a 3% profit? (Assume that the overhead is fixed at 7%.)

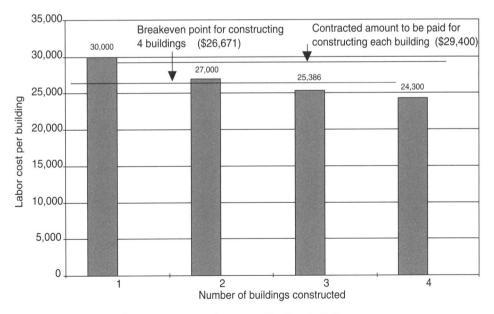

Figure 11.3   Learning Curve Data for Four Similar Buildings

From Figure 11.3, it can be determined that the average cost of constructing three buildings is $27,462 per building ($30,000 + $27,000 + $25,386)/3). Allowing for a 10% margin, the contractor will want to seek compensation of $30,208 per building (27,462 × 1.1). If the contractor is paid only $29,400 per building, the learning advantage gained in the fourth building is lost. The margin realized with a payment of $29,400 per building if only three buildings are constructed is 7.06% (computed as $29,400/$27,462). Allowing 7% for overhead, the remainder would be profit. Thus, the contractor will essentially break even with the payment of $29,400, earning only .06% in profit or essentially no profit.

Example:

A contractor made an agreement with a firm to weld ten special vessels for them. After six vessels were completed, the owner suspended the contractor's work. The reason for the delay was never disclosed to the contractor, but it was apparently not directly associated with the contractor's operation. When the work was permitted to resume, the contractor noticed that the labor cost per unit was much higher than it had been previously. What was the cost of the delay to the contractor? The analysis of the impact of the delay is summarized in Figure 11.4.

If the delay had not occurred, the learning-curve prediction was that the average labor cost per unit would be $7,047. Since the realized cost was $7,760, it can be computed that the average added labor per unit was $713 per unit, or $7,130 for the ten units. The contractor would probably make a claim to recover $7,130 from the owner to compensate for the impact of the delay.

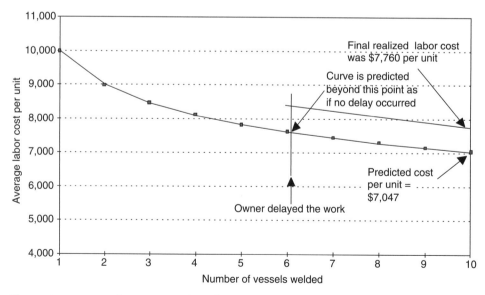

Figure 11.4 Learning Curve Principle Applied to a Delay

OTHER SOURCES OF LOST PRODUCTIVITY

There are various other sources of lost productivity, and these are often even more difficult to quantify. A reduction in worker morale can be devastating to worker productivity. Morale can decline significantly for a number of reasons, including disregard for worker safety, excess changes that necessitate rework, poor relationships with other site personnel, poor site conditions, and so on. Changes, delays, and rework are often sited as causes of production losses. Changes and delays may mandate that workers be reassigned to other tasks or be transferred temporarily to other projects. This will interrupt the smooth flow of work and will compromise worker morale, all of which will cause a decline in productivity.

Many of the losses in productivity will be difficult to quantify without an analysis of the specific circumstances involved. Some might be quantified with relative accuracy. For example, if access is denied to a particular area, it may be impossible to accomplish any work, clearly causing a delay in progress. The delay in the shipment of owner-provided materials may directly impact the project by delaying progress. Similar delays could be caused by changes that require material substitutions, especially if those materials require a long lead time for delivery. Delays may also cause work to be extended into the winter season, when productivity generally declines or when work cannot always be performed; e.g., concrete generally cannot be placed in freezing or near-freezing conditions.

Some situations might not have been anticipated at contract award that will drastically impact productivity. For example, a change could be made late in the construction phase. This could occur at a stage of project completion such that the use of a crane is no longer feasible and the work might have to be done through arduous manual labor, or the owner may have already occupied the facility prior to final completion and the contractor will be forced to finish the work in a facility that is partially furnished and occupied. The nature of these situations will impact productivity, but the extent of that impact will depend on the nature of the work that must be performed.

Whether a change results in productivity losses or a change in the manner of performing the work, in the final analysis, a delay will generally result. While formulas cannot adequately compute many of these types of losses, the use of schedules can be very helpful and often compelling in convincing others of the extent of the impact of a change, suspension, or directive that changes the way the work is performed. The accuracy of the schedule will be instrumental in determining the viability of using the schedule to make a convincing argument.

In some cases, judgment will need to be exercised to quantify the anticipated productivity losses. In fact, this is almost always the case. For example, suppose that a specialty contractor has a contract to put up 200,000 square feet of decking on a large project. The intent was to use five workers, and it was estimated that the crew could put up 250 square feet of decking in one hour. In an eight-hour day, the crew would put up 2,000 square feet of decking. Thus, the crew could be completed with the work in 100 working days. Before the work begins, the specialty contractor is told that completing the job in 100 days is not acceptable. The contractor needs to speed up the operation. It is decided to double the crew size and work five ten-hour days. What is the expected duration?

By doubling the crew size, the productivity would be at 85% of that of a five-worker crew. Thus, the crew productivity of a five-worker crew is reduced to 212.5 square feet per hour and, with two crews, the productivity would be 425 square feet per hour. Note that this is the production rate that is achieved simply by doubling the crew size. The issue of working overtime must now be addressed. With five ten-hour days, the production rate is 90% of that of a work schedule involving five eight-hour days. With overtime, the production rate of the ten workers is reduced to 382.5 square feet per hour, or 3,825 square feet in a ten-hour workday. Now it can be determined that the work can be completed in 52.3 days. The specialty contractor must now evaluate the reasonableness of the computations. Will both crews be served by a single crane? Will they work effectively as two crews or will they work as one large crew? Will safety be adversely impacted with the additional workers or the additional hours, or both? Once these aspects have been considered, an informed decision can generally be made. While the duration of the task is cut nearly in half, the financial impact will be considerable. It must be remembered that every day includes two hours of overtime for each worker.

FINAL COMMENTS

Questions concerning productivity generally relate to efforts to determine the expected level of productivity if some given conditions are imposed or if certain conditions do not occur. In many instances, these questions are asked after given events have already occurred. If work is still to be performed, managers may base their decisions about the work on models that indicate the impact that certain job conditions have on overall performance.

Retrospective use of productivity models is often done to establish grounds for a claim of some type or to determine if adequate grounds for a formal claim exist. The results of the models are always conjecture. Decisions have to be made based on the best information available, but one may never be able to establish quantitatively the value of the decision. The productivity models have been presented in this chapter to help highlight the essence of the implications of performing work under differing conditions. These models are generic in nature and their use may require modification under changing circumstances.

REVIEW PROBLEMS

1. The third unit of an aircraft hanger project was completed in 2,700 hours. The fifteenth unit was built in 2,100 hours. What is the learning rate (based on doubled units)? How long would it take to build the hundredth unit?
2. A tower (the first one) was installed in 900 hours. The improvement rate for doubled units is expected to be 90%. What will be the time required to complete the fifth tower? The thirtieth tower?
3. A contractor was in the business of building storage tanks. The learning rate was 0.80 on doubled averages. He broke even on the twelfth unit, which brought the average

of all the units to 1,500 hours per unit with a labor cost of $18,000 per unit. What was his profit margin on a contract for 30 such units? Assume that this was the first contract and that he was paid the same price per storage tank. Give the profit margin in dollars and in percent.

4. The A value for 5 sluice-gate installations was 160 hours (learning was 0.90 for doubled averages). Find the average time to install 25 sluice gates. What is the time required to install the twentieth sluice gate?

5. A tower was erected in 7,000 hours. The company feels that the learning rate per tower will be 0.80 on doubled units. How many towers will have to be erected before the time per tower is reduced to 3,000 hours? How much time will be consumed on the fiftieth tower?

6. The cumulative average time for constructing 30 electrical assemblies is 170 hours. The rate of learning for doubled averages is about 0.95. How much time was consumed in building the fiftieth unit? What was the total time required for all units after 100 units were completed? What total hours were needed to assemble units 100 through 150?

7. The learning rate of 0.90 on cumulative averages is realized when constructing metal storage tanks. The first unit took 1,400 hours to construct. After six units were constructed, the hunting season opened and the entire work crew took off a total of five days. Ignoring the impact of weekends, determine the time required to complete units 7 through 12. If wages are $10 per hour, what was the cost of the hunting season in terms of lost productivity?

8. A welder and a carpenter decided to get out of the construction industry and build farm trailers instead. From building a few trailers on weekends, they estimated that the first trailer would take about $700 of their own labor to build and that an 85% learning rate can be anticipated on the cumulative average time as each trailer is built. (Note: They decided that their hourly wages should be no less than those they received in the construction trades.) The material costs for each trailer will be about $500, and the craftsmen do not see any way that this can be reduced. They estimate that each trailer can be sold for $1,000. In addition to making their wages on labor, they want to make 15% profit on the trailer materials. How many trailers must be built before this rate of profit can be realized? What is the labor cost on the fiftieth trailer?

9. A learning rate of 92% on cumulative averages is realized when constructing metal storage tanks. The first tank took 1,200 hours to construct. After six tanks were constructed, the work was stopped for two days, after which the work resumed. Find the time required to complete units 7 through 10. If wages are $15 per hour, what was the cost of the delay from lost productivity? What was the total cost of constructing 15 tanks?

10. Study the following data:

Unit #	Labor Cost/Unit
1	$1,000
5	900
25	810

Based on the trend of production costs on these units, develop a formula that will be usable to predict future costs; that is, find the value of *s*. Find the cost for unit 100.

11. An offshore drilling operation requires a large number of welders to build a platform. The contractor is paying welders $25 per hour on a 40-hour (8 hr./day) workweek. The contractor has never had the same welders return to the platform after they have been on the job for one week. He is now considering a seven-day workweek (56 hours) so that he can take advantage of the "experience" gained in the previous five days of work. Of course he would have to pay double time for anything beyond 40 hours. He figures that all other expenses would equalize (5 trips for 35 workdays with 7-day work weeks, or 7 trips for 35 workdays with 5-day workweeks). He incurs an outfitting cost of $2,000 per welder. What type of learning rate must take place for the "scheme" to work? State any assumptions that are made.

12. A contractor has kept accurate records on the installation of pressure vessels in a refinery. He had the following data:

Unit #	A
3	1,530
4	1,450
7	1,370
10	1,325
12	1,300
15	1,275

    A strike delayed further installations for two weeks (16 days) before the sixteenth unit could be installed. The contractor had a crew size of 11 workers who receive an average wage of $18 per hour. The rent on the equipment and the company overhead is $500 per day. The contractor noted that he "broke even" after having completed the eighth unit. Find the theoretical time to complete the first unit. Find the learning rate (in percent). When work resumes, how many units must be completed before the crew is working at the same pace it had when it completed the fifteenth unit? How long will it take to complete the sixteenth unit? What is the cost of the delay on units 16 through 20?

13. A project was expected to be completed in ten weeks by a crew of five workers. The work was to be completed without overtime, namely, with 40-hour workweeks. Workers were to be paid $20 per hour, which included all fringes. Before construction began, the owner decided that the project schedule was to be accelerated by having the workers work eight hours on Saturdays. Assume wages are paid at time-and-a-half on Saturdays.

    a. What is the revised expected duration of the project?
    b. What is the added cost due to the acceleration?

14. A four-worker crew was assigned to install plumbing in a house. To accelerate the work, it was suggested that two additional crews (consisting of 4 workers each) be assigned to do the work.

    a. If the initial estimate was that the four-worker crew could complete the work in 12 days, what is the best estimate of the completion time with the additional crews?

b. What is the estimated labor cost of the change in worker assignments if all workers are paid $20 per hour?
15. A high-rise project was scheduled to be painted in 30 working days when a single crew was employed, beginning its task at the bottom floor. What would be the expected duration of the painting work if two additional crews were employed on the project? One crew was assigned to start at the top of the building and work downward. The other crew was assigned to start at the one-third point in the building and work upward.
16. The finishing work on a building project was to be completed by an ideally sized crew of five workers in 45 days (40-hour workweeks). To accelerate the work, it was decided that the crew size would be increased to eight workers and that a second eight-worker crew would be added to the project. In addition, all workers were scheduled to work 60-hour workweeks (10 hours per day from Monday through Saturday). Assume workers are paid $18 per hour.
    a. What is the expected duration of the project with the suggested changes?
    b. What is the cost of the acceleration in terms of labor cost alone?

12

CPM in Dispute Resolution and Litigation

The best way to avoid litigation is to prepare for it.

In many cases it may only be in the "final analysis," during the resolution of disputes, that the critical path method (CPM) is considered to be worth the time and effort required to use it to its fullest extent. It is at this point that the CPM technique is particularly useful because it provides a model of the project. The model illustrates how a project was originally planned and how it changed as the project progressed. By introducing changes into the model, it is possible to show how other portions of the model are impacted; it thereby furnishes a basis upon which assessments of time and money awards can be made.

GOING TO COURT

Disputes normally come down to assessing three aspects of delays: (1) who was at fault or who caused the delay, (2) how much of a time delay occurred (project delay), and, consequently, (3) what monetary awards should be made. The critical path method is primarily useful in addressing the second of these conditions and in some cases can be used to assist in determining at least a portion of the monetary awards. The only item it does not normally address is that of determining the party at fault, as this will be difficult to assess without considering specific wording in the contract. However, CPM networks should enable anyone to evaluate the impacts of the action of any party (owner, contractor, subcontractor, or others). Most important, the assessment can be considered as being objective (unbiased) provided the model (network) used is valid.

There are numerous court cases involving disputes in which the project duration has allegedly been compromised by change orders, differing site conditions, excusable delays, nonexcusable delays, suspensions of work, and other causes. Many of these cases concern contractors who felt they had been delayed as a result of some occurrence for which there was entitlement to additional time and/or money. The full extent of these cases will not be presented here because numerous other construction textbooks address this topic adequately. The cases presented here relate primarily to those in which the schedule itself became an elemental component of the court's decision.

The objectivity of a CPM network is one aspect that allows it to be used effectively in court and other arenas in which disputes are adjudicated. Once validated, an analysis of the network can be used to provide evidence as to the extent of delays, if any, caused by actions, or "inactions," of any party. As such, the CPM can be an effective evidence-seeking tool. An example is the case of Blackhawk Heating & Plumbing Co., Inc.'s, claim against the U.S. government, as described in the Board of Contract Appeals Decision (75-1 BCA [P11, 261] GSBCA No. 2432. April 30, 1975. Contract No. GS-05BC-4963).

In the Blackhawk case the Board of Contract Appeals based its decision on a network prepared by a government-hired consultant four years after the project was completed. The after-the-fact CPM was developed with heavy reliance placed on Blackhawk's daily work logs. The decision stated that "Because of the after-the-fact nature of the . . . CPMs and the fact that they were prepared at the government's request . . ., we feel compelled to take a long, hard look at the basis on which the CPMs were constructed."

The consultant had started with a CPM submitted by the contractor. This had not been formally accepted by the government, but the third CPM submitted was felt to be a reasonable plan for meeting the specified project duration. The consultant modified the "as-planned" network only to the point of breaking down enough activities to accurately depict the point at which disputed delays impacted the project. Next, an "as-built" network was developed. The activity sequence was basically the same as planned, except that while work was to be performed on a floor-by-floor basis, the contractor's actual work logs showed that work was in fact taking place on multiple floors simultaneously. Thus, floor-to-floor restraints were removed from the network.

After examining results of computer runs using the network and considering challenges to it by the contractor, the Board found that "Network 4 and Run 1 thereunder are a sound basis upon which to evaluate various project delays." Included in the contractor's challenges to the network were that the consultant had used "dummies" to depict delays and had assigned durations to the dummies. It was argued that this was contrary to the consultant's own definition of a dummy. The Board agreed, but found it did not invalidate the legitimacy of the model. What the consultant had called dummies were really activities that consumed time alone and no other resource. Thus, even though the network was constructed well after the fact and had technical problems, it was still held to be a valid depiction of the project and fit for providing objective analysis of delays.

A portion of the decision involved the finding that a government-caused problem did, in fact, delay the contractor's work. The delayed work did show up on the critical path of the "as-planned" network. However, at the time of the delay the path was not critical, as the contractor had been largely responsible for another unrelated delay that had caused the critical path to shift. To support his argument, the contractor presented a network model representing only a portion of the project. This was disregarded in the decision as being incomplete and thereby not acceptable as an evaluation tool.

In the Blackhawk case several delays occurred at the same time, referred to as *concurrent delays*, and a decision about the separable impacts of each delay could be determined. It is probably accurate to conclude that it was the availability of the CPM that permitted the Board to resolve the relative impacts.

Such was not the situation in Minmar Builders, Inc.'s claim against the government (72-2 BCA [P9599] GSBCA No. 3430. July 28, 1972. Contract No. GS-03B-15477). In this case the two construction schedules submitted for the addition of a classroom and laboratory at Gallaudet College were nothing more "than a bar chart showing the duration and projected calendar dates for the performance of the various contractual tasks."

The fact that bar charts are not acceptable as evidence to show the cause of project delays is clearly stated in this decision: "Since no interrelationship was shown as between the tasks, the charts cannot show what project activities were dependent on the prior performance of the [activities], much less whether overall project completion was thereby affected. In short, the schedules were not prepared by the Critical Path Method (CPM) and hence are not probative as to whether any particular activity or group of activities was on the critical path or constituted the pacing element for the project."

A detailed CPM analysis was utilized by Pathman in its claim for delay damages in *Pathman Construction Co. v. Hi-Way Electric Co.*, 382 N.E. 2d 453 (1978). A professional analyst was hired by Pathman to help develop support for its delay claim for damages. Pathman claimed it was delayed by Hi-Way through its practice of submitting shop drawings and material lists to the General Services Administration in a form that was not acceptable. Other delays by Hi-Way caused some of Pathman's work to be shifted to the winter months, necessitating the need to winterize the project. While the court stated that it would not apportion damages based on speculation, the evidence presented through the CPM schedule was quite compelling. Without a CPM schedule, Pathman would not have been as persuasive in court or succeeded in its claim for damages.

Beyond the simple use of a CPM network, it is imperative that the CPM be used properly if decisions are to be based upon it. It has already been shown that a CPM may be accepted as being persuasive evidence in deciding the extent of project delays. This was predicated, however, on it having been properly prepared and documented. Thus, supporting documentation is necessary, and if this information is collected in a manner that relates it directly to an existing CPM, it will be all the more powerful. Finally, for a CPM to be totally effective, it must be analyzed properly. This requires that it represent the project accurately to begin with and that as the project changes, the assessment of these changes is correctly and fully integrated into the network.

In the case of *Brooks Towers Corp. v. Hunkin-Conkey Construction Co.*, 454 F. 2d 1203 (1972), the court used CPM networks as the basis on which it decided a claim for delay damages. In this case Hunkin-Conkey entered into a contract to build a commercial and apartment building. A portion of the project was to be completed within 15 months, and the entire project was to be completed in 18 months. The contractor had prepared a CPM study prior to bidding the project because the schedule was considered tight. According to the contractor, the schedule duration was extended by numerous changes issued by the owner.

It was the nature of how those changes were processed that required an interpretation by the court. On many changes, the contractor stipulated the price and the additional time associated with each change. Each change quotation from the contractor was submitted on the owner's form, which included space for the owner to indicate the approval of the price that was quoted and a separate space for the approval of the added time requested. On a majority of the quotations, the price was specifically approved but no entry was made for the time extensions. These extensions became the court's central focus.

The owner contended that when no action was noted on the quotation, the request for a time extension was denied. The contractor took the opposing view, stating that no action on the owner's part signified implied approval. This became the issue on which the court had to decide. Numerous court decisions were cited in which inaction was construed as approval. The court stated "... where a duty exists to communicate either an acceptance or rejection, silence will be regarded as an acceptance." The extent of the time extension warranted was then determined by close scrutiny of the CPM schedule, which was quite detailed and took into consideration "all Bulletins, formal Change Orders, Field Change Orders, related correspondence, Daily Progress Reports and Monthly Pay Requests."

Progress schedules were also introduced in the case of *Lichter v. Mellon-Stuart Company*, 193 F. Supp. 216 (1961), but a twist occurred in the final determination by the court. Lichter, a specialty contractor doing business as Southern Fireproofing Company, entered two subcontracts with Mellon-Stuart on an addition to the Federal Reserve Bank in Pittsburgh. One agreement was for exterior stonework and the other for interior masonry work. The time of completion of the project was 20 months, but the completion dates of the stone and masonry work were not specified. Instead, the subcontract stated that the subcontractor was to perform "... as required by the progress of the work and as directed by the Contractor...." When it prepared its bid, Southern had anticipated performing the stonework in the summer months and that the masonry work would be done in an enclosed building.

When work began, the stonework was delayed to the point where work was performed during the winter months and the masonry work was done in a building that was not yet fully enclosed. This problem arose through delays caused by the issuance of 192 change orders, the postponement of the stonework by three months, the late delivery of materials, failure of other subcontractors to complete their work pursuant to the job schedule, strikes, winter weather, failure of the owner's representative to make timely decisions, and various other occurrences. The court concluded that, because of the contract provision related to the subcontractor's obligation to perform "as directed by the Contractor," Southern could file a valid claim only if a delay constituted a breach of the contract. The court also stated that it was incumbent on Southern to proceed to complete the work and that it was a mistake for it to abandon the work. While Southern was awarded some damages caused by a breach on the part of Mellon-Stuart, the contract provision as noted above limited its liability on other portions of the suit. In some cases, no matter how well it can be shown that the delay was caused by one party, the contract provisions may preclude any recovery.

CPM can also be used effectively to show how a project is expected to be completed in less than the specified contract time. By so doing at the outset of the project, the contractor can establish the grounds for claiming costs for delay if the owner delays the project from its scheduled, albeit early, finish. An example of this was presented in *Owen L. Schwam Construction Co., Inc. v. U.S.* (79-2 BCA [P13,919] ASBCA No. 22407. May 31, 1979. Contract No. DACA 51-75-C-0051).

The essence of part of this case was that the contractor provided a plan to construct a theater at Fort Devens, Massachusetts, in less than the contract duration of 360 days by avoiding a winter shutdown. The plan required that structural steel be in place by about mid-November, allowing close-in of the structure before prohibitive winter weather arrived. Government-caused delays through defective specifications and the extended approval of certain shop drawings delayed foundation work, making it impossible for the contractor to complete closing on the theater as intended. It was concluded that a consequence of foundation problems was that "the forming and pouring of auditorium grade walls, a critical path activity . . . was delayed. . . . It caused a 35-calendar day delay with respect to [contractor's] early completion plan." It was noted that the contracting officer's final decision stated that the structural steel was the cause of the delays. However, the board found that "the structural steel work was not a prerequisite activity to completion of the foundation. . . ." Both of the preceding quotes base decisions on CPM techniques and terms, specifically "critical path activity" and "prerequisite." It is also interesting to note that in this case the contractor was able to claim the salary cost of the superintendent during the winter shutdown period (during which he took his vacation) because his salary continued and the delay was attributed to the actions of the owner.

A final useful observation from this case is how the credibility of witnesses and their testimony is assessed. It was noted that the government's chief witnesses testified that the delaying effects of the events identified as causing construction delays had been exaggerated. They did not deny that they had occurred, but they indicated that they had been "blown out of reality" and had no effect on timely project completion: ". . . neither of them supported their general assertions . . . with any specifics. Moreover, the area engineer frequently pleaded an inability to recall the details of past occurrences and understandably, since the instant contract was only one of 25 projects assigned to him."

In contrast were the testimony and credentials of the contractor's president ("a one-person operation"). His "testimony demonstrated a comprehensive and specific grasp of the technical details and the events . . ., his constant attention on the job was required." His credentials included graduate work in civil engineering, contract administration work for the government, and supervising and managing construction projects for private owners as well as the U.S. Navy and U.S. Army. Consequently, the Board record for this case states that:

> In view of the above findings concerning the generality of the testimony of the Government's primary witnesses as contrasted with the specific and comprehensive testimony of appellant's president, . . . we are inclined to afford greater credibility to the latter's testimony with respect to our findings. . . ."

Must a contractor be delayed beyond the contractual project duration in order to successfully file a claim for delay damages? Not according to the decision rendered in *D'Angelo v. State of New York*, 362 N.Y.S. 2d 283 (1974). D'Angelo, operating under the business name of Triple Cities Construction Company, entered a contract with the state of New York for the reconstruction and realignment of 8.2 miles of highway, including a cattle pass and a number of culverts scattered along the length of the project. Triple Cities completed the project on time but still filed a claim for delay damages from the state. A detailed schedule or detailed records were never produced by Triple Cities, but heavy reliance was placed on the recollection and memory of Mr. D'Angelo, who presented himself as being thoroughly familiar with project operations.

The court found that the state could not delay the project and then defend its actions on the premise that the project had still been completed within the contractually stipulated duration. The court poignantly stated that Triple Cities "could rightfully expect to operate free from needless interference by the State, and, therefore, they are entitled to compensation where, as here, they could have completed their work ahead of schedule and thereby saving substantial sums of money, absent the delays caused by the State." While no specific information about network schedules was introduced, the case makes it clear that the difference between the contractor's planned duration and the contract duration does not constitute float to be utilized by the owner.

A different conclusion was reached in *Callanan Industries v. Glens Falls Urban Renewal Agency*, 403 N.Y.S. 2d 594 (1978). Callanan entered two contracts with Glens Falls. On one project, work could not commence until a utility line was relocated, and on the other the work was stalled until a house was demolished. Both projects were started by Callanan several months after the contracts were signed. For example, the house was removed three months after the preconstruction meeting was held, considerably later than anticipated by Callanan. The court denied the claim for damages by Callanan, stating ". . .time was plainly not of the essence is amply demonstrated by plaintiff's completion of its actual work in the same time as originally planned. . . ." It further stated that the owner had not "needlessly delayed" the contractor. Essentially, the court said that since the project was completed within the originally contracted duration, the delay must have been anticipated.

Judge P. J. Kane of the New York Supreme Court took strong exception to the court's decision in the Callanan case. He noted that the parties to the contract were in a meeting shortly after contract award in which ". . . it was fully expected that [house demolition] would be completed by June 1, 1972, at the latest. . . ." The house was not removed until August 24, 1972, the very day on which Callanan began work in earnest. Clearly, the Callanan decision is controversial and in strong contrast to the decision of the D'Angelo case.

The case of *Haas and Hanie v. the U.S. Government* (GSBCA Nos. 5530, 6224, 6638, 6919, 6920, June 8, 1984. Contract No. GS-09B-C-7003-SF) stressed the value of network schedules. Haas and Hanie entered a contract to construct a multistory building in Hawaii. The contract documents required the contractor to submit a CPM schedule to the owner and to make regular reports regarding this schedule. A network schedule was developed, but it was never updated or even utilized to any great extent. The contractor requested that the CPM requirement be waived and that the owner accept reports based on the bar chart. This request was accepted. As construction progress continued, the owner made 166 changes to the project. The contractor finished the project late and was assessed liquidated damages. The contractor sought an extension of time based on the impact that the changes had on the normally scheduled duration.

When Haas and Hanie presented its case, it used a bar chart to show its planned sequence of activities and their durations. They backed up this information by showing construction progress as measured by the amount of money spent. While the Board conceded that Haas and Hanie might have a legitimate claim for added compensation caused by the changes, the Board stated that this was not made apparent from the testimony that was offered. First, the bar chart did not show how one activity related to the other activities. Second, measuring progress by the amount of money spent has no basis in fact. The case for Haas and Hanie was also jeopardized by several subcontractors who testified that there was no concerted effort expended by the general contractor to coordinate the various activities on the site. There were several issues that did not help the case for Haas and Hanie, but the implication from the Board was that the failure to use a network schedule was a major error in judgment on the part of Haas and Hanie.

The contract language is very relevant in the settlement of many legal issues. This was certainly true in *Marriott Corporation v. Dasta Construction Company* (26 F. 3d 1057). In 1984, The Marriott Orlando World Resort contracted with Dasta Construction Company of Kansas City, Missouri, to perform a majority of the exterior skin work on the guest tower of a large hotel complex including affixing layers of stucco, plaster, and waterproofing onto the cement block walls. With direct input from Dasta, a CPM schedule was prepared for the project. Dasta's work was to begin in July 1984 and be completed in 15 months. In pricing its bid, Dasta placed considerable reliance on the CPM schedule that had been prepared, showing a steady flow of work for the Dasta workers.

When Dasta's team arrived on the project, they discovered that the project was running behind schedule by at least five months, primarily due to the poor workmanship of the concrete placement that directly impacted Dasta's work. Marriott made numerous changes as construction progressed, often resulting in the inability of Dasta to fully utilize the workers it had on site. This inefficiency was costly. In addition, costs were increased when Marriott failed to provide adequate vertical transportation and safety measures. Furthermore, Marriott refused to allow Dasta to bring in its own hoist to alleviate the vertical transportation problem. For this and various other reasons, Dasta sought added compensation on the contract. While Dasta felt justified in its claim against Marriott, the court relied almost entirely on the contract verbiage which stated that Marriott "specifically reserves the right to modify the progress schedule as required by the conditions of work." Intentions and perceptions can be deceptive. Marriott was simply exercising its contractual rights and was deemed justified in making the changes to the CPM schedule.

TYPES OF SCHEDULES

When the schedule is the subject of a claim, different types of schedules might be considered. Of course, the first schedule to consider is the original or "as-planned" schedule that presents information on the schedule as envisioned, generally by the contractor, at the beginning of the construction phase. On some projects, the owner approves the schedule submitted by the contractor. This approval has the project beginning with both the contractor and the owner having the same understanding of the project schedule. The as-planned schedule may be the only schedule that exists in a delay dispute, especially if no updating of the schedule occurred during the construction of the project. In that instance, the strength of the lawsuit will be dependent on the accuracy of the schedule in depicting the actual timing of the activities.

In a delay dispute, the claim by the contractor may be based on a "what if" schedule. Using a what if schedule, the assumption is made that the original schedule represents an accurate depiction of how the project was expected to be completed. Another schedule is then developed that is a modification of the as-planned schedule. The what if schedule is similar to the as-planned schedule with the introduction or insertion of the event or events that resulted in the disputed delay. The difference in the project duration between the what if schedule and the original or as-planned schedule is then assumed to be directly attributable to the delay for which a claim is being made. This type of schedule analysis assumes that the contractor would have constructed the project precisely as it was planned had it not been for the disputed delay. This assumption is rarely reasonable, especially on large or complex projects. As a result, it is difficult to make a compelling argument about the true impact of a delay using the what if schedule.

A stronger argument can be made to support a delay claim when it is based on an "as-built" schedule. The as-built schedule is one that depicts the actual occurrence times for the various activities. As described earlier, the as-built schedule will provide information on the actual start and completion dates of each of the activities contained in the schedule. The as-built schedule should be accurate as it is based on historical information. Of course, if the project had been constructed exactly as shown in the as-planned schedule, the as-built schedule will not appear to be different in any respect. This is not a realistic expectation. In a delay claim, the as-built schedule forms the basis for the "but for" schedule. This is done by modifying the as-built schedule to represent the project schedule "but for" the disputed delay. The but for schedule is constructed by modifying the as-built schedule by deleting the event or events that resulted in the delay. The difference in duration between the as-built schedule and the but for schedule is then assumed to be attributed to the delay. The weakness of this approach is that, on large or complex projects, it is not always clear which activities actually were controlling the pace of work. Thus, two schedulers may not make the same conclusions given the same information. Nonetheless, this is more convincing than an analysis based on the as-planned schedule.

A more compelling argument to support a delay claim is possible when the schedule is updated on a periodic basis and the analysis is made at the time of the delay. An analysis of the impact of a delay is made at the point of realization that a delay has occurred. This forms the basis for contemporaneous period analysis. If the impact of a delay is evaluated at the time that the delay occurs, it will be clear which activities are critical and, as a result, it will be easier to determine the impact of a specific delay.

Some differences in activity durations might actually be due to the variability that exists in the ability to estimate durations with accuracy. The duration of an activity might be adversely impacted by a heavy rainfall. It may also be possible that the duration of an activity is shorter than anticipated because the winter-scheduled activity took place during unseasonably mild weather. In a claim situation, the focus may be on the impact of perhaps a single delay or a single change made to the original contract. For example, if the owner issues a change order, the impact on the schedule might be considerable. In preparing the claim, it will be necessary to somehow isolate the impact of that single change. To show the impact of this change, it will be desirable to use the contemporaneous period analysis approach. The schedule will show the project duration as it would have been expected to be completed had it not been for this one change. This schedule will show the project status at a given cutoff date. Once the project is updated, the revised project completion date can be observed. At this point, another schedule is prepared in which the delay from the change being proposed is inserted in the schedule. This contemporaneous period analysis will give a reasonably accurate view of the potential impact of the schedule. The impact of the change will simply be the difference in the project duration as shown in the initial updated schedule and the schedule that shows the change.

If the contractor maintained accurate scheduling records, it may be a simple task to show how the schedule would have been completed without a particular change. The strength of the claim will weigh heavily on the accuracy of the updated schedule, the credibility of any witnesses employed, and how convincingly the schedule shows the impact of a single change on the entire schedule.

IMPACT OF CHANGES

Changes to the original contract are a common and readily accepted reality on virtually all construction projects. Most contracts grant the owner the power to make changes to the contract, and the contractor is similarly contractually mandated to perform such work. A change in which the owner requests the purchase of a better-quality item can be easily priced if the item has not yet been purchased and is readily available without a long lead time. Changes that impact activity duration are difficult to quantify in monetary terms because a single change can impact the rest of the project in many different ways. Some changes will result in increased costs that will result in a loss in productivity as described in Chapter 11. Still others will be difficult to quantify. Some changes will have more far-reaching impacts. One type of impact is referred to as the ripple effect in which a single change might influence other activities that are seemingly not related to the change in question. This may occur when the change results in a delay in one activity that has an impact on several "downstream" activities.

Various impacts of changes are identified in Table 12.1. It will be noted that most of these costs will be difficult to estimate with great accuracy. There is often no clear line of distinction between one type of cost and another. Many of these costs may not

Table 12.1
POTENTIAL IMPACTS OF CHANGES ON PROJECT COSTS

Type of Cost	Cause of the Added Cost or Loss
Lost opportunities	Project delays resulted in other projects not being pursued
Increased overhead costs	Duration increase will increase various overhead costs
Stacking of trades	Changes result in more workers working in congestion
Increased overtime	To maintain the schedule, the contractor worked overtime
Worker fatigue	Extended overtime will break the rhythm of worker production
Work acceleration	To maintain the schedule, the contractor added crews/equipment
Learning-curve effect	Delay resulted in adverse impact on learning
Increased administration time	Specific changes had to be communicated to all relevant parties
Impact of weather	Project extension results in activities taking place in cold climate
Lost morale	Worker morale will suffer if excessive changes occur
Supervisory impact	Supervisory staff must study the change to adjust the schedule
Additional workers	Adding workers to the crews will decrease productivity
Rework	If changes are not communicated, mistakes will be made
Materials delivery	Changing delivery dates may add to project costs
Ripple effect	Changes in one area may impact crews in other areas
Price increases	Late purchases mandated by a change could be costly
Work flow interruption	Excessive changes will upset the smooth flow of work
Remobilizing equipment	Bringing equipment back to the project to perform a change
Impact on other projects	Delay in assigning workers and equipment to other projects
Cooperative impact	Excessive changes will adversely impact the contractor/owner roles

be recognized by the contractor and, as a result, there may be no compensation. For this reason, many contractors will state that they do not make a profit on changed work, but at the same time many owners accuse contractors of using changes as a means of making their profits. While the loss in productivity of a single change may not be great, the loss in productivity associated with a large number of changes (hundreds or thousands of changes may occur on a single project) could be devastating to the contractor. These cumulative impacts are often the source of litigation when the contracting parties cannot agree on the appropriate payment for changes to compensate the contractor for performing the work. A review of the types of costs associated with changes will reveal that the schedule is often impacted. It should also be evident that the schedule may be one of the mechanisms by which the contractor attempts to justify the request for the added costs associated with changes.

IMPACT OF DELAYS

For a variety of reasons, delays may occur on a construction project. These might be caused by late material delivery, productivity being below that which was anticipated, delays associated with changes, delays requested by the owner for a number of reasons (to evaluate site conditions, design changes, etc.), delays caused by weather conditions, and so on. Delays, regardless of the source, will impact the schedule. By contract, some delays may be compensable while others may not. The compensable delays will certainly be ones that the contractor will try to accurately evaluate to quantify the payment to be made for the delay(s). To justify these added costs, the schedule is usually a crucial evaluative tool. Without an accurate schedule, a compelling argument cannot be made to show the true costs of delays. Consequently, litigation involving delays is generally supported with information contained in the construction schedule.

FINAL COMMENTS

The review of these few cases is intended to leave the reader with an appreciation for the fact that CPM project schedules represent a unique tool in the resolution of contract disputes with regard to evaluating the impacts of delays. However, one must recognize that, as with any tool, the schedule must be prepared and analyzed correctly in order to be put to effective use. This preparation starts with the initial development of the network activities, continues through the project monitoring phase to provide adequate documentation of the facts revolving around delays, and is completed with the proper interpretation of the delays/changes encountered, the state of the project at the time of the delays, and the consequent impacts resulting from them.

REVIEW QUESTIONS

1. In the Blackhawk case, what role did construction schedules play in deciding the merits of the delay claim? Describe briefly how the schedules were prepared.
2. In the Minmar case, what role did construction schedules play in deciding the merits of the delay claim? Describe briefly how the schedules were prepared.
3. Discuss the role that the first or initial schedule for a project can play in litigation involving delays. What assumptions must be valid for this role to be justified?
4. What is the role of as-built schedules in litigation involving delays?
5. To minimize the chance of losses, suggest a methodology to follow when a delay that has the potential of resulting in litigation occurs on a construction project.
6. Give an example of how the "but for" schedule might be used in a contractor's claim for added compensation.
7. Explain one of the added costs that might be incurred by the contractor as a result of a change issued by the owner.

13

Short-Interval Schedules

Even the most complex projects are constructed one day at a time.

The importance of the role of planning and scheduling on construction projects is well emphasized in the construction literature and in academia. However, most of the methods of planning and scheduling relate to overall project schedules, those that provide a global view of an entire project. It is generally recognized that these schedules serve as excellent vehicles by which owners can monitor the progress of contractors on their projects. It is also an efficient means of providing guidance for overall project coordination by project managers and job superintendents.

While project schedules or networks address an essential need for owners and top-level project managers and superintendents, they generally do not provide the necessary detail by which project activities can be scheduled on a day-to-day or even a task-to-task basis. Project networks are not the only, and certainly not the final, mechanism for providing guidance for project scheduling. In essence, a more detailed plan must be developed in order to bridge the gap from the overall project schedule to the organization of the tasks performed at the crew level. This can be done effectively through the use of schedules that cover short periods of time, such as one, two, or three weeks. Such schedules for organizing the activities of crews are short-interval schedules, also referred to as *look-ahead schedules, window schedules, "roll-up" schedules, two-week schedules,* or *short-range plans*.

Short-interval schedules are of particular value to first-line supervisors or craft foremen. They assist in organizing such resources as time, materials, equipment, workers, and information.

Short-interval schedules are a vital link between the master project schedule and the execution of specific tasks. It is important to recognize that supervisors must observe long-lead-time items as identified on the master schedule, they must anticipate the short-term resource requirements as identified in short-interval schedules, and they must provide effective direct supervision for tasks. With this link between these tasks, the short-interval schedule should be tied to the master schedule. In fact, the short-interval schedule might be viewed as an additional level of detail that overlays the master schedule. It would be too cumbersome to include this level of detail in the master schedule. Furthermore, the management of such detail on the master schedule would be an excessively arduous task.

The general objective of most short-interval schedules is simple. It is to provide a means of coordinating the various resources on the project over a short period of time. The successful utilization of short-interval schedules is dependent on the thoroughness of the information included in them. Many items not necessarily included in the master schedule should be included in the short-interval schedule. Items to be covered in short-interval schedules include the following:

All "work in place" activities
Material and equipment deliveries
Quantities of selected materials (concrete, water, fuel)
Work performed by subcontractors
Equipment maintenance
Safety enhancements
Quality control (inspections)
Issuance of permits
Surveying functions (setting grades, alignments, control points)
Planned delays (shutdowns, holidays, company picnics)
Regulatory constraints
Owner functions involving the project (meetings, inspections)
Project events involving third parties

Care must be exercised to avoid assigning two different crews to perform work in the same area. Production rates should also be reviewed to ensure that they are realistic and achievable. In general, the limited resources should be examined to be sure that they are not overextended.

In relation to the master project schedule, the short-interval schedule is typically more accurate. This is because this schedule must be current. While the master schedule may be permitted to slip considerably before it is updated, short-interval schedules must be current and must therefore reflect conditions as they actually exist. Because of the variances that are permitted in master schedules, short-interval schedules cannot be effectively prepared directly from the master schedule. Short-interval schedules must reflect exactly the conditions on the construction site in terms of what tasks have been done and which activities are yet to be performed. Short-interval schedules can be prepared with accuracy only when clear lines of communication exist between the various parties involved in the construction process. If the short-interval schedule is to be both current and accurate, specific time or scheduling constraints of the various parties must be known.

SHORT-INTERVAL SCHEDULES IN THE LITERATURE

An examination of more than 15 textbooks on planning and scheduling disclosed an interesting phenomenon: textbooks tend to provide very little information on short-interval schedules. In fact, the few sources that addressed the topic did little more than provide a definition of them. No scheduling textbook devoted more than two paragraphs to the topic. The lack of a serious treatment of short-interval schedules in major scheduling texts should in no way be construed as meaning that such schedules are not important. The reverse is true, as the effective use of short-interval schedules is frequently vital to the successful completion of construction projects.

HOW CONTRACTORS USE SHORT-INTERVAL SCHEDULES

With so little information about short-interval schedules in the literature, it was necessary to seek additional information directly from construction practitioners. Approximately 30 construction firms that had construction projects of significant size, generally in the tens of millions of dollars, provided examples of the types of short-interval schedules that they used on their projects. The most meaningful information gleaned from the various contractors was the actual format used for making the short-interval schedules. These were typically presented as bar charts structured to cover a period of two, three, or four weeks. The two-week schedules were organized to cover the upcoming two-week period (see Figure 13.1). The three-week schedules were either organized to cover the upcoming three-week period or they were designed to

TWO-WEEK SHORT-INTERVAL SCHEDULE

Date →																	
Activity	Respon.	F	S	S	M	Tu	W	Th	F	S	S	M	Tu	W	Th	F	Comments

Project: ________________ Date Prepared: ____________ Prepared By: ____________

Figure 13.1 Sample Two-Week Short-Interval Schedule

THREE-WEEK SHORT-INTERVAL SCHEDULE

Date →																								
Activity	Respon.	F	S	S	M	T	W	T	F	S	S	M	T	W	T	F	S	S	M	T	W	T	F	Comments

Project: ________________ Date Prepared: ____________ Prepared By: ____________

Figure 13.2 Sample Three-Week Short-Interval Schedule

cover the upcoming two-week period with a look back that recorded the accomplishments of the previous week (see Figures 13.2 and 13.3). (Figure 13.3 shows slightly more detail than Figure 13.2.) Basically, short-interval schedules show the days of the week and the list of activities that are to be scheduled during the selected time period. Since they are used on widely varying projects, short-interval schedules are structured in an open-ended fashion. This allows a particular type of format to be utilized on many different projects without making any modifications to the general format as the projects change. Some firms used four-week schedules in which one week documented the previous week's work and the remainder planned the activities for the upcoming three weeks.

The accuracy of a short-interval schedule is directly affected by the timing of its preparation. Assume that a particular short-interval schedule covers two full weeks of construction activity. The ideal time in which to prepare this schedule would be the Thursday or Friday of the week immediately preceding the scheduling period. This will result in an accurate schedule, as only the work actually accomplished on Friday of that week must be estimated. If prepared a week earlier, the short-interval schedule may not be current because of the variability of the work actually performed between the time of schedule preparation and the beginning point of the short-interval schedule.

Short-interval schedules are prepared each week. Although the focus of discussions at meetings involving the short-interval schedules tend to emphasize the activities and events to take place during the upcoming week, looking ahead two or three weeks forces all parties to begin to plan for future developments on the project. If a crew was to produce at a faster-than-planned rate, the schedule for the second week would be activated early. At any rate, looking a few weeks into the future serves a valuable function in getting all parties to begin devising their own work plans.

THREE-WEEK SHORT-INTERVAL SCHEDULE

Contractor:__________ Schedule Period: ___/___/___ to ___/___/___ Page ___/___
Project Title:__________ Date Prepared:__________ Prepared by:__________
Job No.__________ Month ⟶

Line	Master Activity	Work Item	Day of Month ⟶ / Crew ↓	M	T	W	T	F	S	S	M	T	W	T	F	S	S	M	T	W	T	F	Special Requirements or Comments on Delay

KEY: ▱ Scheduled Activity ☒ Completed Activity

Figure 13.3 Sample Detailed Three-Week Short-Interval Schedule

As noted, many short-interval schedules include a look back of one week. These provide a basis for establishing continuity from a week past to the upcoming week. This link between the two weeks (the past week and its following week) can help clarify the nature of the tasks to be performed and the rate at which they are to take place. In addition, this look back constitutes valuable documentation that could be useful in preparing an as-built schedule. Since short-interval schedules contain considerable detail, this historical record can be very valuable (see Figure 13.4).

While the information in short-interval schedules may be accurate, it is important to also establish a forum at the project level in which the relevant information can be communicated to those parties—subcontractors or suppliers—identified in the schedule. One common means of communicating this information is through weekly project planning meetings.

Some construction contract documents address or mandate the use of short-interval schedules. The 90-day schedule is one in which sufficient detail is required in order to show the nature of the activities that will occur on a week-by-week basis. At the same time, this schedule covers a long time period during which progress can be effectively monitored (see Figure 13.5). One provision in a contract on a large subway project included the following provision concerning a 90-day work schedule:

> A schedule covering the first ninety days of the Contract shall be submitted within fifteen days following the date of the Notice to Proceed. The schedule shall be time-scaled and resource loaded for workforce requirements and shall be submitted in either bar chart or critical path method (CPM) format. Work items

286   Chapter 13

Two-Week Short-Interval Schedule

Date Activity	Respon.	M	Tu	W	Th	F	S	S	M	Tu	W	Th	F	Comments
Main Bldg.														
Struct. Steel	STI													
West wing		Stair		Deck	Deck	Insp.								
East wing			Deliver Steel						Columns		Beams	Deck		
Conc. Deck	R-mix								(Pour)					Order 3 days early
Fireproofing	ISC	1st floor	2nd floor									3rd floor		
Sprinkler	Jackson		piping									piping		Verify by phone

Project: __Rec. Center__   Date prepared: __June 6, 2003__

Prepared by: __J. Hazden__

Figure 13.4 Example of a Working Two-Week Short-Interval Schedule

90-DAY WORK SCHEDULE

Activity	Month Week	1	2	3	4	1	2	3	4	1	2	3	4	Comments

Project: _____ Firm: _____ Date Prepared: _____ Prepared By: _____

Figure 13.5 Sample 90-Day Work Schedule

defined in the schedule shall not exceed fourteen days in duration. The submittal shall be accompanied by a written narrative that describes the schedule and the approach to the work that the Contractor intends to employ during the initial ninety-day period of the Contract.

The U.S. Army Corps of Engineers employed the following provision in one of its contracts. Note that greater detail is required in this provision, which addresses the "general approach" for completing the entire project.

A preliminary network defining the Contractor's planned operation during the first 90 calendar days after notice to proceed will be submitted within 25 days after notice of award or prior to the start of work, whichever is earlier. The Contractor's general approach for the balance of the project shall be indicated. Costs of activities expected to be completed or partially completed before submission and approval of the whole schedule should be included.

The U.S. Bureau of Reclamation uses a modification of the 90-day schedule that covers 60 days: "The Contractor shall submit a 60-day look ahead report that shall be a listing of all activities or milestones that are scheduled to start, finish, or continue during the next 60-day period."

One public agency was so enamored by a general contractor's use of short-interval schedules that it decided to include a contract provision that addressed the two-week short-interval schedule. Unfortunately, the provision was not one that was embraced by the contractor. It stated, "The Contractor shall submit a written activity schedule to the Owner's Representative. The schedule shall indicate the Contractor's proposed activities for the forthcoming 2-week period. This schedule shall be submitted to the Owner's Representative no less than 7 calendar days prior to the first working day of the schedule." The provision stipulated that the contractor had to submit the short-interval schedule seven days prior to the start of the week being scheduled. The contractor on the project contended that too much had changed after seven days and that the early preparation was not helpful for project coordination.

In the contractor's view the requirement was essentially that the short-interval schedule from the previous week was to be submitted. Unfortunately, the contractor saw no value in using such an "old" schedule. In practice, most contractors employ short-interval schedules in roughly the same way. This was clear from the copies of short-interval schedules provided by different contractors. From these, a composite short-interval form was developed that shows all of the information that was contained in the different short-interval schedules (see Figure 13.3). Other examples of short-interval schedules are more typical of what most contractors use. Most tend to be on large (14″ × 17″ sheets) pieces of paper that lend themselves to large handwriting and have ample space to make entries and adjustments.

In essentially all cases, the CPM schedule is used as the basis for developing short-interval schedules. The CPM or master schedule is important, but the short-interval schedule is an acknowledgment that limitations of the master schedule are severe when trying to plan activities for a short duration. Short-interval schedules address that limitation. The daily work plans that are prepared to describe the duties and tasks of individual workers are developed from the short-interval schedules. The master schedule is typically prepared by the project manager (hopefully with input from the job superintendent) and the short-interval schedules are ideally prepared with input from the various craft supervisors.

The format of most short-interval schedules is some variation of a bar chart placed on a two- or three-week calendar. The concrete placements and large crane utilization are commonly highlighted because of the importance of these activities. The responsible parties for the various activities are clearly noted on some of these schedules. When short-interval schedules are prepared, it is a good idea to use a checklist to verify completeness of the schedules. Specific elements that may be included in the short-interval schedule, as noted earlier, include the following:

All "work in place" activities expected to take place
Material and equipment deliveries
Work performed by subcontractors
Equipment maintenance
Safety enhancements
Quantities for selected tasks (as for concrete pours)
Delays
Third parties
Surveying
Owner responsibilities
Testing

Before distributing the short-interval schedule to the appropriate parties, it may be a good idea to verify some data, primarily to ensure that the schedule is realistic. Some of these quick checks include the following:

- The schedule should accurately reflect realistic production rates from past or completed projects.
- Ensure that two crews are not scheduled for work in the same area at the same time.
- Verify that major pieces of equipment are not assigned to two different work items that are to take place concurrently.

OTHER SHORT-INTERVAL SCHEDULES

While the two- or three-week short-interval schedules are in wide use, they are by no means the only forms of schedules in use that cover a short time interval. One type of short-term schedule is the punchlist. The punchlist is a form that is generally prepared by the owner's representative at the point of substantial completion. The punchlist is used to document the remaining shortcomings in the project before it can be declared to have reached final completion. The punchlist tends to be a detailed listing of all noted deficiencies in a project. The punchlist itself can take a variety of forms. Figure 13.6 is a punchlist that might be used on a building project. Naturally, civil projects or industrial projects utilize different lists of items.

The punchlist is generally not a time-scaled schedule. It is simply a listing of all minor work items that remain to be performed (see Figure 13.6). Since these are often of a minor nature, it is not prudent to try to schedule the actual time at which each punchlist item will be addressed. On a three-story office building, the punchlist might consist of hundreds of deficiencies that must be addressed. The major task for the general contractor is to determine which first-line supervisors or which subcontractors are

PUNCHLIST

Project: _____ Prepared by: _____
Room #: _____ Date: _____

Feature	Spec No.	North	South	East	West	Comments
1. Walls						
-Paint						
-Wall Board						
-Base						
-Electric						
2. Ceiling						
-Acoustic						
-Soffit/Coves						
-Wall Board						
-Electric						
-Mechanical						
3. Floors						
-Carpet						
-Tile						
-Fin. Concrete						
-Wood						
4. Doors						
-Frame						
-Jamb						
-Hardware						
-Operation						
-Seals						
5. Windows/Skylights						
-Frame						
-Glass						
-Hardware						
-Operation						
-Seals						
6. Casework						
-Finish						
-Electrical						
-Mechanical						
7. Other						
-Built-ins						
-Furr Downs						

Figure 13.6   Sample Punchlist

responsible for the respective items. Once this has been accomplished, the general contractor must communicate this information to the appropriate parties.

The daily project schedule is on an even shorter time frame than the punchlist. It provides a good picture of the specific tasks to be performed by the employees of the general contractor and the various subcontractors. In addition, it includes the tasks to be performed by engineering, with separate tasks spelled out for the surveying crew. Even the large equipment on the site is scheduled for the day on an hour-by-hour basis. The project schedule can take many forms (see Figures 13.7 and 13.8). The important

DAILY PROJECT SCHEDULE
Tasks to Be Performed

Project: _____   Prepared by: _____
                             Date: _____

Work Items Today	Foreman/Crew
1.	
2.	
3.	
4.	
5.	
6.	
7.	
8.	
9.	
10.	

Work Items Today	Subcontractor
1.	
2.	
3.	
4.	
5.	
6.	
7.	
8.	
9.	
10.	

Work Items Tomorrow	
1.	
2.	
3.	

Work Items Tomorrow	
1.	
2.	
3.	

Survey Tasks Today	
1.	
2.	
3.	

Engineering Tasks	
1.	
2.	
3.	

Mat Is Needed			
1.		5.	
2.		6.	
3.		7.	
4.		8.	

General Notes	
1.	
2.	
3.	
4.	

Equipment Description	Hrs 7-8	8-9	9-10	10-11	11-12	1-2	2-3	3-4	4-5	Remarks
1.										
2.										
3.										
4.										
5.										
6.										

Figure 13.7   Sample Daily Work Plan

Daily Schedule

Date: _____
Daily Activities
1.
2.
3.
4.
5.
6.
7.
8.
9.
10.
11.
12.
13.
14.
15.
16.
17.
18.
19.
20.
21.
22.
23.
24.
25.

Figure 13.8 Sample Daily Schedule Form

information to be conveyed concerns the overall commitments that are to be met by the various subcontractors and the various other support forces under the control of the general contractor. The responsible parties might also be spelled out.

Another short-term schedule is one that is focused on the crew level (see Figure 13.9). In actual practice, these daily crew planning charts exist in a variety of

DAILY CREW PLANNING CHART
Tasks to Be Performed

Crew: _____
First Line Supervisor: _____
Prepared by: _____
Date: _____

Worker	7-8	8-9	9-10	10-11	11-12	12-1	1-2	2-3	3-4	4-5
1.										
2.										
3.										
4.										
5.										
6.										
7.										
8.										
9.										
10.										

Materials Needed	Major Tools Needed
1.	1.
2.	2.
3.	3.
4.	4.
5	5.

Equip. Plan	7-8	8-9	9-10	10-11	11-12	12-1	1-2	2-3	3-4	4-5
1.										
2.										
3.										

Contingency Tasks to Perform (when time is available)
1.
2.
3.
4.

Comments and Notes
1.
2.
3.
4.

Figure 13.9   Sample Crew Planning Chart

forms. For many foremen or first-line supervisors, the daily crew planning chart is something that is committed to memory and is never reduced to writing. While this can be effective in many routine tasks, it has its drawbacks. First, there is no documentation of the day's events. Second, memory can fail and this can be costly if it impacts other crews. This can happen, for example, if two crews suddenly have a need for the same front-end loader that neither of two foremen anticipated at the beginning of the day.

Had the tasks been reduced to writing, the need for the equipment might have been readily apparent and the conflict between the two crews might have been resolved.

The daily crew planning chart is an example of a form that captures the essence of the information that foremen should be thinking about on a daily basis. The chart shows how each worker will be utilized during the day, the essential materials that will be required, the special tools that will be needed, and the major equipment that will be used to accomplish the work. In addition, there is also space to include contingent tasks in the event that work runs ahead of schedule, the work on the primary task is delayed for some reason, or workers have free time during the day. A comments portion is also included to document any special issues.

Some plans take place more frequently than daily. One such plan is the pretask plan, which is prepared prior to the performance of each task. An example of a pretask plan is shown in Figure 13.10. The pretask plan is implemented at the crew level. Before a new task is performed, the foreman or a designated worker will meet with the crew to discuss how the upcoming task is to be performed. Although work planning is discussed at this meeting, particular emphasis is placed on ensuring that the task will be performed in a safe manner.

When the pretask planning meeting is held, the crew members will discuss the possible ways that an injury might occur when using the proposed approach. The pretask planning form (Figure 13.10) contains some of the common hazards resulting in injury that are listed to help the crew focus on the possible hazards associated with a particular task. As suggestions of potential hazards are offered, means are sought to minimize or eliminate those potential hazards. When the crew has reached a consensus on how the task is to be performed, each member of the crew is asked to sign the form. The pretask plan is generally posted in the work area for inspection by other supervisory or safety personnel. Where pretask plans are implemented, there is rigid adherence to having a pretask plan prepared prior to performing each task. It is possible that a single crew might have four or five pretask planning meetings in a single day. These have been found to be particularly effective in reducing the incidence of injuries on construction projects.

Pretask Plan

Foreman: Date: Time: AM or PM Project: Location: Task:	**Hazard Elimination** **Personal Protective Equipment** ☐ Safety harnesses ☐ Special gloves ☐ Protective clothing ☐ Eye protection ☐ Hearing protection ☐ Face shields ☐ Safety footwear ☐ Respirators ☐ Other:
Potential Hazards Present ☐ Pinch points ☐ Chemical burn ☐ Materials/scrap ☐ Thermal burn ☐ Elevated work ☐ Slips, trips, falls ☐ Cave-in ☐ Electrical shock ☐ Cuts ☐ Fire ☐ Heat stress ☐ Noise ☐ Rigging ☐ Overhead work ☐ Strains, sprains ☐ Harmful vapors ☐ Abrasions ☐ Overextension	**Specific Safety Measures** ☐ Toeboards, netting ☐ Eyewash, shower ☐ Use monitor ☐ Housekeeping ☐ Slope, shore, t-box ☐ Scaffold ☐ Set safe position ☐ Contain sparks ☐ Fire extinguisher ☐ Erect barricades ☐ Use proper tools ☐ Get help ☐ Change procedure ☐ Contain materials ☐ Other: **Items Verified** ☐ Permits obtained ☐ Task reviewed

Figure 13.10   Example of a Pretask Planning Form

Pretask Plan

Potential Hazards Present (cont'd)	Items Verified (cont'd)
☐ Particles in eye ☐ Confined space	☐ Lines drained/purged ☐ Others informed
☐ Lockout/tagout ☐ Dropped items	☐ Close drains/vents ☐ Chemical burns
☐ Spill hazard ☐ Open holes	☐ Safety gear present ☐ MSDS reviewed
☐ Overhead work ☐ Asbestos/lead/etc.	☐ Check for low points ☐ LOTO in place
☐ Other: _____	☐ Other: _____

Description of procedure to perform task: _____

_____

_____

_____

_____

Signatures of persons performing the work

(By signing this form, crew members indicate that they understand the task plan and feel fully prepared to proceed to perform the work safely.)

Figure 13.10 *(continued)*

FINAL COMMENTS

One of the components of most construction contracts is that the project be delivered within a specified period of time. This means that the proposed method of achieving that objective is conveyed through the use of CPM schedules. While these serve the needs of both project owners and project managers, they lack the necessary detail to provide meaningful guidance about the specific work activities to be performed on a daily basis. The daily activities can be effectively managed with the necessary focus on details with the use of short-interval schedules. Short-interval schedules are generally prepared weekly and show the scheduling of activities for a two- to three-week period. Fashioned in the form of bar charts, they can readily convey information to most personnel. While no sophisticated formulas are employed in short-interval schedules, their success depends instead on developing the schedules from accurate information concerning the current status of project activities. When properly prepared, the short-interval schedule can be an effective tool to organize the activities of specialty contractors and various trades. Resource utilization can also be effectively managed. The success of most large construction projects relies heavily on the proper use of short-interval schedules.

REVIEW QUESTIONS

1. Discuss the suggestion of preparing short-interval schedules one week prior to discussing them at a project meeting in order for all personnel to have ample time to review the information.
2. Some short-interval schedules are prepared with one week of information looking back (work performed the prior week) and two or three weeks looking forward. Discuss the value or use that the information contained in the "looking-back" portion of the short-interval schedule can serve.
3. How does the detail in a short-interval schedule generally compare to the level of detail contained in the project schedule?
4. Suggest who might be the appropriate personnel in a construction firm to prepare a short-interval schedule.
5. It was noted that concrete placement activities are generally highlighted in short-interval schedules. There might be some obvious reasons for this. What other activities might warrant highlighting?

14

Linear Scheduling

Only one activity can take place at a location at a point in time.

Some types of construction projects involve a great number of similar activities, occurring in succession, to be repeated throughout the project duration. On such projects, contractors often stipulate using a fixed allocation of resources for the identical tasks occurring in succession. This constraint mandates that there can be no overlap in these activities. While these can be scheduled with traditional networking techniques, they may be more effectively modeled using linear scheduling. In order to model a project with linear scheduling, the project must be able to be presented in a linear fashion. Linear scheduling is a convenient way to portray repetitive activities and offers the additional feature of helping to identify activities that might result in conflicts.

Examples of projects with sequences of repetitive activities include large highway projects, high-rise buildings, levies, seafront walls, sewer lines, and the like. The more traditionally accepted scheduling methods in the construction industry do not always lend themselves to effective management of these types of projects. Bar charts become complex mazes of activities and create complications in readily ascertaining space-time relationships. The critical path method (CPM) becomes very convoluted in assessing the start-to-start relationships and finish-to-finish relationships of activities. These shortcomings of traditional scheduling methods have resulted in the development of or resurgence of the use of linear scheduling.

Linear schedules are especially applicable where similar activities are repeated. The repetitions might describe the installation of a pipeline in which every 100 feet are considered to be repeated. A track of 50 homes under construction might also be described with the use of a linear schedule. On a large concrete structure, the repetitions might be based on lifts or concrete pours. On high-rise structures, the repetition might be on a floor-by-floor basis, as shown in Figure 14.1. Note that Figure 14.1(a) shows this information in the form of a precedence diagram; 14.1(b) depicts the same information in the form of a linear schedule. This comparison shows the simplicity of the linear schedule without compromising any of the detail in the information being conveyed.

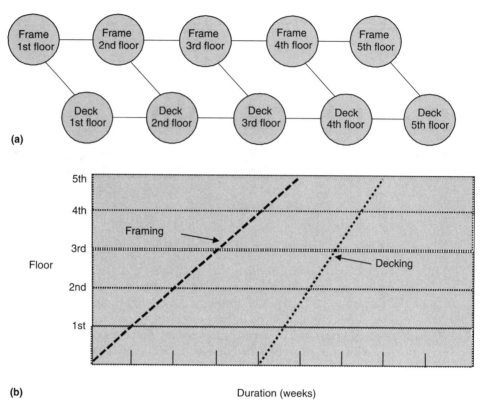

Figure 14.1 Example Showing How Linear Schedules Depict Information Contained in Precedence Diagrams

Another comparison is shown in Figure 14.2, where a simple highway construction project is presented as a precedence diagram and also as a linear schedule. The precedence diagram shows a finish-to-start relationship between activities, while the linear schedule does not exhibit such a clear distinction between the completion of one activity and the start of the next. It should become clear that the linear schedule is a more accurate portrayal of the actual construction of the highway. For example, there will be a point in the schedule as the last portion of the fill work is being done that finishing grading is well underway and paving work will have started. There are other differences between precedence diagrams and linear scheduling. While the precedence diagram may use the beginning-of-day convention, this has no meaning in the linear schedule. Since the project duration is shown as a continuum in the linear schedule, there is no need to think in terms of beginning-of-day or end-of-day conventions as the entire day is portrayed in the linear schedule.

Where is the critical path in the linear schedule? This is not an easy question to answer. The linear schedule itself does not identify the critical activities. In linear schedules, the level of detail is such that most of the activities will be critical. In the example, only two activities are not critical, namely, striping the pavement and performing the signage work. These activities will take place as the landscape work is performed. Once

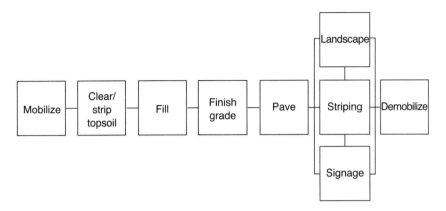

(a) Example of a precedence diagram for a road construction project

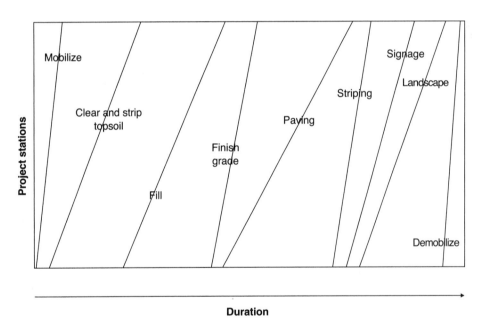

(b) Example of a linear schedule for a road construction project

Figure 14.2   Example to Compare Precedence Schedules with Linear Schedules

these three activities are completed, demobilization will complete the project. Note that the landscape work has a lower slope than that shown for the striping and paving activities. With three sequential activities, the activity that is critical is the one that has the lowest slope. Rather than thinking about which activity is critical, it may be better to think about which activity could more easily be shortened in order to reduce the project duration.

WHAT IS LINEAR SCHEDULING?

Linear scheduling is a unique means of resource leveling or allocation with a simple graphic display of time-space interaction. This technique is known by a number of other titles, including the *vertical production method, time-space scheduling method,* and *repetitive-unit construction.* The term *linear scheduling* has become more widely accepted in recent years.

The exact origin of linear scheduling is unknown, but it stems from efforts in manufacturing work to prevent delays or bottlenecks. In the manufacturing industry a related scheduling technique is more widely known as the *line of balance (LOB).*

The development of a linear schedule for a project is similar to any other scheduling process. The first three steps, familiar to most schedulers, are:

1. Identify activities.
2. Estimate activity production rates.
3. Develop activity sequence.

In accomplishing these three steps, one must determine whether the linear schedule method is the most appropriate. As a rule, the project will work well in linear scheduling if the vast majority of the activities can be grouped as a family of repetitive and nearly identical tasks. The activities should be defined in a level of detail comparable to that found on a bar chart.

The fundamental aspects of the use of linear schedules can be described through the use of some examples. Figure 14.3 shows a simple linear schedule that consists of three activities for the installation of a white board fence. For this operation, Activity A consists of the layout of the centerline of the fence; Activity B consists of augering the post holes, installing the posts, and attaching the boards; and Activity C consists of painting the fence. In Figure 14.3, the vertical axis represents the station position or distance along the fence. Since the horizontal axis is time, the slope of the activities represents the rate of production (distance/time). That is, the steeper the slope, the faster the activity is accomplished. In this example, it should be obvious that erecting the fence itself will take longer than laying out the fence or painting the fence.

As noted earlier, with the linear schedule there is no need to think in terms of an early-start schedule or a late-start schedule. Rather, the linear schedule is generally more akin to the expected schedule, which is some form of compromise between the early and late schedules. Is the linear schedule deterministic or probabilistic? At first glance, the linear schedule would appear to be deterministic; however, a closer examination will reveal that it too is a reflection of the probabilistic approach. Consider the time buffer that exists between activities. The very presence of a time buffer is an acknowledgment that the timing of the occurrence of activities cannot be scheduled with pinpoint accuracy. While an activity in a linear schedule will typically be characterized by a straight line, this is not a realistic portrayal, as a wavy line would be more realistic. This means that productivity over a short duration might be

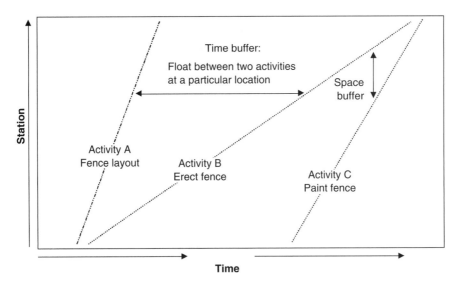

Figure 14.3 Linear Schedule Showing Three Activities

considerably below that which is anticipated and at other times will far exceed the expected production rate. It is practical to portray it as a straight line and include a sufficient time buffer to accommodate for the hour-to-hour or day-to-day fluctuations in productivity.

As shown in Figure 14.3, the horizontal distance between two activities is a graphic representation of the free float of the earlier activity at that location. It is evident that the float for the activity may change for different locations along the fence. The layout work must begin promptly so that the fence installation can commence. At this point in time (start of the layout work), the layout activity has very little float, but as the layout work nears completion, its float is considerable. Thus, the layout work must begin promptly in order to keep the project on schedule, while a delay at layout completion may not have an adverse impact on the project duration.

Once the linear schedule format is understood, it is easy to grasp the nature of a project. As mentioned, the horizontal distance between two contiguous activities is a graphic depiction of the float. What does the vertical distance between two contiguous activities represent? It simply represents the physical distance between two activities at a point in time, also known as a space buffer. It essentially shows how close together two contiguous activities are. A vertical line extended fully through the linear schedule at any selected time represented on the horizontal axis will intersect all activities that are scheduled to occur at that time. It is worthwhile to examine this relationship. Even if an activity has ample float, the operation of one activity might be too close to another operation to effectively accomplish both at the same time.

While an examination of the schedule may reveal potential conflicts if activities are taking place in too-close proximity of each other, it is imperative that activity lines

302 Chapter 14

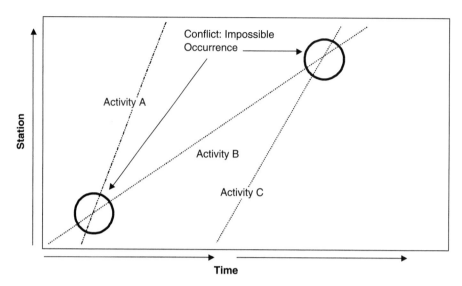

Figure 14.4 Linear Schedule Showing Conflict that Must Be Resolved

not cross. This would indicate that two activities are occurring concurrently at the same location, generally an irreconcilable conflict (see Figure 14.4). It should be evident that there might be some activities that might actually appear to be able to occur at the same place at the same time. For example, one activity might show that the signage is being installed at a particular station or location along a roadway. Another activity, painting the centerline stripe of the roadway, might take place at the same station at the same time. While these activities might actually cross when portrayed on a linear schedule, in reality they are not in conflict. It is only because of the lack of sufficient detail that there is a visual, but no actual, conflict.

One interesting feature of linear schedules is their simplicity. It is easy to grasp the operations that are taking place and it is also easy to see the impact of making modifications to the schedule. For example, suppose it is desirable for the fence project to be completed earlier than originally scheduled. The activity that warrants immediate focus is the fence construction activity, as this activity (Activity B) essentially drives the schedule (see Figure 14.3). One way that the schedule can be reduced is to add a second crew for this activity. This second crew will begin the work later than the first crew and it will begin at a considerable distance from the first crew. The schedule compression is readily observable when this is done (see Figure 14.5). Note that Activity B now appears as two different activities that have the same production rate as seen by the slope of the line.

A comparable compression could also be accomplished by increasing the crew size or by working overtime (see Figure 14.6). Note that Activity B begins and ends at the same time that it did in Figure 14.5. This was achieved by changing the rate of the work pace, which is shown as a steeper line.

Linear Scheduling 303

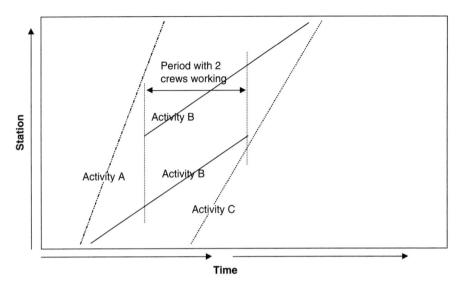

Figure 14.5 Linear Schedule Showing Project Compression by Adding a Second Crew

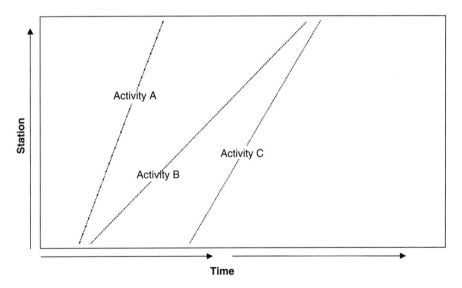

Figure 14.6 Linear Schedule with Project Compression by Accelerating One Activity

EXAMPLE 1: PROJECT TO REPLACE A STATE PARK WALKWAY

This walkway project is for the complete placement of an asphalt pavement walkway that will be 16,000 ft. long and 4 ft. wide. The following activities are used:

Activity	Description
A	Demolition and Removal of Old Asphalt Pavement
B	Level and Roll
C	Place 3-Inch Aggregate Base
D	Pave with Asphalt
E	Install New Signs and Stripping

Production Rate Diagrams

First, a production rate diagram (also called a velocity diagram) is developed that shows the time-space relationship of each activity. Figure 14.7 shows the production rate diagram for Activity A, Demolition and Removal of Damaged Asphalt Pavement. It can be seen that the production rate for this activity is 2,000 ft/wk (10,000 ft/5 wk). Not all production rates will be as nicely linear as it is for Activity A. There may be reasons for

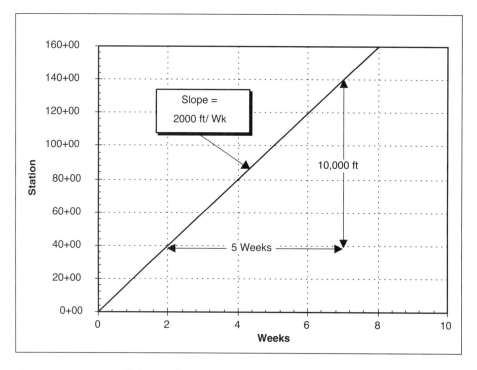

Figure 14.7 Demolition and Removal of Damaged Asphalt Pavement Activity Production Rate

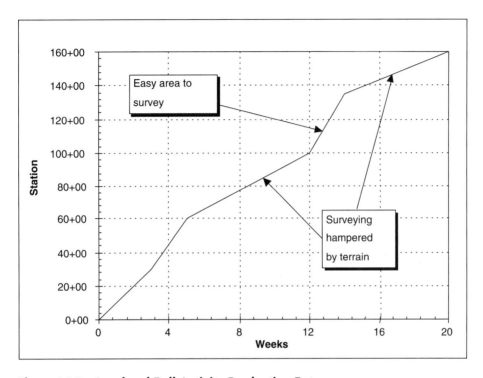

Figure 14.8 Level and Roll Activity Production Rate

altering the production rates for an activity along the time axis or distance axis. This could be a result of resource limitations or physical hindrances.

For Activity B, Level and Roll, the production rate does not remain linear from beginning to end, as shown in Figure 14.8. These differences in production could be attributed to varying ground conditions or restrictions on operations. Steeper sections on the velocity curve indicate higher production rates. Flatter sections on the diagram show areas where the production is expected to be lower. Reasons for this may include:

- Saturated ground conditions
- The importance of leaving existing vegetation undisturbed
- The proximity of the walkway to an embankment or stream
- Learning effects

It should be noted that production rates can also be changed by altering the number of resources assigned to an activity. The linear schedule yields a graphic image that can be useful when leveling resources or when balancing crews. Ideal resource utilization occurs when all activities have the same production rate, or slope. This will provide the shortest project duration.

Buffers

Buffers provide space or lag between activities. The buffer can be in terms of either time or distance, or both. It stands to reason that paving a section cannot begin until after the aggregate base is in place. Figure 14.9 shows this relationship: paving is to follow the laying of the aggregate base by no less than 5,400 ft *and* 3 weeks. When compared to conventional precedence schedules, the horizontal space between two sequential activities is analogous to float. Of course, a portion of this space is also needed to allow for the fluctuations that will occur in the rate of work accomplished. The vertical space between sequential activities is simply needed to permit each activity to be performed without conflict, congestion, or interruption.

Generating the Linear Schedule

After each activity velocity diagram is generated and buffer requirements defined, the linear schedule can be produced. Figure 14.10 shows the linear schedule for the walkway project. Note that a linear schedule is simply the portrayal of all of the activity velocity diagrams for a project included in one illustration. The project duration is 30

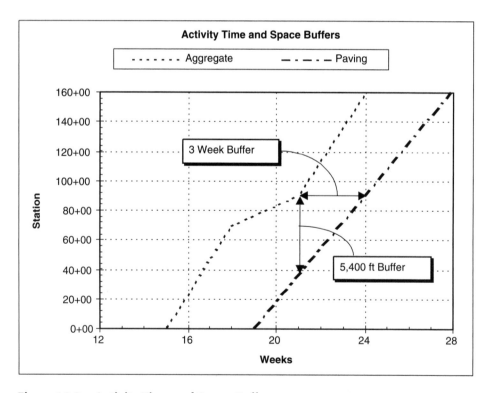

Figure 14.9   Activity Time and Space Buffers

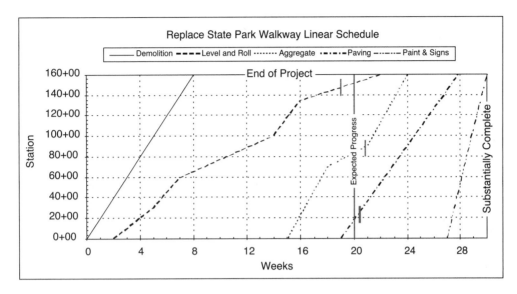

Figure 14.10 Replace State Park Walkway Linear Schedule (Stations on the Vertical)

weeks. Note that the duration could be shortened considerably by increasing the production rate of the level and roll activity. Depending on the magnitude of the time or space buffers incorporated by schedulers, the total project duration is shortened or lengthened. A determination of acceptable risk must be considered to avoid interference between activities. The objective is to allow productivity to continue in an orderly manner without bottlenecks or delays.

As with other forms of project scheduling, a drop string used on the schedule quickly facilitates tracking of actual progress versus the originally submitted schedule projection. From this, management decisions can be made to adjust activity production rates, material flow, and labor crews for separate activities. Some members of the construction industry have found it more useful for grasping the schedule to use the x axis for distance and the y axis for time. This arrangement is shown in Figure 14.11.

Linear scheduling offers a wide latitude of choice in graphic techniques used. The only limitation is the imagination of each scheduler. Color coding, symbols, zones, legends, boxes, and bar charts placed above linear schedules are among the many innovations incorporated in linear scheduling. The creation of unique combinations of scheduling methods occurs on schedules throughout the industry. One department of transportation uses a method that combines the bar chart, CPM, and linear scheduling methods. Finally, computer modeling is rapidly emerging in many developmental stages to meet the needs of those employing this scheduling technique.

A cursory introduction to linear scheduling can leave the impression that this method is limited to projects that traverse physically on-site across a horizontal plane or in a road curve stationing manner. This is not the case, as the following example will show.

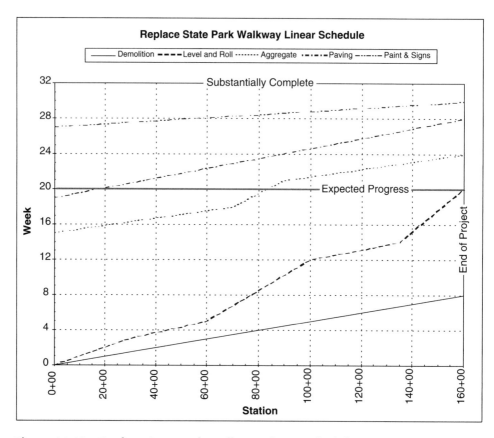

Figure 14.11 Replace State Park Walkway Linear Schedule (Stations on the Horizontal)

EXAMPLE 2: PROJECT TO CONSTRUCT 500 TRACT HOUSING UNITS

This project consists of a developer's housing tract of 500 housing units, each having essentially identical sets of plans and specifications. The entire project duration is one and one-half years, or 78 weeks. Activities for each unit are listed in sequence below:

Activity	Description
A	Survey and Layout
B	Earthwork
C	Foundation
D	Rough Carpentry
E	Roof
F	Rough Utilities

G	Finish Carpentry
H	Finish Utilities
I	Interior Finish
J	Landscaping

Figures 14.12 and 14.13 show two of the activity velocity diagrams for this project. Figure 14.14 shows the linear schedule for this project. Notice that most of the velocity diagrams are not linear. Buffers between activities also vary. The schedule shown was developed before any resource leveling was done. Figure 14.15 shows the linear schedule of a multistory building project.

When monitoring projects with linear schedules, one can readily evaluate the project status directly on the schedule. As such, the as-built schedule information can be recorded directly on the original schedule.

The activities described in the examples have been of a "linear" nature, but all activities need not be of this type. For example, in the installation of a sewer line, it may be necessary to install a lifting station. Such an activity would take place at a single location along the sewer line. On the schedule it would be depicted as a line with a very shallow slope or no slope. Such activities are represented in a fashion that is similar to the bars in a bar chart.

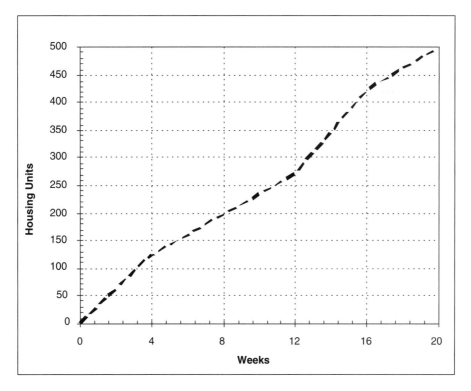

Figure 14.12   Earthwork Activity Production Rate

310   Chapter 14

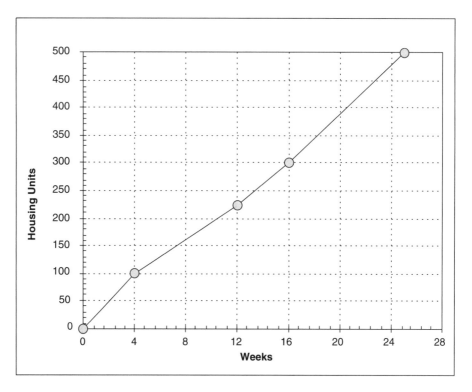

Figure 14.13   Rough Utilities Activity Production Rate

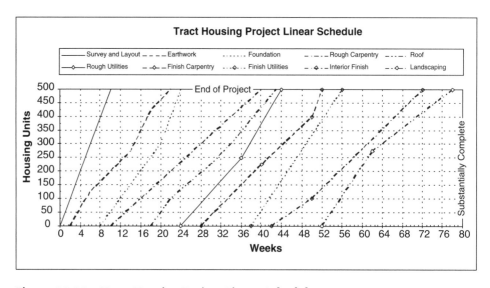

Figure 14.14   Tract Housing Project Linear Schedule

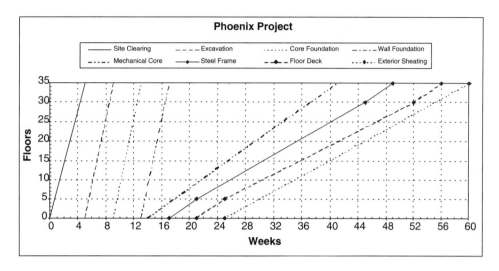

Figure 14.15 Linear Schedule for a High-Rise Building

Figure 14.16 shows the linear schedule for a simple sewer line. This schedule shows only the general activities, but greater detail could be represented. For example, the schedule does not depict an activity to lay out the pipeline or the final above-grade finish work. What is shown, however, is the removal of a utility pole and the setting of a flow meter. Note that these are easily illustrated with a horizontal line as time passes as they are performed, but their location remains unaltered. Even though the pole removal consumes a considerable amount of time, the trenching operation can already begin; the pole must be removed by the end of the third week or the trenching work

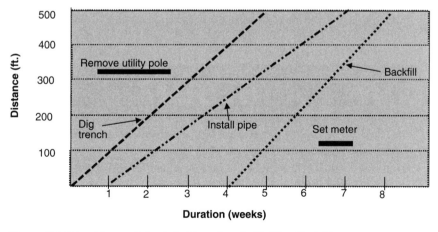

Figure 14.16 Linear Schedule Showing Activities that Occur over Time at a Single Location

will be interrupted. Although the flow meter is scheduled to be set in approximately one week, this work could probably begin a week earlier or start a week later without compromising the project schedule. Conflicts could be readily seen if the pole removal or the setting of the flow meter crossed one of the other activity lines.

FINAL COMMENTS

Linear scheduling techniques have a variety of useful applications. They are most readily adapted for projects in which tasks are repetitive. In general, linear schedules are more readily adapted for projects in which logic is not inordinately complex. Conflicts can be readily identified in linear schedules. When schedule compression is to be achieved, linear schedules assist in identifying those activities that warrant primary consideration. Perhaps the major advantage with using linear schedules is that they are easily understood by most construction personnel, obviating the need to develop additional schedules to communicate with other parties.

REVIEW QUESTIONS

1. A project consists of replacing the track on an existing railroad line, involving 3,000 feet of track. The old track can be removed at a rate of 70 feet per hour. The old railroad ties can be removed at a rate of 200 feet (measured along the track) per hour. The new ties can be installed at a rate of 25 feet per hour and the replacement track can be installed at a rate of 40 feet per hour. Show this operation on a linear schedule. Assume that the minimum float for an activity at any location is one day.
2. A 1,500-foot subdivision street is to be constructed. The layout will be completed in three days, the fill will be brought in over the course of eight days, the compaction/grading will be completed in five days, the asphalt topping will be done in three days, and the striping will take one day to complete. Using a minimum float guideline of one day, develop a linear schedule for this project. What is the project duration if no activities are to be interrupted? Suggest ways that the duration can be shortened.
3. A shallow trench for a 1,000-foot water line installation will take ten days. It will take four days to install the first 400 feet of pipe. A water meter will then be installed during the following two days at the end of the pipe. After the meter is installed, the remaining 600 feet of pipe will be installed in six days. Develop a linear schedule for this project.
4. Describe a project that might not be well suited for linear scheduling.

15

PERT: Program Evaluation and Review Technique

One thing is certain: all planning is associated with uncertainty.

Scheduling is the basis for the management of time on a construction project. The estimates for the various activities in a project form the basis on which the success of the schedule relies. Thus far, it has been assumed that the estimates of activity durations were accurate. While this may be an acceptable premise in most construction schedules, it is essential for the prudent scheduler to have a clear understanding of the uncertainty actually associated with duration estimates. Even if a deterministic approach is used for duration estimates, it is important to constantly recognize that uncertainty is an inherent characteristic of virtually every activity.

Uncertainty is more easily understood by those with a background in statistics. Although a firm knowledge of statistics would be an asset for reading this chapter, the material is presented in such a manner that formal training in statistics should not be required (refer also to Chapter 4). The theoretical basis for the statistics being presented will not be provided, because the applications of statistics to scheduling will be the primary focus.

UNCERTAINTY IN ACTIVITY DURATION ESTIMATES

When an activity is stated as having a given duration, it should be recognized that the duration is not an absolute value. For example, the excavation of a basement may be stated as requiring 30 days to complete. What does this really mean? Will the excavation take exactly 30 days to complete, or will the actual duration vary from the estimated duration? It could mean that, on the average, the duration is 30 days; however, this could mean that, in half of the cases, the duration is longer than 30 days, and in half of the cases, the duration is less than 30 days. With high liquidated damages assessed against the contractor for late completion, the contractor cannot afford to use a duration on which an overrun in time occurs in half the instances. This is where a basic understanding of statistics comes in handy (refer to Chapter 4).

Assume that the duration of the excavation activity is to be described in terms that define the relative uncertainty associated with it. In reality, the contractor would be more descriptive if he or she stated that the shortest imaginable duration for the excavation activity is 12 days, the worst-case scenario for the duration is 60 days, and the most likely duration (the mode) is 18 days. This is typical of the information needed to describe the uncertainty associated with a project duration. With the worst or most pessimistic estimate, the mode, and the shortest or most optimistic estimate, it is possible to estimate the mean that should be used for an activity. Note with the values given above that the distribution of the time estimates is not a normal or bell-shaped distribution. That is, the optimistic estimate is typically closer to the mode and mean than is the pessimistic estimate. As in the example above, the most optimistic guess of the duration is only six days less than the mode, while the most pessimistic estimate is 42 days more than the mode. This is fairly typical in that there is a conceivable minimum duration that can be attained under only the most ideal circumstances. On the other hand, the pessimistic duration may be quite long in order to take into account a variety of occurrences that might delay an activity.

Uncertainty in scheduling is considered in the method referred to as the *Program Evaluation Review Technique (PERT)*. PERT is based on activity estimates derived from a "three time estimate," namely, an optimistic estimate, the estimate of the most likely (mode) duration, and the pessimistic estimate. The mean estimate of the activity duration can then be computed as follows:

$$T_e = \frac{T_o + 4T_m + T_p}{6}$$

Where:

T_e = Mean value of the activity duration (expected)
T_o = Optimistic activity duration
T_m = Most likely duration
T_p = Pessimistic activity duration

In the above equation, note that the estimates are weighted, with the heaviest weight placed on the most likely duration estimate. Although this mean value of the activity duration has some meaning, it does not, in itself, convey any information about the degree of uncertainty. That is, it would be helpful to have a measure to describe the extent to which the duration is expected to vary from the derived mean value. Such a measure is known as the *standard deviation* and it can be derived from two of the time estimates used to calculate the mean. The standard deviation is determined as follows:

$$\text{Standard deviation} = s = \frac{T_p - T_o}{6}$$

PERT: Program Evaluation and Review Technique

The preceding calculation of the standard deviation is an approximation. Rather than explaining the theoretical basis on which the standard deviation is calculated, it is more appropriate to simply be able to use the information. To do this, refer to Table 15.1, which gives information about the probabilities associated with different numbers of standard deviations.

The information in Table 15.1 assumes that the distribution of the time estimates is normally distributed. Normally distributed data implies that the data are symmetrically distributed about the mean, similar to a bell-shaped curve (see the figure contained in Table 15.1). Duration estimates are rarely normally distributed; however, when many activities are included in a schedule, it is acceptable to assume that a normal distribution exists. Table 15.1 gives information about a standard normal curve. This means that probabilities of occurrence of a specific duration can be determined by simply knowing the number of standard deviations that the value in question is from the mean. The table is set up to give information on the probability that a particular duration will be less than some specified value that is given in terms of the number of standard deviations that the value extends beyond the mean.

Consider the duration estimates given in the preceding example. The mean is calculated as being 24 days ($[12 + (4 \times 18) + 60]/6$), and the standard deviation as being eight days ($[60 - 12]/6$). If the standard deviation had been calculated as being two days, it would imply that the estimate of the mean time has little uncertainty, while a standard deviation of 25 days would imply a great deal of uncertainty.

Table 15.1 is set up to provide the probability of occurrence for a variety of standard deviations. The number of standard deviations is measured to the right of the mean. Thus, a value of zero standard deviations (point at the mean) to the right of the mean (probability of .5 noted in the table for 0 standard deviations) implies that half of the durations would be less than the mean. If the value in question is one standard deviation to the right of the mean, the probability is .8413 that the value actually realized will be less than that value.

Let's examine the implications of having a mean duration of 24 days with a standard deviation of eight days. In the example, the contractor estimated that the excavation work would be done in 30 days. What is the probability that the excavation work will actually be done within that estimated time frame? This is where Table 15.1 can be used to good advantage. To use Table 15.1, the information must be converted into "standard" terms. For example, a duration of 30 days is greater than 24 days by 0.75 standard deviations ($[30 - 24]/8$). For 0.75 standard deviations to the right of the mean, the table shows that the probability that the duration will be under 30 days is 77.34%. What is the probability that the duration will be less than 40 days, a duration that is two standard deviations beyond the mean? The probability that the duration will be less than 40 days is 97.72%. Once the probabilities are known, the contractor in the example might be well advised to reassess the 30-day estimate and revise the duration to 40 days. The analysis shows that there is a 77.34% probability of completing the task in less than 30 days and a 22.66% (100% − 77.34%) chance of running over this time, a high risk for the contractor (see Figure 15.1).

Suppose the mean duration for an activity is calculated with a three-time estimate to be 78 days with a standard deviation of 16 days. What is the probability that the activity will take longer than 100 days? First, one must determine the number of

Table 15.1
STANDARD NORMAL CURVE

x = The number of standard deviations to the right of the mean

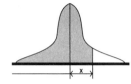

The area under the curve (as shown in the figure) always includes the portion containing the mean.

X	0	1	2	3	4	5	6	7	8	9
0.0	.5000	.5040	.5080	.5120	.5160	.5199	.5239	.5279	.5319	.5359
0.1	.5398	.5438	.5478	.5517	.5557	.5596	.5636	.5675	.5714	.5754
0.2	.5793	.5832	.5871	.5910	.5948	.5987	.6026	.6064	.6103	.6141
0.3	.6179	.6217	.6255	.6293	.6331	.6368	.6406	.6443	.6480	.6517
0.4	.6554	.6591	.6628	.6664	.6700	.6736	.6772	.6808	.6844	.6879
0.5	.6915	.6950	.6985	.7019	.7054	.7088	.7123	.7157	.7190	.7224
0.6	.7258	.7291	.7324	.7357	.7389	.7422	.7454	.7486	.7518	.7549
0.7	.7580	.7612	.7642	.7673	.7704	.7734	.7764	.7794	.7823	.7852
0.8	.7881	.7910	.7939	.7967	.7996	.8023	.8051	.8078	.8106	.8133
0.9	.8159	.8186	.8212	.8238	.8264	.8289	.8315	.8340	.8365	.8389
1.0	.8413	.8438	.8461	.8485	.8508	.8531	.8554	.8577	.8599	.8621
1.1	.8643	.8665	.8686	.8708	.8729	.8749	.8770	.8790	.8810	.8830
1.2	.8849	.8869	.8888	.8907	.8925	.8944	.8962	.8980	.8997	.9015
1.3	.9032	.9049	.9066	.9082	.9099	.9115	.9131	.9147	.9162	.9177
1.4	.9192	.9207	.9222	.9236	.9251	.9265	.9279	.9292	.9306	.9319
1.5	.9332	.9345	.9357	.9370	.9382	.9394	.9406	.9418	.9429	.9441
1.6	.9452	.9463	.9474	.9484	.9495	.9505	.9515	.9525	.9535	.9545
1.7	.9554	.9564	.9573	.9582	.9591	.9599	.9608	.9616	.9625	.9633
1.8	.9641	.9649	.9656	.9664	.9671	.9678	.9686	.9693	.9699	.9706
1.9	.9713	.9719	.9726	.9732	.9738	.9744	.9750	.9756	.9761	.9767
2.0	.9772	.9778	.9783	.9788	.9793	.9798	.9803	.9808	.9812	.9817
2.1	.9821	.9826	.9830	.9834	.9838	.9842	.9846	.9850	.9854	.9857
2.2	.9861	.9864	.9868	.9871	.9875	.9878	.9881	.9884	.9887	.9890
2.3	.9893	.9649	.9898	.9901	.9904	.9906	.9909	.9911	.9913	.9916
2.4	.9919	.9920	.9922	.9925	.9927	.9929	.9931	.9932	.9934	.9936
2.5	.9938	.9940	.9941	.9943	.9945	.9946	.9948	.9949	.9951	.9952
2.6	.9953	.9955	.9956	.9957	.9959	.9960	.9961	.9962	.9963	.9964
2.7	.9965	.9966	.9967	.9968	.9969	.9970	.9971	.9972	.9973	.9974
2.8	.9974	.9975	.9976	.9977	.9977	.9978	.9979	.9979	.9980	.9981
2.9	.9981	.9982	.9982	.9983	.9984	.9984	.9985	.9985	.9986	.9986
3.0	.9987	.9987	.9987	.9988	.9988	.9989	.9989	.9989	.9990	.9990

PERT: Program Evaluation and Review Technique   317

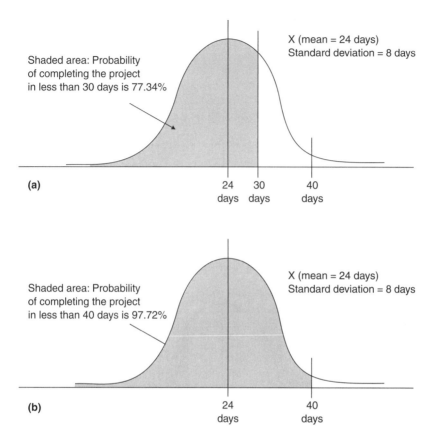

Figure 15.1 Normal Distribution Showing the Probability of Completing the Excavation Work in Less than a Specified Duration

standard deviations that the duration in question is beyond the mean. In this case, the duration of 100 days is 1.375 standard deviations beyond the mean of 78 days. From the table it can be determined that the probability that the duration will be less than 100 days is 91.55 percent. However, this was not the question; what is desired is the probability that the duration will exceed 100 days. Since the probability of completing in less than 100 days and completing in more than 100 days must be 100 percent, it can be quickly determined that there is an 8.45 percent (100% − 91.55%) probability that the duration will exceed 100 days (see Figure 15.2).

Continuing with the same example, what is the probability that the duration will be between 90 and 100 days? First, it can be determined that 90 days is 0.75 standard deviations from the mean. From the table, it is determined that the probability of accomplishing the task in less than 90 days (0.75 standard deviations from the mean) is 77.34 percent. Since it has already been established that the probability of exceeding 100 days is 8.45 percent, the probability of doing the activity in 90 to 100 days is 14.21% (100% − 77.34% − 8.45%) (see Figure 15.3).

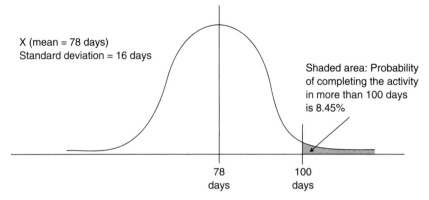

Figure 15.2 Normal Distribution Showing the Probability of Completing an Activity in More than 100 Days

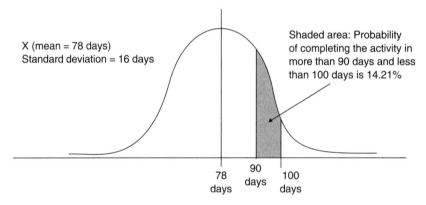

Figure 15.3 Normal Distribution Showing the Probability of Completing an Activity in More than 90 and Less than 100 Days

The preceding examples have all dealt with the durations of single activities. The above explanation should suffice to resolve most questions of probability on those activities. When several activities with differing values of standard deviations are involved, additional assumptions are worth making. This will be explained further.

UNCERTAINTY IN THE DURATION ESTIMATES OF AN ACTIVITY CHAIN

Project schedules may consist of hundreds or even thousands of activities, each with a unique level of uncertainty. The explanation in the following discussion centers on a simplified network consisting of one chain of activities. The chain will be called the *critical path*.

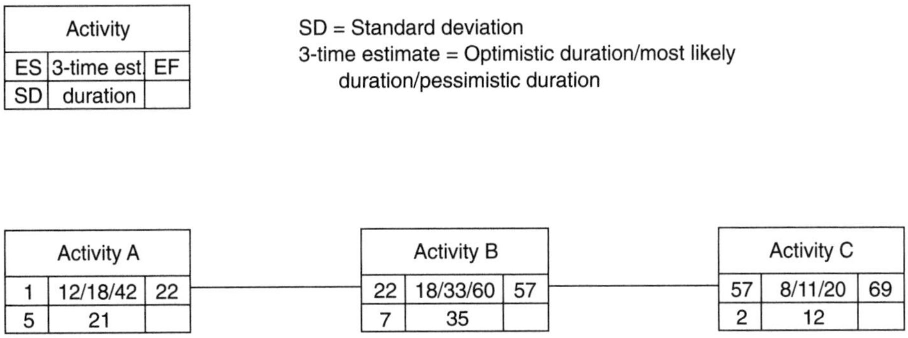

Figure 15.4 Legend

```
    Activity
ES | 3-time est | EF
SD | duration  |
```

SD = Standard deviation
3-time estimate = Optimistic duration/most likely duration/pessimistic duration

Network

Activity A: ES=1, 3-time est=12/18/42, EF=22, SD=5, duration=21

Activity B: ES=22, 3-time est=18/33/60, EF=57, SD=7, duration=35

Activity C: ES=57, 3-time est=8/11/20, EF=69, SD=2, duration=12

Figure 15.4 Simple Precedence Network Using Three-Time Estimates

What is the probability that the critical path activities will be completed within a specified time when uncertainty is associated with each of the activities in the path? For ease of illustration, the critical path will be defined as consisting of three activities, Activities A, B, and C. The information about these activities can be described as follows (see Figure 15.4):

	Mean Duration	Standard Deviation
Activity A	21 days	5 days
Activity B	35 days	7 days
Activity C	12 days	2 days
	68 days	

From this information, it can be established that the mean duration along the critical path is 68 days. Assume that the project duration is contractually established as not to exceed 90 days. What is the probability that the project duration along the critical path will not exceed 90 days? Since the mean value of the duration along the critical path is simply the sum of the means of the activities in the chain (68 days), the only remaining unknown is the standard deviation of the critical path chain. Unfortunately, the standard deviation cannot be derived directly from the standard deviations of the individual activities. Another statistical parameter must first be generated. This statistical parameter is the variance. This is a simple process, as noted:

$$V = s^2$$

Where:

V = Variance of an activity
s = Standard deviation of that same activity

$$V_{cp} = \sum_{start}^{end} s_{cp}^2 = \sum_{start}^{end} V_{cp}$$

Where:

cp = Chain of critical path activities
V_{cp} = The variance for the critical path

Simply stated, the variance of the activities along the critical path is equal to the sum of the variances of the activities comprising the critical path. Since the standard deviation of each critical activity is known, it is a simple matter to square each standard deviation to determine the variance of each activity. Once the variances of the critical activities have been totaled, the variance along the critical path is known. Of course, the statistic that is desired for the critical path is the standard deviation. This is easily done:

$$s_{cp} = V_{cp}^{\frac{1}{2}}$$

After the mean and standard deviation of the critical path have been determined, it is a simple procedure to find the probabilities. In the example, the mean for the critical path has already been determined. The standard deviation must be calculated to find the probability of completing the critical path activities within 90 days. The information of interest at this stage is the mean (sum of the mean durations on the path) and the standard duration (square root of the sum of the activity variances). Thus, the mean duration is 68 with a standard deviation of 8.83 days.

Activity	Duration	Std. Deviation	Variance
A	21 days	5 days	25
B	35 days	7 days	49
C	12 days	2 days	4
Critical Path	68 days	$\sqrt{78}$ = 8.83 days	← 78

Once the mean and standard deviation of the critical path are known, the critical path can be treated as if it is a single activity. For example, to answer the question of the probability of completing within the 90-day period, first one must determine that 90 days is 2.49 standard deviations from the mean of 68 ([90 − 68]/8.83). From Table 15.1, this converts to a 99% (0.9936) probability that the critical path activities will be completed in less than 90 days. Similarly, there is a 91% probability that these activities will be completed in less than 80 days.

UNCERTAINTY IN THE DURATION ESTIMATES OF PROJECTS

Now suppose that the above project is made slightly more complex. Instead of having a single chain, the project network consists of two independent chains. Assume that the chain described in the above example is Chain "A." This second chain of activities is referred to as Chain "B," with the mean duration for this chain being 59 days with a calculated standard deviation of 14.5 days (see Figure 15.5). This information for the standard deviation would be generated in the same fashion as described above for Chain "A." For Chain "B," it is calculated that the probability that the chain will be completed in less than 90 days is 98.4 percent and the probability that it will be completed in less than 80 days is 92.6%. The information can be summarized as follows:

Chain	Duration	Standard Deviation
"A"	68	8.83
"B"	59	14.5

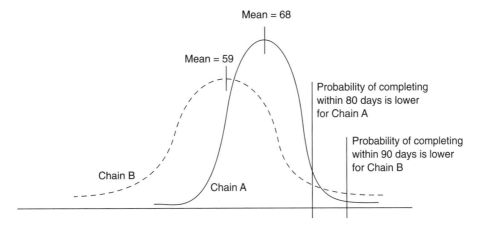

Figure 15.5 Probability as Influenced by the Standard Deviation

There is a slightly greater probability (99.3% versus 98.4%) of completing Chain "A" in less than 90 days than completing Chain "B" in less than 90 days. However, for the 80-day criteria, Chain "B" has a higher probability (92.6% versus 91.3%) of completing in less time. From this it should be observed that it would be improper to refer to Chain "A" as the critical path simply because it has the larger mean duration. For the 90-day limit, Chain "B" is actually more critical, since its probability of finishing on time is less. The reason that Chain "B" is more critical for the longer durations is because of its larger standard deviation.

Obviously, both Chains "A" and "B" must be completed before the project is completed. So what is the probability that the project will be completed in less than 80 days? This is easy to solve as the probability of completing the project within 80 days is simply the product of the probabilities of all the chains making up the network for the project. That is, both chains must be completed for the project to be completed. In this example, there are only two chains, so the probability of completing the project in less than 80 days is 84.5% (.926 × .913).

The above example uses a simple network comprised of only two chains of activities that are independent. This is rarely the nature of networks, as a greater number of activities are generally involved and there is a greater degree of interdependence among the various chains that make up the network. This complexity vastly complicates the process of solving networks by PERT.

Although PERT solutions are rarely used on construction schedules, a scheduler should maintain an awareness of the principles of probability. This will keep schedules more realistic.

MONTE CARLO SIMULATION

PERT is not a practical approach to address the scheduling needs of most projects. It is a worthwhile topic to discuss in terms of construction schedules, however, as it points out the variable nature of the durations assigned to many construction activities. For the most part, PERT is too complex to be understood by most personnel associated with construction projects, as few will have had formal course work in statistics.

While the discussion of PERT may be largely academic, it continues to have merit. There are also those circumstances in which PERT may have real applicability. While no known court cases have utilized PERT, it appears that the concept would have good application in some circumstances.

How can probability and statistics be used effectively? The manual approach is cumbersome even on very simple networks. Thus, the use of PERT almost mandates a computerized solution. One related statistical approach consists of a Monte Carlo simulation. The Monte Carlo simulation simplifies the statistical solution. Assume that a PERT network consists of 500 activities, a modest number of activities for a medium-sized project. Each of these activities will have the duration defined by the three-time estimates. From the three-time estimates for each activity, a probability distribution of the duration for each activity can be developed. With the Monte Carlo simulation, a duration for each activity is randomly selected from its probability distribution. In this way, the probabilities associated with each activity are reduced to a single number through this random selection process. Once this has been done for every activity in the network, a duration for the project has been established. However, this is simply one duration estimate that defines the duration based on the computations that resulted from the random selection process.

The concept of the Monte Carlo simulation approach is one that essentially requires the use of the computer. A brief description will make this clear. Assume the simulation is applied to a 500-activity project. Many different computations will be made on the network. The first duration computation for the project will begin by assigning a duration to each activity that is randomly selected from the duration distribution of each activity. Suppose the first computation results in a project duration of 211 days. The duration for the project is computed a second time, this time using durations for the activities that are again derived from a random selection process. Suppose the duration is now determined to be 205 days. While these computations would be time consuming done manually, they consume little time when performed with the computer. This process can be repeated hundreds of times with little effort on the computer. Once sufficient computations have been made, a profile of the project duration is developed. For example, the durations may range from a low of 198 days to a high of 223 days. From these computations, the mean and standard deviation of the project duration are obtained. From this information informed decisions can be made using the probability of achieving different durations.

Naturally, the solution for the project duration may contain considerable bias if the random selection process chooses extreme values from some of the duration distributions. This potential bias is recognized by the Monte Carlo simulation process, so another pass is made through the network with independent random selections being made of the duration of each activity. This is repeated many times, perhaps thousands of times, to derive a distribution of the duration of the entire project. The concept is simple but it does require the use of the computer.

FINAL COMMENTS

PERT is used very little in the construction industry. Despite the small amount of use, it is worthwhile to have a good understanding of this technique. Virtually all activity durations vary. When activity durations are assigned to activities, a best-guess number is generally used. This single number assignment of a duration makes it easy to perform network calculations. It should always be borne in mind that the duration assignments are guesses and that considerable uncertainty may actually exist in the value used. This approach is generally regarded as viable because all activity durations represent these best guesses, and in the normal computation of many activities in a network, the errors of overestimation and underestimation will cancel out each other. The use of PERT is more accurate in its depiction of a project as the scheduler is made very aware of the uncertainty that is associated with each activity duration estimate. PERT makes it possible for the scheduler to make more informed decisions about the probability of achieving stated project durations.

REVIEW PROBLEMS

1. After measuring the distance on a map between two cities several times, it was decided that the distance was 35 miles with a standard deviation of three miles. What is the probability (expressed in percent) that another measurement of the distance would be more than 38 miles? What is the probability (expressed in percent) that the distance is actually more than 33.5 miles?

2. In driving from Columbia to Tulsa, several different routes can be taken:
   - Route A was calculated as taking seven hours with a standard deviation of one-half hour.
   - Route B was calculated as taking seven and one-half hours with a standard deviation of one-quarter hour.

   a. What is the probability (expressed in percent) that the trip can be completed in less than seven and one-half hours on route A?

   b. What is the probability (expressed in percent) that the trip can be completed in less than seven and one-half hours on route B?

   c. What is the probability (expressed in percent) that the trip will take more than seven and three-quarters hours on Route A? on Route B?

3. The weather person said the high for the next day will be 85° F. What was meant by the forecaster was that the predicted high will be 85° F with a standard deviation of 4 degrees. That is, the mean probability of reaching 85° F is 50%.
   - What is the chance of having a high of 90°? _____ %
   - What is the chance of a high of only 74°? _____ %

324   Chapter 15

4. If the contract for a project is 120 days, what is the chance (in percent) of meeting the schedule? Assume the owner delays the job start by five days. What is the probability (in percent) that the project will still finish at the originally scheduled completion date?

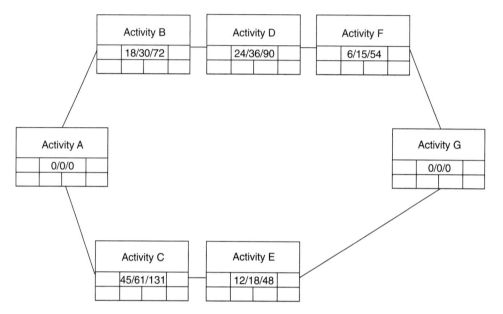

5. What is the probability (in percent) of completing the project in 55 days? In 60 days? In 65 days?

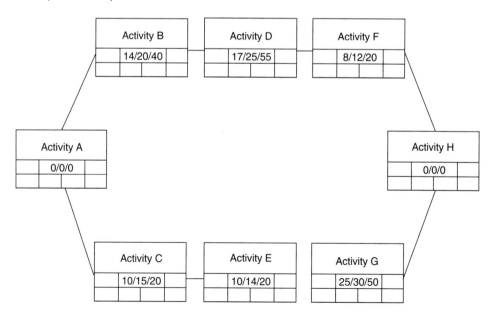

6. PERT data have been computed for the network below. Determine the probabilities of meeting the scheduled durations of 24 days and 32 days.

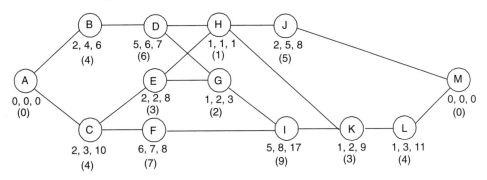

Critical Path Activities	Duration	Std. Dev. (σ)	Variance σ²
A	0	0	0
B	4	.67	.44
D	6	.33	.11
G	2	.33	.11
I	9	2	4
K	3	1.33	1.78
L	4	1.67	2.78
M	0	0	0
	28		9.22

CRITICAL PATH INFO:		DURATION = 28 (MEAN)		STD. DEV. = 3.04

Scheduled Duration	Mean	Dur-Mean	Dur-Mean/S	Prob. Of Meeting Schedule
28	28	0	0	.50
30	28	2	.66	.74
25	28	−3	−.987	.16
35	28	7	2.3	.99
24	28	−4	___	___
32	28	4	___	___

16

Arrow Diagrams

Failing to plan is planning to fail.

Precedence diagrams currently dominate the industry, and it is doubtful that there is a great need for new schedulers to work with arrow diagrams. There may be some instances, however, where arrow diagrams will be encountered, and it is in those situations that an understanding of arrow diagrams can be valuable. For this reason, this chapter is devoted to an explanation of arrow diagrams.

Arrow diagrams were widely used in the construction industry in the 1960s and 1970s. Their use has steadily declined since that time. This chapter will present some of the nuances associated with drawing arrow diagrams and reasons why arrow diagrams are not in wide use today. The traditional form of arrow diagram is also called an Activity-on-Arrow (also called A-on-A) network or ADM (Arrow Diagram Method). In arrow diagrams, the activities are denoted as arrows. The nodes that precede and follow the arrows or activities are referred to as events and these consume no time. Thus, the arrow diagram consists of two basic elements, events and activities that are denoted by nodes and arrows, respectively.

Nodes, termed events, are typically drawn as small circles, ovals, or rectangles and represent a point in time in an arrow diagram. They are used to signify the start and completion of activities. Events are not associated with the utilization of any resources, whether time, money, equipment, or materials. Events noted on a diagram can provide valuable information and can highlight milestone occurrences in a project. A common example of a significant event or milestone on most building projects is the point in time at which a building has been "dried in," when the exterior enclosure is sufficiently complete so interior finish work can be performed without the threat of the elements. Once a building has been dried in, the work activities generally go through a major change. On a smaller scale, the starting point and the finishing point of each activity are also events.

Arrows, which are symbolic of activities, represent the performance of operations or tasks that consume time. In some cases they represent only the passage of time. They are drawn as lines, generally with arrowheads on the right end to designate the direction of progress within the time-oriented diagrams; thus the term arrow diagram. The tail of the arrow represents the starting point of the activity and the head of the arrow represents its finish. Activities consume time and generally other resources as well. The curing of concrete is an example of an activity that ostensibly consumes time but no other resources. While the consumption of other resources may be associated with the curing of concrete, they are generally indirect or minor in nature.

In an arrow diagram, the relationships between events and activities illustrate the logic of the process (plan) used to complete the project. Each activity occurs between two events, namely the start and finish times for the activity. As soon as the beginning event (often termed the i node) is reached, the activity can begin. Its end event (the j node) cannot occur until the activity has been completed. An event actually "occurs" only when all incoming or preceding activities have been completed.

ACTIVITY RELATIONSHIPS

There are no strict requirements for how an arrow diagram network should be drawn, but there are a few simple conventions that tend to be followed. Most important, arrows should be drawn showing progress from left to right. Loops can therefore not be created and, in fact, are illogical. Arrowheads are not required but are strongly recommended, particularly when arrows are drawn in a near-vertical direction, as it may be difficult to otherwise determine the direction of progress or the relationship between activities. The length of the line need not have any meaning; it is simply drawn long enough to connect its beginning and ending events. Note that arrow diagrams can be drawn as time-scaled models. Ideally, each network should be drawn so that all activities are tied into a single starting event at its beginning and a common ending event at the end. Some schedulers even advise that the arrow diagram should start and end with a single activity. In some scheduling software programs, this is a requirement. From an aesthetic point of view, arrow diagrams are best drawn so that the crossing of lines is kept to a minimum. The solution of arrow diagrams, however, is not impacted by the crossing of lines; i.e., arrows can cross. In the past, several software programs were available that solved scheduling problems as arrow diagrams. Almost all of these programs no longer exist or the option of solving the schedule as an arrow diagram is no longer available.

Several basic types of relations are found in arrow diagrams. Figure 16.1 illustrates the most simple and basic relationship that can exist between two activities. In this example, Activity G (Install mirror) cannot begin until Activity B (Hang wallpaper on wall) has been completed. Note that it can also be stated that wallpaper hanging can begin once the schedule has reached event 1 and that mirror installation can begin once the schedule has reached event 5. In this example, event 5 coincides with the completion of wallpaper hanging. Hanging wallpaper, or Activity B, can also be described as Activity 1–5 (1 is the number of the i node for activity B and 5 is the number of the j node for this activity) and Activity G, mirror installation, can be referred to as Activity 5–7.

Figure 16.2 illustrates an example in which three activities are dependent on the completion of a single preceding activity. If multiple activities (e.g., Activities F, G, and H) commence at a common event, it signifies only that the activities could all start at

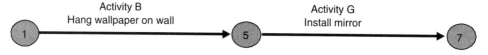

Figure 16.1 One-on-One Relationship Between Two Activities

the same time. The installations of the vanity light, mirror, and duplex coverplate are not necessarily required to actually start at exactly the same time (their start times are independent of the other start times), although none of them can start before the immediately preceding event (hanging wallpaper) has occurred. That is, the wallpapering work must be completed before any of these tasks can be performed.

Just as several activities may have their start times dictated by the completion of a single activity, so can a single activity have its start time dictated by the completion of several preceding activities. This type of relationship is shown in Figure 16.3. Note that "Install plasterboard" cannot begin until "Install wall plumbing," "Route electrical wiring in walls," and "Install insulation" have been completed. Note that the convergence of multiple activities, ending at a common node, does not necessarily mean that the preceding activities will all end at the same time, but that they must all be completed before the following activity can begin. In this example, the installation of wall plumbing, the routing of electrical wiring in the walls, and the installation of insulation must all be completed prior to attaching plasterboard to the studs. The plumbing, electrical, and insulation activities might ideally be performed in some preferred sequence, but they can theoretically take place simultaneously.

A more generalized relationship among several activities is shown in Figure 16.4. In this example, three activities must be completed before three subsequent activities can begin. In this example, Node 5 might be considered a milestone due to its significance. Note that Node 5 follows the completion of Activities A, B, and C and that it precedes the

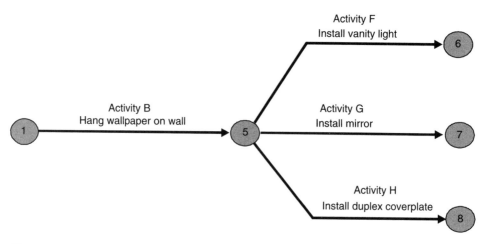

Figure 16.2 Many-on-One Relationship Among Several Activities

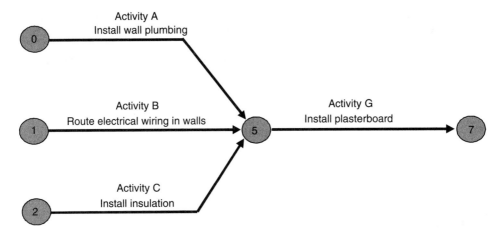

Figure 16.3 One-on-Many Relationship Among Several Activities

start of Activities F, G, and H. As already mentioned, one typical example of a milestone event in building construction is when the building has been "dried in," as the interior finish work can begin only after the building has been properly enclosed so that the elements will no longer jeopardize the work. A building can be considered dried in only after the roof is completed, the siding or masonry is installed, the windows are installed, all exterior doors are hung, all exterior door thresholds are installed, all exterior caulking is completed, and so on. After the building has been dried in, many interior tasks can begin. Another milestone might be the completion of work by a general contractor's own forces. After that point in a project, the remainder of the work will be performed by subcontractors.

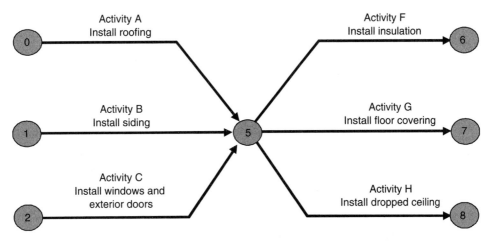

Figure 16.4 Many-on-Many Relationship Among Several Activities

Some activities may not be linked logically to other activities in a network. For example, in a network for a construction project, many procurement activities may be shown as having a possible start time at project start, meaning they have no preceding activities. It is highly unlikely that all of the procurement activities will actually occur at one time at project start. Naturally, procurement activities are focused on acquiring materials for which an eminent need exists on a project or on those materials for which a long lead time is required for delivery. The network may show that all procurement activities can take place at the project start, but this will probably not reflect when all of these activities will actually take place. In fact, when all the activities are actually scheduled later in the process, an appropriate scheduled start time will be selected that will ensure that the various materials will arrive in a timely manner for the various scheduled activities.

THE i–j NOTATION OF ACTIVITIES

As mentioned earlier, arrows or activities are frequently designated by their starting and ending nodes, commonly referred to as their "i" and "j" nodes. Algorithms used to do calculations for an arrow diagram require that all events (nodes) in the network be numbered. This is especially true of arrow diagrams that will be input into computer programs. Activities are then often identified by the numbers of the two events or nodes that precede and follow them. Node numbering should be done systematically following such guides as:

- Make each activity's i node smaller than its j node (sometimes required by computer programs).
- Leave gaps in numbers to allow for future network changes and additions (e.g., 5, 10, 15, 20, 25, etc.).
- Make each activity's i–j node combination unique (generally required by computer programs; see Figure 16.5).

Dummies

The relationships that exist among activities cannot always be shown as simply as those demonstrated in Figures 16.1 to 16.4; it may be necessary to include in the diagram a "pseudo-activity" called a dummy. A dummy is treated as an activity (normally drawn as a dotted line), but it is assigned no duration, meaning it does not consume time or

Figure 16.5 Typical Means of Designating Activities by Their Starting and Ending Nodes

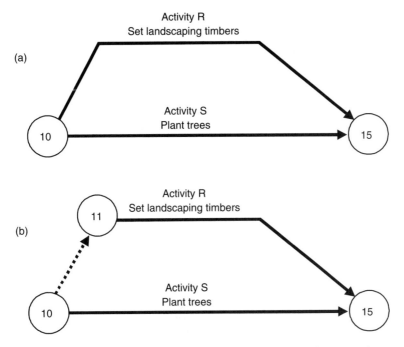

Figure 16.6 Unique Activity Designation Through the Use of a Dummy

any other resources. Some schedulers use dummies to designate delays; however, this use is inappropriate, as time is obviously a key component of a delay. It would be more appropriate to create a delay activity, one that has a specified duration assigned to it. As illustrated in the following examples, dummies may be used to ensure unique i–j node numbering for all activities, permit proper logic (activity ordering) to be displayed, or, in rare cases, serve as a beginning or ending "activity" for the network.

In Figure 16.6, the dummy is required to avoid having two activities with the same i–j designation of 10–15. For computer programs to function, unique i–j node numbering for an activity is absolutely essential. As a general rule, it is best to add the dummy to the beginning, rather than the end, of an activity. This avoids computational errors later.

Most dummies are required to maintain proper logic of various construction activities. For example, Figure 16.7 shows portions of a network involving tasks associated with constructing a garage. In example (a), one can conclude that Activities C and D can begin only after both Activities A and B have been completed. But this is not what was intended. Suppose Activity C (Install prefinished shop cabinets) can begin only after Activities A (Place concrete slab in garage) and B (Install garage door) have been completed, but Activity D (Install garage door opener) can begin as soon as the garage door is installed. The corrected logical sequence is accomplished with the use of a dummy, as shown in example (b) of Figure 16.7. Although the garage door opener might be installed with greater ease with the concrete slab in place, this is not essential.

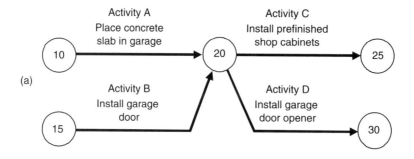

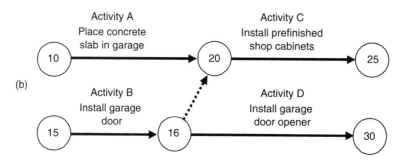

Figure 16.7 The Use of a Dummy to Show Proper Logic

One must understand the use of dummies to become an accomplished scheduler with arrow diagrams. As with all arrow diagrams, an activity cannot begin until the immediately preceding event has been reached. It is very important to recognize that every activity that precedes an event or node must be completed before the subsequent activity or activities can begin. Figure 16.8 includes two scenarios designed to help explain this fundamental issue, especially as it relates to the use of dummies. Of particular interest are the required preceding activities for Activity F (Install floor registers). In illustration (a), for instance, the floor registers can be installed as soon as the wallboard is installed, the vinyl flooring is installed, and the carpet is installed. Note that since Node 30 is preceded by Activity B and the dummy that links its start to Activity A, Activity F can start only after Activities A, B, and C are completed. But on this project, the floor registers can be installed whether or not the wallboard is installed. The proper logic is shown in illustration (b), where the addition of dummy 30–31 changes the predecessors for installing floor registers. The floor registers can be installed as soon as the vinyl flooring and carpet are in place.

Other uses of dummies are to establish a single starting activity and a single end activity for a project network. These are not crucial issues and do not pose a problem in most project schedules. Many computer programs in use today can easily handle a project that has more than one starting activity. It is often a matter of convenience that

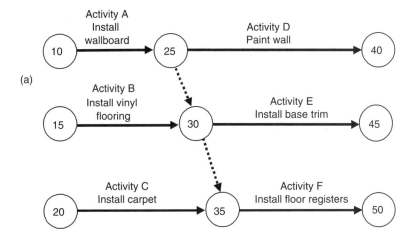

(a)

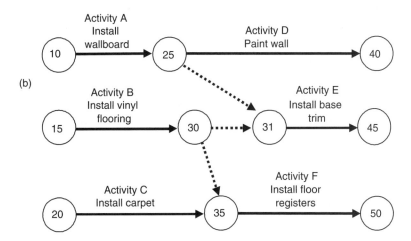

(b)

Figure 16.8 The Use of a Dummy to Change Activity Predecessors

there be a single starting node. Nonetheless, a single starting and/or ending activity can be achieved by the use of dummies, as illustrated in Figures 16.9 and 16.10.

The various uses of dummies in a network are illustrated in Figure 16.11. Note that, in some instances, a dummy may be required both to show proper logic and to establish unique i–j numbering (e.g., dummy 30–40). A review of Figure 16.11 should be conducted to evaluate the purpose of each dummy.

Arrow diagrams are favored by some people primarily because they became familiar with arrow diagrams early in their career. Proponents of the technique also

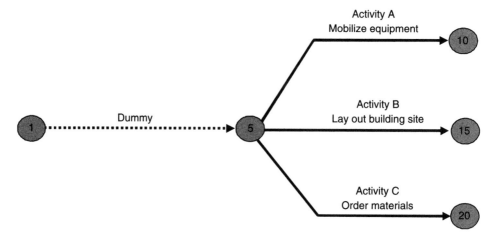

Figure 16.9 The Use of a Dummy to Create a Single Starting Activity

argue that the use of arrows is a natural way to illustrate that time is passing and that it is easy to graphically time-scale the network by making the length of each arrow proportional to the duration of the activities they represent. Another argument is that the arrow diagram automatically generates events at every point of interaction between activity groups.

There are critics of arrow diagrams who contend that the approach only represents finish-to-start relationships between activities. Of course, this is not entirely true, but to the extent that it is applicable, it is also true of precedence diagramming. Second, it is also claimed that drawing the network is a time-consuming process.

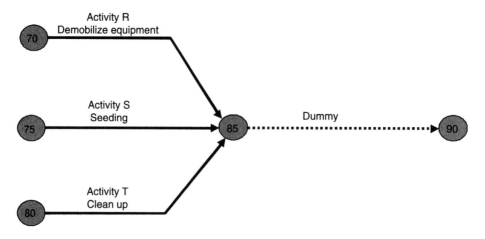

Figure 16.10 The Use of a Dummy to Create a Single Ending Activity

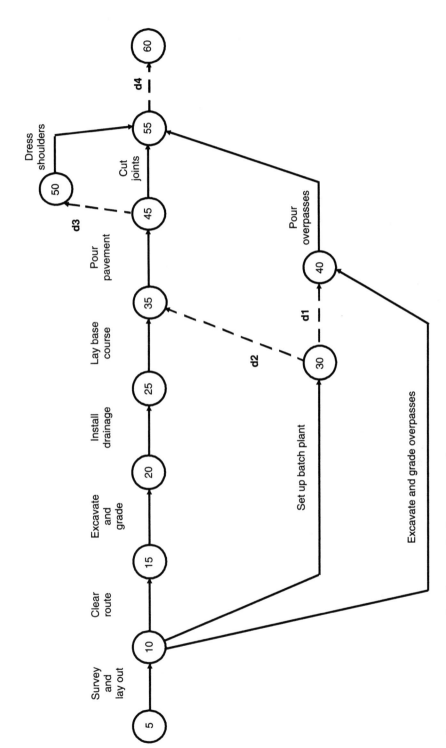

Figure 16.11 Example of Dummies in a Network

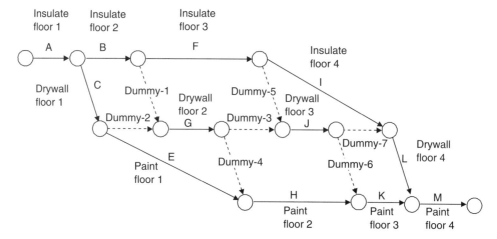

Figure 16.12 Arrow Diagram with Dummy Activities to Maintain Proper Logic

Errors in logic can go undetected when complex relationships requiring numerous dummies are represented. This argument has some merit, especially when activities are added to an existing network. Third, the network generally must be developed manually prior to entering information in a computer program, but this is also applicable to the development of precedence networks in many instances. Fourth, the number of "activities" that must be dealt with is increased by the fact that dummies, unique types of activities, must be used to correctly construct the project logic in the arrow diagram. Last, arrow diagrams are more difficult to modify with the introduction of new activities.

The arrow diagram in Figure 16.12 shows how the complexities of a project can introduce the need for dummy activities. The work being depicted consists of the sequential tasks of insulation installation, drywall installation, and painting. As shown, the insulation on each floor or story of the building must be installed prior to beginning the drywall installation. After the drywall is in place, the wall can be painted. In addition, the insulation installation must be completed on one floor prior to performing that work on the next floor. The dummies in the network are required to show this logical sequencing of activities. When examining the network, consider how the logic is actually altered by removing any of the dummy activities. Each of the dummies is required to maintain proper logic in the network.

PERFORMING TIME CALCULATIONS WITH ARROW DIAGRAMS

The development of a clear and accurate network diagram is an important step in the planning process. Once this diagram has been completed and the durations for all activities have been determined, the network will yield valuable information. This requires the performance of some simple calculations that are fundamental to the use of network diagrams in properly scheduling the activities required to complete a project.

As with precedence diagrams, the values for the start and finish dates for activities can be computed within a network by making what are called "forward" and "backward" passes through the arrow network. These are performed as described in Figure 16.13.

The units of time are often stated in terms of days, but the units could just as easily be hours, shifts, crew-days, weeks, or even months. Since the occurrence time for the first node was set at 1, it can be said that the "beginning-of-day" convention is being used here. Note that this means that the finish date of the activities is defined in terms of the beginning of the following day. For example, if the first activity in a network has a duration of two days and if its start date is set at the beginning of Day 1, the following calculations would be made:

The activity will end on the beginning of Day 3

(early start time of Day 1 + 2-day duration = early start time of Day 3)

Obviously, the activity will actually be completed at the end of Day 2, but hand calculations are easier to make if the beginning-of-day convention is used throughout. Note: Typically, computer programs automatically present information in the form of

---

Forward Pass Calculations*
1. Assign Day 1 as the early event time (ET) of the beginning node.
2. Select an activity starting at this beginning node.
3. Assign the node's early event time as the early start (ES) time for the activity.
4. Calculate the early finish (EF) time for the activity by adding its duration to its early start day.
5. Compare the activity's early (EF) time to the current early event time (ET) of the activity's end node:
   A. If the end node of an activity does not yet have an assigned ET, set the ET for the node as the EF of that activity. At nodes where two or more activities (both real and dummy activities) merge, the latest of the early finishes is to be assigned as the ET of the node.
   B. If the activity's EF is greater than the current ET for the node, assign ET the value of EF.
   C. If the activity's EF is not greater than the current ET for the node, leave the current value for ET unchanged.
6. Determine if any more activities begin at the node currently being used as a beginning node:
   A. If more activities do start there, go back to Step 3.
   B. If no more activities start at this node, select the next node that appears in the network. (If no more nodes exist, go to Step 7.)
7. The early event time of the last node is the estimated duration of the project.

*This process assumes that the network begins and ends with a single node.

Figure 16.13 Steps for Calculating Early Start Dates and Early Finish Dates

the beginning-of-day convention for the start times of activities and the end-of-day convention for the finish times.

The basic steps for calculating the early start and early finish times for activities are illustrated in the following example. The network that will be used is shown in Figure 16.14. While the format for presenting this information on a network may vary, the information of interest will essentially remain the same. The network shows the relationships that exist among the activities and the duration of each activity. Note that the network shown as an arrow diagram in Figure 16.14 is the same as the network shown as a precedence diagram in Figure 3.22.

The forward pass begins with the starting event in the network (Node 1). As discussed, this time has to be set by the scheduler. Since we are using the beginning-of-day convention, the first event would occur at the beginning of Day 1. The early event time of the first event node will also define the early start time of any activity that originates from that node. In this example, "Mobilize" will be assigned an early start time of 1. The early finish time of an activity is calculated by adding the activity duration to the early start time of that activity. Since the duration of "Mobilize" is 1, the early finish time for "Mobilize" is 2 (see Figure 16.15). Note that "Mobilize" will be completed on the beginning of Day 2. Of course, completing on the beginning of Day 2 is the same as completing the activity at the end of Day 1. It is only in this manual approach that this distinction must be kept in mind. (No further comment will be made about this point, as all times will be given with the beginning-of-day convention.)

Since the early finish time for Activity 1–2 is 2, the early event time for Event 2 is also 2. The early start time for Activities 2–3 and 2–4 is then determined to be 2, the early event time for Event 2 (see Figure 16.16). The early finish times of Activities 2–3 and 2–4 can be easily determined by adding their respective durations to the early start time. With an early finish time of 7 for Activity 2–3, the early event time for Event 3 is noted as being 7. Following the same procedure, the early finish time for Activity 2–4 is then determined to be 11. But what is the early event time for Event 4? The early finish time for Activity 2–4 is 11 and the early finish time for Activity 2–3 is 7. Because a dummy activity precedes Event 4, the early occurrence time of Event 3 must be taken into consideration. Naturally, Event 4 occurs after Activities 2–3 and 2–4 are completed. Therefore, both of the preceding activities must be completed before the early event time of Event 4 will occur. Logic should tell us that the controlling activity is Activity 2–4. Thus, the early event time for Event 4 is 11. As a general rule, when more than one activity precedes a node, its early event time will be the latest (largest value) of the early finish times of the immediately preceding activities.

The procedure outlined above is followed for the remaining activities in the network. Remember that the early event time for a node becomes the early start time for those activities that start after that node. The early finish times for the activities are always determined by adding their durations to the early start times. The early finish time for an activity then becomes the early event time for the node or event following that activity. If more than one activity precedes a particular node, the early event time for the node is established as being the latest of the early finish times of the preceding activities. The calculations in Figure 16.16 show that the early event time of the last event, Event 8, is 21. Thus, the project can be completed by the beginning of Day 21.

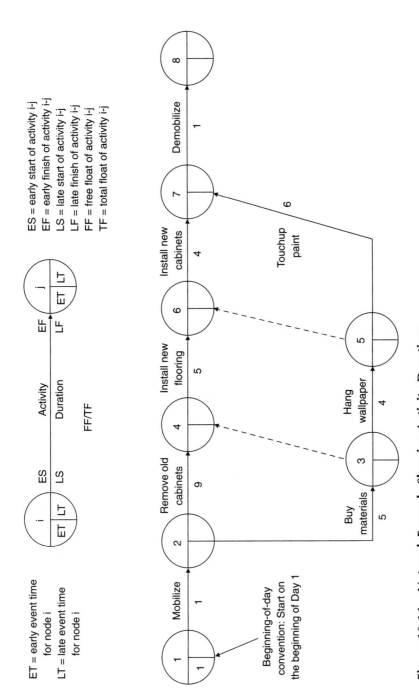

Figure 16.14   Network Example Showing Activity Durations

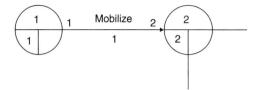

Early occurrence time of node 1 becomes the early start time for activity 1-2

Early finish time of activity 1-2 is computed as ES of activity 1-2 plus the duration of activity 1-2

$$EF_{1\text{-}2} = ES_{1\text{-}2} + Duration_{1\text{-}2}$$

Figure 16.15 Determining Early Event Times, Early Start Times, and Early Finish Times

The network in Figure 16.16 illustrates the early times of all event occurrences, early activity starts, and early activity finishes. While this is valuable information, additional information can also be gleaned from the network. While it is helpful to know the earliest time at which each event will occur and the earliest time that each activity can start and finish, the scheduler also needs to know the latest times that these events and activities can start and finish without delaying the completion of the project. This information is obtained when a backward pass is conducted.

Essentially, each of the early occurrence times computed with the forward pass will be complemented by late occurrence times. These late occurrence times are computed in a manner that is similar to the forward pass, a backward pass. The steps for calculating the late event times and the late start and finish times are described in Figure 16.17. Figure 16.18 shows the basic components of the network of interest when conducting a backward pass. The term *backward pass* comes from the fact that the calculations begin at the last node or event and end with the first node or event.

In conducting the backward pass, the focus of attention begins with the last node in the network. In this example, the last node describes Event 8, which has already been determined to have an early event time of 21. To establish the late occurrence times in a network, it is common to set the late event time of the last node as being equal to its early event time (in this case, 21). The late event time of an event then becomes the late finish time of those activities that immediately precede that event. In this case, the late finish of Activity 7–8 becomes 21. By subtracting the duration of Activity 7–8 from its late finish time, we can compute the late start time for the activity. In this example, the late start time for Activity 7–8 is 20. The late start times of the activities are then assigned as the late event times of the immediately preceding events. Thus, the late occurrence time for Event 7 is 20. This procedure is followed consistently as the backward pass is performed. Care must be exercised when more than one activity follows a particular event. In Figure 16.19, this situation is encountered at Event 2. Note that Activity 2–4 has a late start date of 2 and that Activity 2–3 has a late start date of 5. Which of these values should be assigned as the late event time for Event 2? Logic tells us that if an activity cannot start later than a given date, then the late event must occur prior to or on that date. In other words, if more than one activity follows an event, the late event time for the event will be determined by the earliest late start times of those following activities; i.e., use the smallest late

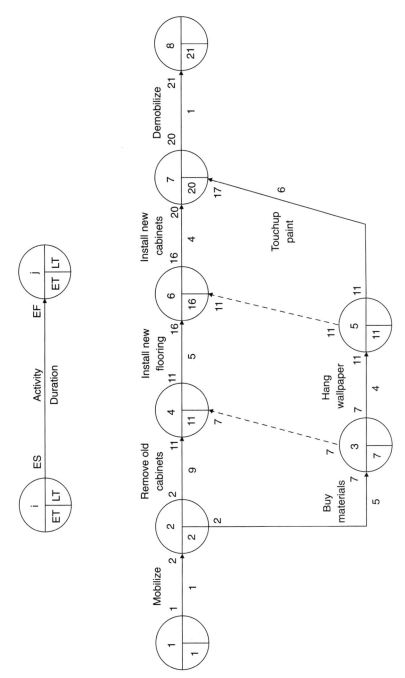

Figure 16.16 Arrow Diagram Showing Early Event Times, Early Start Times, and Early Finish Times

341

342   Chapter 16

> **Backward Pass Calculations***
> 1. Assign the last node a late event time (LT) (normally this is set as being equal to its early event time [ET]). Thus, generally for the last node, ET = LT.
> 2. Select an activity ending at this node.
> 3. Use the j-node's late event time (LT) and assign this value as the activity's late finish (LF) day.
> 4. Calculate the late start (LS) for the activity by subtracting its duration from its late finish.
> 5. Compare the activity's late start time to the late event time of the activity's beginning node:
>   A. If the i-node of the activity does not yet have an assigned LT, set LT equal to LS.
>   B. If the activity's LS is less than the current LT for the node, assign LT the value of LS.
>   C. If the activity's LS is not less than the current LT for the node, leave the current value of LT unchanged.
> 6. Determine if any more activities end at the node currently being used as the j-node:
>   A. If more activities do end there, go back to Step 2.
>   B. If no more activities end at this node, select the next preceding node and go back to Step 2. (If no more nodes exist, go to Step 7.)
> 7. The late event time of the first node is the day on which the project must begin. (It should always be 1 if the LT of the last node was set equal to its ET.)
>
> *This process assumes that the network begins and ends with a single node.*

Figure 16.17   Steps for Calculating Late Start Dates and Late Finish Dates

Beginning the backward pass:
Late occurrence time of the last node in the network is set
equal to the early occurrence time of the last node,
e.g., LT = ET = 21

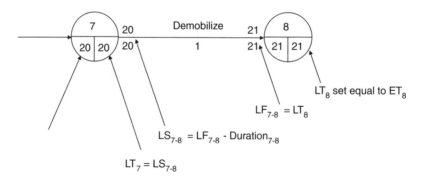

Figure 16.18   Steps Outlining the Backward Pass Calculations

start time of the subsequent activities to set the late event time for a node. The late event time (LT) for Event 2 is then determined to be 2.

The backward pass is relatively simple to perform. If the basic rules are understood, there should be no difficulty in making the calculations. When dummies occur in the network, the early event times of the events they follow are carried forward as the early occurrence times of the dummy activity. In the same fashion, when performing the backward pass, the "late finish dates" of the dummies are carried backward as the late start dates of the dummy activities. Figure 16.19 shows the backward pass calculations for the network, along with all of the early occurrence times and the late occurrence times for the events and the activity starts and finishes.

FLOAT VALUES

With the forward and backward passes done, additional information can be determined by calculating the activity float (sometimes termed *slack*) values for each of the activities. As the term implies, *float* is a way to describe the flexibility associated with the scheduling of a particular activity. The most common types of float are free float and total float:

Total Float: The maximum amount of time that an activity can be delayed (beyond its designated early start time) without delaying the project completion. The total float of an activity can be calculated by taking the difference between an activity's early finish and its late finish or the difference between the early start and the late start. Some texts refer to the total float as *total slack*.

Free Float: The maximum amount of time that an activity can be delayed (beyond its designated early start time) without delaying the early start of any other activity, specifically those activities immediately following that activity. It is calculated by subtracting the activity's early finish from the early event time (ET) of its ending event or j node.

When the total float of an activity is zero, there is no flexibility in the scheduling of that activity. To avoid delaying project completion, any activity with zero total float must start on the scheduled date and must be completed within the stipulated duration. This is true of all activities with zero total float. Regarding activities with zero total float, the delay of any activity by a single day will delay project completion by one day.

It is important for the scheduler to know the float values associated with activities. Of the float values, perhaps the most important is total float (TF), as it addresses the status of the entire project. Free float is more closely associated with the relationships of activities to each other. The total float is measured in terms of the degree of impact that the delay of the starting time or the alteration (lengthening) of the designated duration of an activity would have on the project duration. An activity's total float is directly related to the core concept of the "critical path method." The following definitions explain the importance of total float.

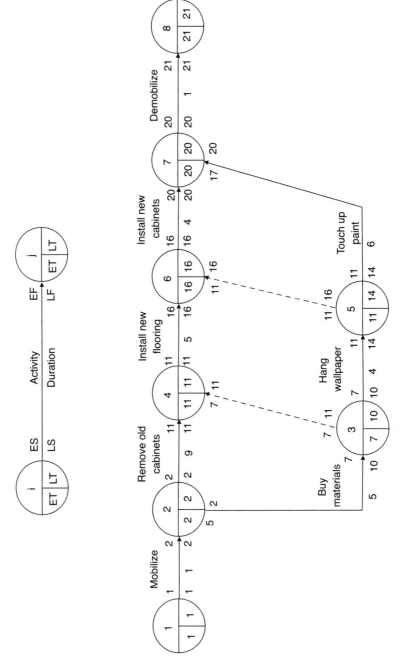

Figure 16.19 Arrow Diagram Showing Late Event Times, Late Start Times, and Late Finish Times

Critical: A term that specifies that an activity cannot be delayed without extending the project duration. Activities with no total float (TF = 0) are defined as critical. These activities have early and late starts that are equal, indicating that they *must* be started at the calculated time and completed within the stipulated duration in order for the project to be completed in the estimated amount of time.

Critical Path: The path (or paths) from the first activity to the last activity in the network that passes through only those activities that have a total float of zero. Each network has at least one critical path connecting the first and last nodes in the network, provided the early and late project completion times are set equal to each other.

The project scheduler has flexibility in specifying when any activity with float will actually be started, as long as it is started by its late start date. If an activity has a total float of ten days, it could be started up to ten days after its calculated early start date without impacting the project duration. However, scheduling earlier activities at times other than their early starts may restrict the start time options for any following activities to times later than their early starts. If an activity has a total float of ten days and a free float of five days, the start date could be delayed ten days without impacting the project duration, but the start date can be delayed only five days if the start times of no other activities are to be impacted.

The float values are not difficult to compute if careful attention is paid to their determination. The total float of an activity is simply the difference between the late start time of an activity and its early start time or, conversely, the difference between its late finish time and early finish time.

Total float of Activity i–j = TF_{ij} = Late finish of Activity i–j − Early finish of Activity i–j

Free float of Activity i–j = FF_{ij} = Early occurrence of Event j − Early finish of Activity i–j

Once the early start and finish times and the late start and finish times have been determined, the computation of the free and total float values is a simple matter. These values are shown in Figure 16.20. As mentioned earlier, the critical path consists of a sequence of activities between the first and last node in the network for which the total float is 0. In the network shown in Figure 16.20, the critical path can quickly be identified as consisting of Activities 1–2, 2–4, 4–6, 6–7, and 7–8. The activities with free float values of 0 are also of interest, as these cannot be delayed beyond their scheduled early completion date without impacting at least one other succeeding activity. In the network, it is apparent that only one activity has a positive free float value, namely Activity 5–7. The results of the computations shown in Figure 16.20 are the same as those derived for the same schedule solved as a precedence diagram in Chapter 3. Also compare the results shown in Figure 16.21 with the tennis court project presented in Chapter 3.

The precedence diagram shown in Figure 3.26 has been converted to an arrow diagram in Figure 16.21. First examine the diagram to see why the dummy variables are necessary. Next, study the computations to see how the various values of ES, EF, LS, LF, FF, and TF have been determined. Finally, compare the results with the values derived in Figure 3.26 and notice that the same values are determined by both methods.

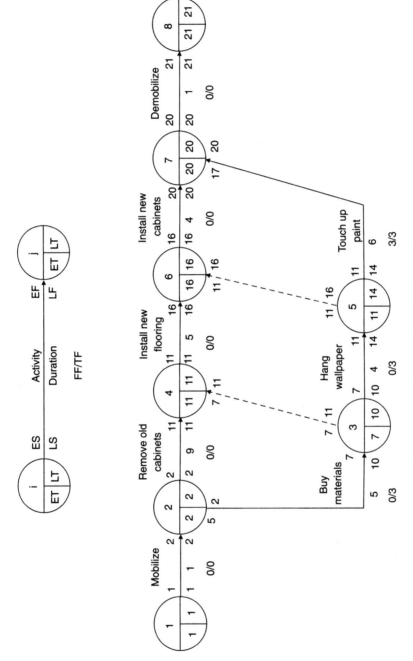

Figure 16.20 Arrow Diagram Showing Free Float and Total Float Values

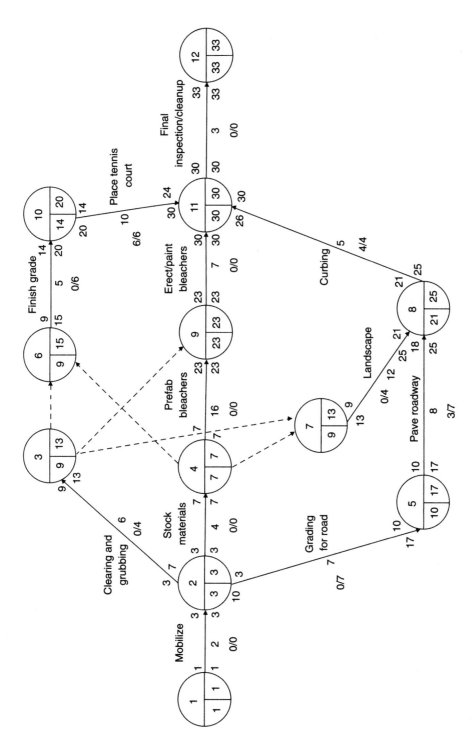

Figure 16.21 Arrow Diagram Showing Scheduling Computations for the Tennis Court Project

The total float can be viewed as presenting information about the flexibility of an activity on a macro level, while the free float presents information about the flexibility of an activity on a micro level. Knowing the free float and total float values of activities can be very helpful to a scheduler. The float values help the scheduler identify those activities that provide the greatest flexibility in altering the start times. This flexibility in start times can then be used to:

- Implement additional activity constraints
- Level resources
- Control contingency time
- Change or delay impact mitigation

As stated earlier, the critical path lies along that continuous sequence of activities from the first activity to the last activity for which the total float is 0. The value of total float along the critical path is 0 only because the late event time for the last event in the network was assigned the same value as the early event time for the last event. It is entirely possible for the critical path to consist of that sequence of activities from the first activity to the last for which the total float is minimal. Thus, the critical path activities could conceivably have total float values that are either positive or negative (when the early and late event times of the end event are not the same), but normally the critical path follows along those activities that have a total float of 0.

Assume that a network has been prepared for the construction of a reinforced concrete bridge. The forward pass computations showed that the duration of the project would be 332 days. In the contract, the duration was stipulated as being 360 days. The scheduler could use the time of 360 for the late finish date for the last activity and then perform the backward pass computations. This would result in the critical path activities having 28 days (360–332) of total float. The assignment of the contract duration as the late finish date for the last activity is not a good one. There is no penalty for completing early on most construction projects, and on many projects there is even a bonus for early project delivery. Utilizing 360 days to finish a project that could be realistically completed in 332 days will generally result in an additional cost to the contractor because of added overhead and other costs.

Now, suppose the contract for this project stipulated that the contract duration was 320 days. Since the forward pass computations resulted in an ET for the last node as 332 days, assigning a value of 320 as the LT for the last node of the critical path activities or the worst value for total float results in a negative value of -12 days. Since it is contractually stated, it would be logical to utilize this value (320 days) as the late finish time for the last activity. The critical path for this network will now consist of activities with total float values that are negative (-12 days). Negative float flags a problem area for a scheduler, as it implies that the early start schedule will result in a completion date that is unacceptable; negative float means that the project will be completed late or possibly that the project must begin at an earlier date. The schedule will generally not be considered complete if any activities contain negative float. It is the scheduler's task to then determine ways in which the logic can be modified or durations for selected activities can be reduced to eliminate the negative float.

UNDERSTANDING FREE FLOAT AND TOTAL FLOAT

A scheduler needs to fully understand the definitions of *free float* and *total float*, and this understanding should be such that a good intuitive feel for these concepts is established. Once this is accomplished, it will be easy to critically assess the reasonableness of the computed values for float. While it is not easy to intuitively evaluate each value of float, there are specific rules that should always apply that make it easier to determine float values. For example, these rules apply specifically when an arrow diagram has at least one critical path along which the total float is 0. The nuances of some rules will be impacted by the placement of dummies in certain portions of the network, but these impacts rarely occur. Note that dummies do not possess float. It is assumed that the late finish date and the early finish date of the last activity are the same. These general rules are as follows:

GENERAL FREE FLOAT AND TOTAL FLOAT RULES

1. There is a minimum value of total float among the activities following each node, and this minimum value is also the minimum value of the total float values among those activities that precede the node.

2. Prior to each node, there is at least one activity that has a free float of 0.

3. The free float of a noncritical activity intersecting the critical path equals its total float.

4. In a chain of activities (a series of activities consisting of only one activity preceding the node and only one activity following each node), the total float for all activities in that chain is the same. Also, except for possibly the last activity in the chain, the free floats of the activities are 0.

5. In any path from the first activity to the last activity in a network, the sums of the durations and the free floats will be equal to the sums of the durations and the free floats of any other path of activities from the first to last activity.

6. When more than one activity precedes a node, first identify the activity following the node that has the smallest total float. This value of total float will also be the smallest total float among the activities that precede the node. The free float of the preceding activity with the smallest total float will be 0. The free float of the other activities can be computed by subtracting the minimum total float value (the total float of the activity preceding the node that has 0 free float) from each activity's total float. For activities preceding a node:

$$FF_{Act} = TF_{Act} - TF_{\text{smallest TF value prior to node}}$$
$$TF_{Act} = FF_{Act} + TF_{\text{smallest TF value prior to node}}$$

7. For all critical path activities, TF = 0 and FF = 0.

350   Chapter 16

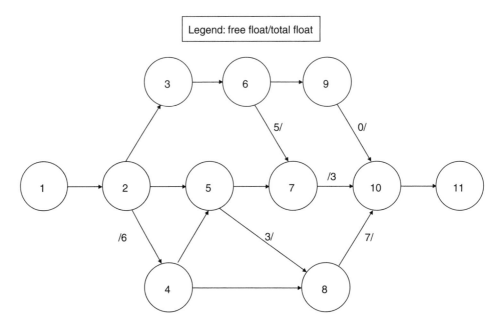

Figure 16.22   Network Showing Limited Free Float And Total Float Values

The foregoing rules will be utilized in an example in which only limited information about free float and total float values is known. As before, the early and late event times for the last event are assumed to be the same. With the limited amount of information provided in Figure 16.22, the remaining free float and total float values can be determined.

In deriving the missing float information for Figure 16.22, first consider Activity 9–10, the only activity preceding Event 10 that can be critical (Rule 7: Critical activities must have free float and total float values of 0). Thus, the total float of Activity 9–10 is 0. Also, since all activities on the critical path have 0 float values, the critical path can be traced. First, by inspection, Activity 10–11 must be critical. Since Activity 9–10 has been identified as being critical, Activity 6–9 must also be critical (Rule 1). By similar reasoning, Activities 3–6, 2–3, and 1–2 are also identified as being critical. Thus, the critical path passes through Events 1, 2, 3, 6, 9, 10, and 11. Since each node must be preceded by at least one activity that has a free float of 0 (Rule 2), the free float is 0 for Activities 2–4, 4–8, and 5–7. The information known about the float values at this stage is shown in Figure 16.23.

Since Activities 7–10 and 8–10 intersect the critical path, their float values can be readily determined since the free float and total float values for these activities are the same (Rule 3). Using Rule 6, the total float for Activity 4–8 is determined to be 7, the same as Activity 8–10, and the total float for Activity 5–8 is determined to be 10 (see Figure 16.24).

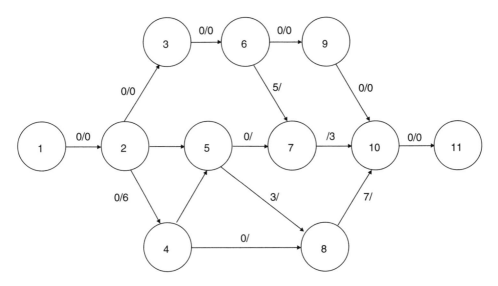

Figure 16.23 Network Identifying the Critical Path

By using Rule 6, the total float values for Activities 5–7 and 6–7 can also be determined. With Rule 1, the total float of Activity 4–5 can be determined as 6. In like manner, the total float of Activity 2–5 is determined to be 3. Using Rule 6, one can derive the remaining free float values for Activities 2–5 and 4–5 (see Figure 16.25).

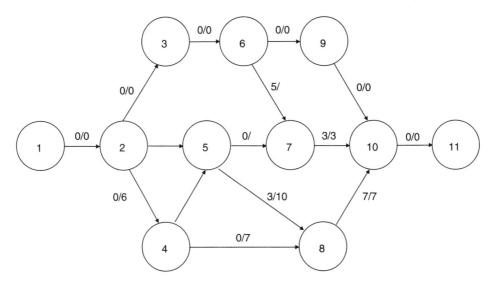

Figure 16.24 Network Showing Float Values for Activities Involving Event 8

352 Chapter 16

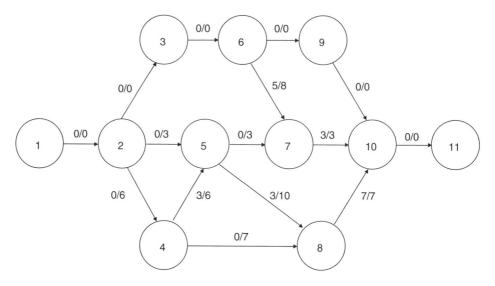

Figure 16.25 Network with All Free Float and Total Float Values Shown

COMPUTATIONS FOR DIFFERENT ACTIVITY RELATIONSHIPS

In Chapter 3, the various activity relationships that can be encountered in a network were introduced. In addition, the issue of lag was described, with lag having either positive or negative values. While it is customary to solve arrow diagrams with activities described with the traditional finish-to-start relationships, arrow diagram computations can be solved with finish-to-start relationships, start-to-start relationships, finish-to-finish relationships, and start-to-finish relationships. All of these relationships can have lag (or lead) times associated with them as well, and the lag values can be either positive or negative. This concept will be illustrated with a simple network consisting of three activities, namely activities A, B, and C.

First it will be assumed that the relationship that exists between the activities is a finish-to-start relationship. This is shown in Figure 16.26. When there are no lag values, the computations are as follows:

$ES_B = EET_J$ where EET_J is the early event time of node J
$EF_B = ES_B + duration_B$
$LS_B = LF_B - duration_B$
$LF_B = LET_K$ where LET_K is the late event time of node K

As described earlier, the early start times and early finish times are computed by conducting a forward pass through the network, and late start times and late finish times are computed by conducting a backward pass through the network. If lag time

Figure 16.26 Finish-to-Start Relationship on an Arrow Diagram

(whether positive or negative) is introduced between activity A and activity B, the computations would be as follows:

$$ES_B = EET_J + lag_{A-B} = EF_A + lag_{A-B}$$
$$EF_B = ES_B + duration_B$$
$$LS_B = LF_B - duration_B$$
$$LF_B = LET_K - lag_{B-C} = LS_C - lag_{B-C}$$

If activities were described by start-to-start relationships, the activities could be depicted as shown in Figure 16.27. Lag values may also exist, whether positive or negative in value. The lag notations on the figures (direction of the dotted lines with arrows) indicate the conditions for which the lag values would be positive. The computations would be as follows:

$$ES_B = EET_I + lag_{A-B} = ES_A + lag_{A-B}$$
$$EF_B = ES_B + duration_B$$
$$LS_B = LET_M - lag_{B-C} = LS_C - lag_{B-C}$$
$$LF_B = LS_B + duration_B$$

If finish-to-finish relationships existed between the activities, the relationships would be as shown in Figure 16.28. Note that the lag values could be positive, negative or zero. The primary computations for the scheduling variables would be as follows:

$$ES_B = EET_J + lag_{A-B} - duration_B = EF_A + lag_{A-B} - duration_B$$
$$EF_B = EET_J + lag_{A-B} = EF_A + lag_{A-B}$$
$$LS_B = LET_N - lag_{B-C} - duration_B = LF_C - lag_{B-C} - duration_B$$
$$LF_B = LET_N - lag_{B-C} = LF_C - lag_{B-C}$$

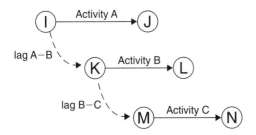

Figure 16.27 Start-to-Start Relationship on an Arrow Diagram

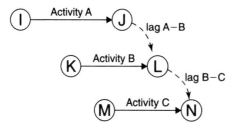

Figure 16.28 Finish-to-Finish Relationship on an Arrow Diagram

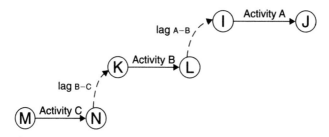

Figure 16.29 Start-to-Finish Relationship on an Arrow Diagram

For start-to-finish activity relationships, the activities would be shown as depicted in Figure 16.29. The basic computations for the scheduling information would be as follows:

$$ES_B = EET_I - lag_{A-B} - duration_B$$
$$EF_B = EET_I - lag_{A-B} = ES_B + duration_B$$
$$LS_B = LET_N + lag_{B-C}$$
$$LF_B = LET_N + lag_{B-C} + duration_B = LS_B + duration_B$$

FINAL COMMENTS

Arrow diagrams are no longer as popular as they once were, but this should not deter students of planning and scheduling from learning this technique. The focus of arrow diagrams on events, in addition to activities, sets them apart from other methods in use. Regardless of the technique used, the same basic information about activities is still determined. For scheduling purposes, it is important to know information about all activities concerning the early and late start dates, the early and late finish dates, and the free and total float values. With this information, it is easy to discern the extent of flexibility of the scheduling of each activity. This is important in order for managers of construction projects to make informed decisions about necessary modifications in the sequencing of project activities.

REVIEW PROBLEMS

Draw an arrow diagram for each of the following activities. Draw all arrows in a left-to-right direction. Assume the data will be computer-input in standard i–j notation. Show all necessary dummies, but do not include any redundant dummies. The activities and their dependencies are as follows:

1.

Activity	Depends on	Activity	Depends on
A	—	H	B, C, E
B	A	I	F, G
C	A	J	B
D	A	K	H, I
E	D	L	H, J
F	D	M	I
G	D	N	K, L, M

2.

Activity	Depends on	Activity	Depends on
A	—	F	B, C
B	A	G	C, D
C	A	H	E
D	A	I	E, F
E	B	J	G, H, I

3.

Activity	Depends on	Activity	Depends on
A	—	G	B, C, D
B	A	H	E, G
C	A	I	E, F
D	A	J	F
E	B	K	H, I, J
F	B, D		

4.

Activity	Depends on	Activity	Depends on
A	—	I	E
B	A	J	E, H
C	A	K	B, D, G
D	A	L	F, K
E	A	M	I, J, K
F	B	N	I
G	C	O	L, M, N
H	D		

5.

Activity	Depends on	Activity	Depends on
A	—	J	F, G
B	A	K	G
C	A	L	K
D	B	M	H, I
E	B	N	E, J, K
F	B	O	J, L
G	C	P	M
H	D	Q	I, N, O
I	E	R	P, Q

6.

Activity	Depends on	Activity	Depends on
A	—	H	B, D
B	A	I	E
C	A	J	E, F, G
D	A	K	G
E	B	L	H
F	C, D	M	I, J, K, L
G	C		

7.

Activity	Depends on	Activity	Depends on
A	—	F	B, C
B	A	G	C, D
C	A	H	E
D	A	I	E, F
E	B	J	G, H, I

8.

Activity	Depends on	Activity	Depends on
A	—	L	D, F, G
B	A	M	G, H
C	A	N	K
D	A	P	L
E	B	Q	M
F	B	R	J
G	C	S	M, N, R
H	D	T	P, Q
I	E	U	S, T
J	F, I	V	S, T
K	D, F, G	W	U, V

9. Draw an arrow diagram that depicts the proper logic for the construction of a pole barn, as described below:

Activity	Activity Description	Dependent on Activity
A	Mobilize	—
B	Order poles	A
C	Clear site	A
D	Order lumber	A
E	Order roofing materials	A
F	Grade site	C
G	Set poles	B, F
H	Trench for gas lines	F
I	Trench for water lines	F
J	Set joists	D, G
K	Install gas lines	H
L	Install water lines	I
M	Install roofing	E, J
N	Install siding	H, J
O	Set doors	M, N
P	Install electrical wiring and fixtures	K, L, M, N
Q	Paint	O, P
R	Clean up	Q
S	Demobilize	R

10. Draw an arrow diagram that depicts the proper logic for the remodeling of a school, as described below:

Activity	Activity Description	Dependent on Activity
A	Mobilize	—
B	Build new cafeteria	A
C	Remodel classroom A	A
D	Remodel theater	A
E	Convert old cafeteria into a lab	B
F	Temporarily move library books	C, D
G	Install security system	F
H	Remodel library	E, F
I	Remodel corridor	E
J	Return books to library	H
K	Clean up and move out	G, I, J

11. Draw an arrow diagram that depicts the logic for finishing work on a house project, as described below:

Activity	Activity Description	Dependent on Activity
A	Mobilize	—
B	Paint walls and ceiling	A
C	Vinyl floor in kitchen and laundry room	B
D	Ceramic floor in bathroom	B
E	Wallpaper in bathroom and kitchen	D
F	Set kitchen appliances	C
G	Install wall-hung mirror	E
H	Hang prefab kitchen cabinets	C, E
I	Set bath fixtures	E
J	Install base trim	H, I
K	Lay carpet in living room, etc.	F, J
L	Touch up paint	G, K
M	Demobilize	L

12. Compute the values of ES, EF, LS, LF, FF, and TF for the activities in the following network.

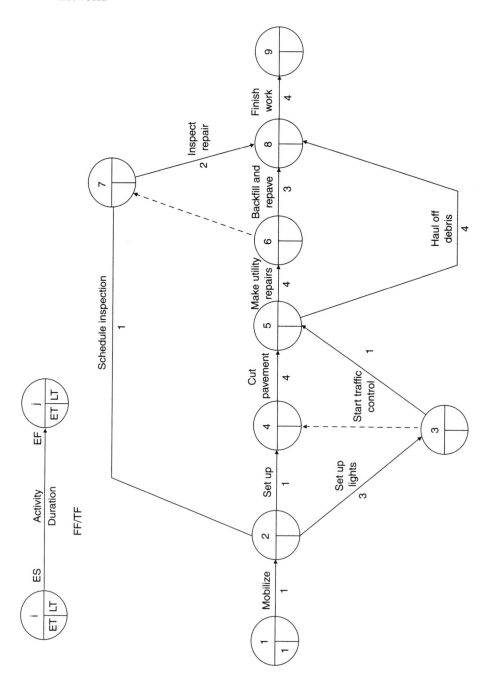

13. Compute the values of ES, EF, LS, LF, FF, and TF for the activities in the following network.

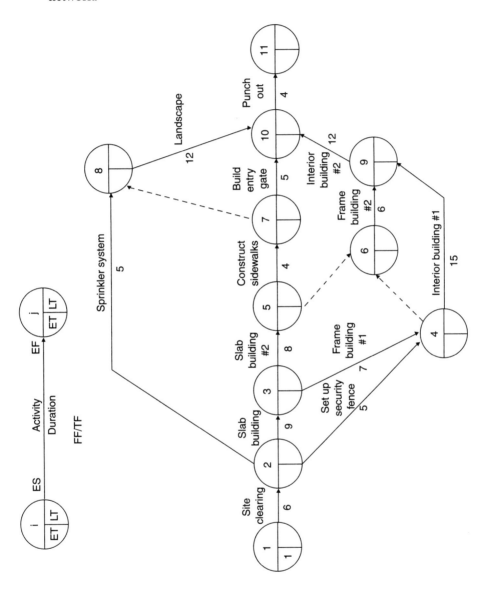

360 Chapter 16

14. Compute the values of ES, EF, LS, LF, FF, and TF for the activities in the following network.

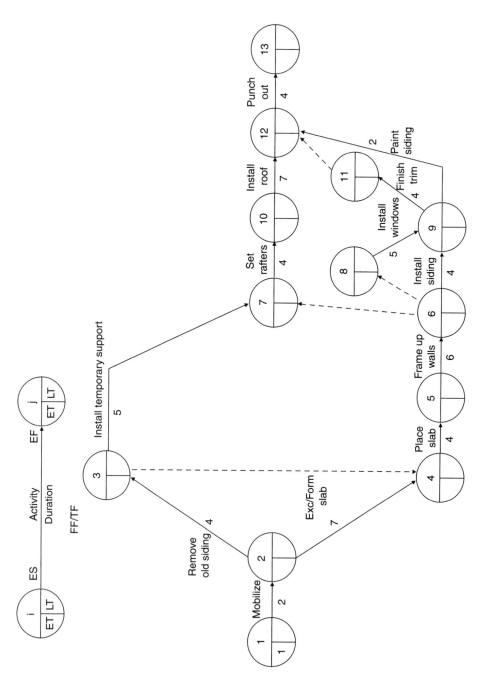

15. Compute the values of ES, EF, LS, LF, FF, and TF for the activities in the following network.

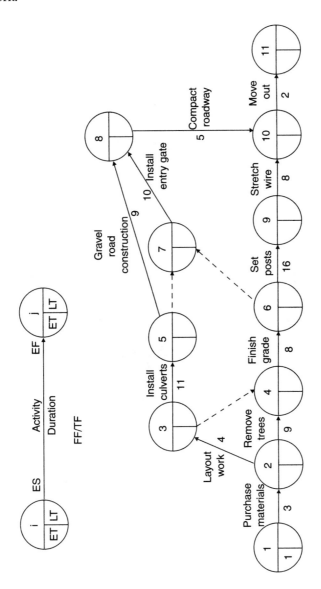

16. Compute the values of ES, EF, LS, LF, FF, and TF for the activities in the following network.

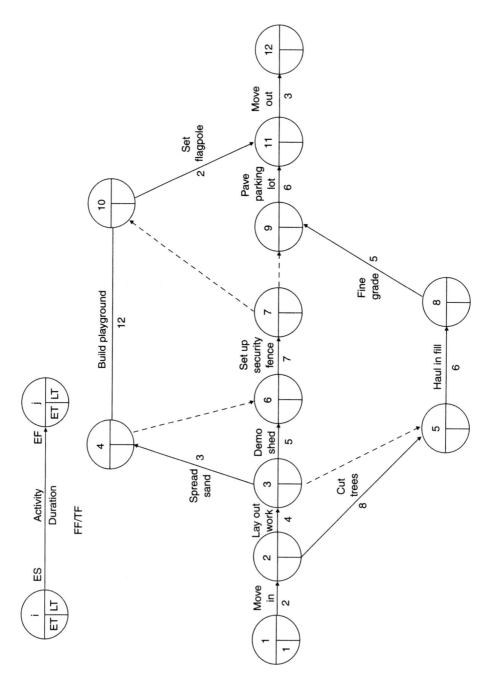

17. Compute the values of ES, EF, LS, LF, FF, and TF for the activities in the following network.

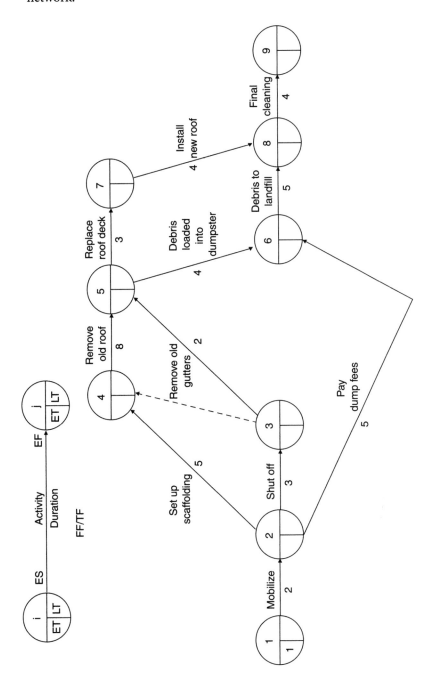

References

Ahuja, Hira N. *Project Management: Techniques in Planning and Controlling Construction Projects.* New York: John Wiley & Sons, 1984.

American Society of Civil Engineers. *Manual of Professional Practice: Quality in the Constructed Project: A Guideline for Owners, Designers and Constructors,* vol. 1. New York: American Society of Civil Engineers, 1988.

Blackhawk Heating & Plumbing v. the United States Government. 75-1 BCA (P11,261) GSBCA No. 2432. April 30, 1975. Contract No. GS-OSBC-4963.

Brooks Towers Corp. v. Hunkin-Conkey Construction Co., 454 F. 2d 1203 (1972).

Callahan, Michael T., Daniel G. Quackenbush, and James E. Rowings. *Construction Project Scheduling.* New York: McGraw-Hill, 1992.

Callanan Industries v. Glens Falls Urban Renewal Agency, 403 N.Y.S. 2d 594 (1978).

Clough, Richard H., and Glenn A. Sears. *Construction Project Management,* 3d ed. New York: John Wiley & Sons, 1991.

Coombes, Jim. *Real World Construction Scheduling.* Seattle, Wash.: Construction Planning and Management, 1990.

D'Angelo v. State of New York, 362 N.Y.S. 2d 283 (1974).

Feldman, William, and Patti Feldman. *Construction & Computers.* New York: McGraw-Hill, 1996.

Fisk, Edward R. *Construction Project Administration,* 3d ed. New York: John Wiley & Sons, 1988.

Gates, M., and A. Scarpa. "Learning and Experience Curves." *Journal of the Construction Division,* 98, no. 1 (1972): 70–101.

Gido, Jack. *An Introduction to Project Planning,* 2d ed. New York: Industrial Press, 1985.

Haas and Hanie v. the United States Government, GSBCA Nos. 5530, 6224, 6638, 6919, 6920. June 8, 1994. Contract No. GS-09B-C-1003-SF.

Harris, Robert B. *Precedence and Arrow Networking Techniques for Construction.* New York: John Wiley & Sons, 1978.

Horowitz, Joseph. *Critical Path Scheduling.* New York: The Ronald Press Company, 1967.

Iannone, A. L., and A. M. Civitello, Jr. *Construction Scheduling Simplified.* Englewood Cliffs, N.J.: Prentice-Hall, 1985.

Lester, Albert. *Project Planning and Control.* London: Butterworth & Company (Publishers) Ltd., 1982.

Lichter v. Mellon-Stuart Company, 193 F. Supp. 216 (1961).

Lu, F. P. S. *The Critical Path Method of Construction Management.* Christchurch, New Zealand: The Caxton Press, 1965.

Marriott v. Dasta Construction Company, 26 F. 3d 1057 (1984).

Means Building Construction Cost Data. Kingston, Mass.: R. S. Means Company, issued yearly.

Minmar Builders, Inc. v. the United States Government, 72-2 BCA (19599) GSBCA No. 3430. July 28, 1972. Contract No. GS-03B-15477.

Moder, Joseph J., and Cecil R. Phillips. *Project Management with CPM and PERT,* 2d ed. New York: Van Nostrand Reinhold Company, 1970.

Mueller, Frederick William. *Integrated Cost and Schedule Control for Construction Projects.* New York: Van Nostrand Reinhold Company, 1986.

Neale, Richard H., and David E. Neale. *Construction Planning.* London: Thomas Telford Ltd., 1989.

O'Brien, James J. *CPM in Construction Management,* 3d ed. New York: McGraw-Hill, 1984.

Owen L. Schwam Construction Co. v. the United States Government, 79-2 BCA (P13, 919) ASBCA No. 22407. May 31, 1979. Contract No. DACA 51-75-C-0051.

Pathman Construction Co. v. Hi-Way Electric Co., 382 N.E. 2d 453 (1978).

Pilcher, Roy. *Principles of Construction Management,* 3d ed. New York: McGraw-Hill, 1992.

Stepman, Kenneth. *A Buyer's Guide to Project Management Software.* Milwaukee, Wis.: New Issues, 1986.

Stevens, James D. *Techniques for Construction Network Scheduling.* New York: McGraw-Hill, 1990.

Wiest, Jerome D., and Ferdinand K. Levy. *A Management Guide to PERT/CPM,* 2d ed. Englewood Cliffs, N.J.: Prentice-Hall, 1977.

Willis, Edward M. *Scheduling Construction Projects.* New York: John Wiley & Sons, 1986.

Additional References

AbouRizk, S., and D. Halpin. "Probabilistic Simulation Studies for Repetitive Construction Processes." *Journal of Construction Engineering and Management*, 116, no. 4 (1990): 575–594.

Al-Harbi, K., S. Selim, and M. Al-Sinan. "A Multiobjective Linear Program for Scheduling Repetitive Projects." *Cost Engineering*, 38, no. 12 (1996): 41–45.

Allam, S. I. G. "Multi-project Scheduling: A New Categorization for Heuristic Scheduling Rules in Construction Scheduling Problems." *Construction Management and Economics*, 6, no. 2 (1988): 93–115.

Arditi, D., and H. Koseoglu. "Factors Affecting Success in Network Applications in a Developing Country." *Construction Management and Economics*, 1, no. 1 (1983): 3–16.

Arni, R. "A Systematic Approach to Stochastic Scheduling." *Construction Management and Economics*, 5, no. 3 (1987): 187–198.

The Associated General Contractors of America. *Cost Control and CPM in Construction: A Manual for General Contractors*. Washington, D.C.: The Associated General Contractors of America, 1968.

———. *CPM in Construction: A Manual for General Contractors*. Washington, D.C.: The Associated General Contractors of America, 1965.

Ayers, C. *Specifications for Architecture, Engineering and Construction*, 2d ed. New York: McGraw-Hill, 1984.

Barrie, Donald E., and Boyd Paulsen. *Professional Construction Management*, 2d ed. New York: McGraw-Hill, Inc., 1984.

Baxendale, T. "Construction Resource Models by Monte Carlo Simulation." *Construction Management and Economics*, 2, no. 3 (1984): 201–217.

Bockrath, J. *Dunham and Young's Contracts, Specifications, and Law for Engineers*, 4th ed. New York: McGraw-Hill, 1986.

Bush, V. *Construction Management*. Reston, Va.: Reston, 1973.

Choate, P. *As Time Goes By: The Costs and Consequences of Delay*. Columbus, Oh.: Academy of Contemporary Problems, 1980.

Christian, J., and D. Hachey. "Effects of Delay Times on Production Rates in Construction." *Journal of Construction Engineering and Management*, 121, no. 1 (1993): 20–26.

Civitello, A., Jr. *Contractor's Guide to Change Orders*. Englewood Cliffs, N.J.: Prentice-Hall, 1987.

Clough, R. *Construction Contracting*, 4th ed. New York: John Wiley & Sons, 1981.

Collier, K. *Construction Contracts*, 2d ed. Englewood Cliffs, N.J.: Prentice-Hall, 1987.

Coombes, Jim. *Real World Construction Scheduling*. Seattle, Wash.: Construction Planning and Management, Inc., 1990.

Coşkunoglu, O. "Optimal Probabilistic Compression of PERT Networks." *Journal of Construction Engineering and Management*, 110, no. 4 (1984): 437–446.

Cusack, M. M. "A Simplified Approach to the Planning and Control of Cost and Project Duration." *Construction Management and Economics*, 3, no. 3 (1985): 183–198.

Cushman, R. *The McGraw-Hill Construction Business Handbook*. New York: McGraw-Hill, 1978.

Easa, S. "Resource Leveling in Construction By Optimization." *Journal of Construction Engineering and Management*, 115, no. 2 (1989): 302–316.

Echeverry, D., C. W. Ibbs, and S. Kim. "Sequencing Knowledge for Construction Scheduling." *Journal of Construction Engineering and Management*, 117, no. 1 (1991): 118–130.

Fazio, P., P. Moselhi, P. Théberge, and S. Revay. "Design Impact of Construction Fast-Track." *Construction Management and Economics*, 6, no. 3 (1988):195–208.

Feldman, William, and Patti Feldman. *Construction & Computers*. New York: McGraw-Hill, Inc., 1996.

Fisk, Edward R. *Construction Project Administration*. 4th ed. Englewood Cliffs, N.J.: Prentice-Hall, Inc., 1992.

Fondahl, John W. *A Non-Computer Approach to the Critical Path Method for the Construction Industry*. Technical Report No. 9, 2d ed. Stanford University, Stanford, Calif.: The Construction Institute, Department of Civil Engineering, 1962.

———. *Some Problem Areas in Current Network Planning Practices and Related Comments on Legal Applications*. Technical Report No. 193. Stanford, Calif.: The Construction Institute, Department of Civil Engineering, Stanford University, 1975.

Fulkerson, D. R. "A Network Flow Computation for Project Cost Curves." *Management Science*, 7, no. 2 (1961): 167–179.

Gould, C. "Rolling Fours: Novel Work Schedule." *Journal of Construction Engineering and Management*, 114, no. 4 (1988): 477–593.

Gould, F. *Managing the Construction Process*. Upper Saddle River, N.J.: Prentice-Hall, 1997.

Halpin, D., and R. Woodhead. *Construction Management*. New York: John Wiley & Sons, 1980.

Handa, V. K., and R. M. Barcia. "Construction Production Planning." *Journal of Construction Engineering and Management*, 112, no. 2 (1986): 163–177.

Hillebrandt, P. M., and J. L. Meikle. "Resource Planning for Construction." *Construction Management and Economics*, 3, no. 3 (1985): 249–263.

Hinze, J. *Construction Contracts*, 2d ed. New York: McGraw-Hill, 2001.

Hohns, M. *Preventing and Solving Construction Contract Disputes*. New York: Van Nostrand Reinhold, 1979.

Jervis, B., and P. Levin. *Construction Law: Principles and Practice*. New York: McGraw-Hill, 1988.

Johnston, D. W. "Linear Scheduling Method for Highway Construction." *Journal of the Construction Division*, 107, no. 21981 (1981): 241–261.

Lambert, J., and L. White. *Handbook of Modern Construction Law*. Englewood Cliffs, N.J.: Prentice-Hall, 1982.

Latterner, C. G., D. M. Dresdner, J. A. Spiech, and G. M. Uslan. *A Programmed Introduction to PERT.* New York: John Wiley & Sons, 1963.

Laufer, A., and R. L. Tucker. "Is Construction Project Planning Really Doing Its Job? A Critical Examination of Focus, Role and Process." *Construction Management and Economics,* 5, no. 3 (1987): 185–186.

Levin, P. "Claims and Changes." *Handbook for Construction Contract Management,* Silver Spring, Md.: WPL, 1978.

Levitt, R., and J. Kunz. "Using Knowledge of Construction and Project Management for Automated Schedule Updating." *Project Management Journal,* 16, no. 5 (1985).

Lorterapong, P., and O. Maselhi. "Project-Network Analysis Using Fuzzy Sets Theory." *Journal of Construction Engineering and Management,* 122, no. 4 (1996): 308–318.

Lumsden, Philip. *The Line-of-Balance Method.* London: Pergamon Press Ltd., 1968.

McDonald, P., and G. Baldwin. *Builder's and Contractor's Handbook of Construction Claims,* Englewood Cliffs, N.J.: Prentice-Hall, 1989.

Moselhi, O., and K. El-Rayes. "Scheduling of Repetitive Projects With Cost Optimization." *Journal of Construction Engineering and Management,* 119, no. 4 (1993): 681–695.

Muth, John R., and G. L. Thompson. *Industrial Scheduling.* Englewood Cliffs, N.J.: Prentice-Hall, 1963.

O'Brien, J. J. "CPM Scheduling for High-Rise Buildings." *Journal of the Construction Division,* 101, no. 41975 (1975): 895–905.

Peterson, R. "Critical-Path Scheduling for Construction Jobs." *Civil Engineering,* 32, no. 8 (1962): 44–47.

Popescu, Calin. *How to Use CPM in Practice, Part II: Resources.* Austin, Tex.: Department of Civil Engineering, The University of Texas at Austin, 1975.

Popescu, Calin, and John Borcherding. *How to Use CPM in Practice, Part I: Time,* 2d ed. Austin, Tex.: Department of Civil Engineering, The Univeristy of Texas at Austin, 1975.

Pultar, M. "Progress-Based Construction Scheduling." *Journal of Construction Engineering and Management,* 116, no. 4 (1990): 670–688.

Rasdorf, W., and O. Abudayyeh. "Cost- and Schedule-Control Integration: Issues and Needs." *Journal of Construction Engineering and Management,* 117, no. 3 (1991): 486–502.

Richter, I., and R. Mitchell. *Handbook of Construction Law & Claims,* Reston, Va.: Reston, 1982.

Royer, K. *The Construction Manager in the 80's,* Englewood Cliffs, N.J.: Prentice-Hall, 1981.

———. "The Federal Government and the Critical Path." *Journal of Construction Engineering and Management,* 112, no. 2 (1986): 220–225.

Rubin, R., S. Guy, A. Maevis, and V. Fairweather. *Construction Claims.* New York: Van Nostrand Reinhold, 1983.

Schlick, H. "Schedule and Resources of Fast Track Renovation Work." *Journal of the Construction Division,* 107, no. 4 (1981): 559–574.

Scott, N., and L. Murphree, Jr. "Project Planning and Control By Time-Sharing Computer." *Journal of the Construction Division,* 98, no. 1 (1972): 37–48.

Sears, Glenn A. *A CPM-Based Cost Control System.* Technical Report No. 199. Stanford, Calif.: The Construction Institute, Department of Civil Engineering, Stanford University, 1975.

Seibert, J., and G. Evans. "Time-Constrained Resource Leveling." *Journal of Construction Engineering and Management*, 117, no. 3 (1991): 503–520.

Shaffer, L. R., J. B. Ritter, and M. L. Meyer. *Critical Path Method I & II*. Urbana, Ill.: University of Illinois, 1964.

Simon, M. *Construction Contracts & Claims*. New York: McGraw-Hill, 1979.

Stepman, Kenneth. *A Buyer's Guide to Project Management Software*. Milwaukee, Wis.: New Issues, Inc., 1986.

Stokes, M., and J. Finuf. *Construction Law for Owners and Builders*. New York: McGraw-Hill, 1986.

Sweet, J. *Legal Aspects of Architecture, Engineering and the Construction Process*, 2d ed. St. Paul, Minn.: West, 1977.

Tenah, K. *The Construction Management Process*. Reston, Va.: Reston, 1985.

Thomas, H. "Effects of Scheduled Overtime on Labor Productivity." *Journal of Construction Engineering and Management*, 118, no. 1 (1992): 60–76.

Thomas, H. Randolph, and Karl Raynar. *Effects of Scheduled Overtime on Labor Productivity: A Quantitative Analysis*. Source Document 98. Austin, Tex.: Construction Industry Institute, The University of Texas at Austin, 1994.

Trimble, G. "Resource-Oriented Scheduling." *International Journal of Project Management*, 2, no. 2 (1984): 70–74.

Turlington, Richard G. *NUSC PERT/Time/Cost Program User's Manual, vol. IV*. NUSC Technical Report Document 4627. Newport, R.I.: Program and Financial Management Plan Resources Information Systems, Naval Underwater Systems, August 1973.

"The Use of CPM and PERT in Construction Management." Reprinted from *The Constructor, The Management Magazine of the Construction Industry*. New York: The Reuben H. Donnelley Corporation, 1963.

White, A. "The Critical Path Method and Construction Contracts: A Polemic." *Construction Management and Economics*, 3, no. 1 (1985): 15–24.

Whiteman, W., and H. Irwig. "Disturbance Scheduling Technique for Managing Renovation Work." *Journal of Construction Engineering and Management*, 114, no. 2 (1988): 191–213.

Index

A

ACTIM value, 129, 130, 131
Activity codes, 224
Activity-on-Arrow (A-on-A) technique, 10, 326
Activity-on-Node (A-on-N) technique, 10, 35
Acts of God, 109
Actual cost of work performed (ACWP), 229, 234, 235, 237, 238, 243
Allocation, resource, 83, 118–119, 120, 123, 129, 138, 216
    Brooks method of, 118, 129–136
    computer scheduling and, 5, 7, 28, 218, 220–221, 226, 228
    linear scheduling and, 297, 298, 300, 307, 312
    parallel method of, 119–120, 123, 125, 129–130, 149
Arrow diagrams, 326–363
As-built schedules, 211–214, 277
At completion variance (ACV), 237
Avoidable delays, 108

B

Backward pass calculations, 52, 54, 58–59, 120, 139, 140–143, 216, 229, 340–343, 348, 352
Bar charts, 2–5, 7, 9–10, 21, 29, 35, 78, 93–94, 120, 138, 201, 204, 210–211, 213–214, 242–245, 272, 276, 283, 285, 288, 296–297, 300, 307, 309
    for job progress, 204–205
    shortcomings of, 3–7, 9, 82, 136, 200, 288, 297
    value of, 7–9, 204, 296
Base calendar, 218, 220, 224
Baseline schedule, 226
Beginning-of-day convention, 50, 53, 200, 337–338
Bell-shaped curve, 85, 315
Bids, 73
Billings, front-end loading of, 112
Blackhawk Heating & Plumbing Co., Inc. v. U.S., 271–272, 280
Blueprints, 1
Brooks method of resource allocation, 130
Brooks Towers Corp. v. Hunkin-Conkey Construction Co., 272
Budgeted cost at completion (BAC), 233–234, 236–238
Budgeted cost of work performed (BCWP), 229, 233–238, 243
Budgeted cost of work scheduled (BCWS), 221, 233–237
Buffers, 306

C

Calendar days, 102–104
Callanan Industries v. Glens Falls Urban Renewal Agency, 275
Cash-flow analysis, 165–176
Change orders, 111, 278–279

Completion, 101, 103, 105, 107, 113, 230, 234, 237
  at completion variance (ACV), 237
  budgeted cost at, 229, 233–235, 240–241, 243
  estimated cost at, 89, 172, 234, 237–238
  liquidated damages and, 91, 97, 101, 105, 106, 109, 116, 136, 178, 183, 208–209, 248–251, 253, 276, 314
Compression, network, 189
Computer aided design programs (CAD), 230
Computer scheduling, 216–231
Conceptual estimates, 73
Constraints, 26–27, 219
Contract provisions, 90–116
Cost and Schedule Control System Criteria (C/SCSC) technique (U.S. Dept. of Defense), 233
Cost code system, 203
Cost compression, 185–190, 192
Cost performance index (CPI), 235, 237–238
Cost variance (CV), 229, 234, 238
Costs, 32, 73, 172, 177, 232, 240
  actual cost of work performed, 229, 243
  budgeted cost of work performed, 229, 243
  budgeted cost of work scheduled, 234, 243
  direct, 177–178, 180, 209–210
  indirect, 177–181, 209–210
Crash cost, 184
Crew planning, 290–293
Crew stacking, 250
Critical activities, 350
Critical path, 32–33, 54, 345–349
Critical Path Method (CPM), 9–10, 20–21, 32, 94, 98, 164, 176–177, 208, 210, 216, 218, 222, 225–226, 232, 240, 270–276, 280, 285, 287, 296–297, 307
Crowding, 249–251

D

Delays, 45, 108–109, 253–254, 265, 279–280, 288
  forgetting function and, 262
Direct costs, 181
Dummies, in arrow diagrams, 330–336
Duration of activities, 31–32, 72, 79, 85–88, 339

E

Early start/finish time 32
Earned value, 232–245

Editing, global, 229
Environmental constraints, 26
Estimated cost at completion (EAC), 234, 237–238
Estimating costs, 72–79
Event times, 340
Events, 28, 326, 350
Extension of time 109–111

F

Filtering, 219, 228–229
Float, 50–52, 55–58, 62, 82, 98–99, 343, 346, 349–352
  back, 139, 142, 144, 229
  free, 46, 48, 50–52, 54–57, 61, 89, 137–144, 147, 201, 208, 216, 225, 301, 343, 345, 348–351
  independent, 53–58
  interfering, 48, 53–58
  ownership of, 98–100, 116
  total, 46, 48, 50, 52–57, 61, 89, 99–100, 119–120, 130, 201, 208–209, 343, 345, 348–351, 354
Forgetting function, 262
Forward pass, 337–339
Four-dimensional computer-aided design programs (4D-CAD), 230
Fragnet (network fragment), 210

G

Gallaudet College, 272
Gantt charts, 227
Gantt, Henry, 2
General Services Administration, 272
Global changes, 221
Global editing, 229
Goodwill, 253

I

Immediately preceding activities (IPAs), 25, 27, 47
Improvement factor, 141
Independent float, 55
Indirect costs, 177, 179
Interest, 165–166
Interfering float, 56

K

Kane, Judge P.J., 275

L

Lag values, 50, 58–61, 353
"Lazy S" cash flow curve 171, 175–176
Lead time, 51
Learning curve, 254, 256–264, 279
Lichter v. Mellon-Stuart Company, 273
Liens, 167
Line of balance (LOB) scheduling method, 300
Linear scheduling, 297–312
    buffers and, 307
    examples of, 300, 309
Lines, 294
    zero-value link, 184
Link lag, 48, 52–53, 183
Link lines, 45, 47
Linking, 219
Liquidated damages, 91, 97, 101, 105–106, 109, 136, 178, 183, 208–209, 248–251, 253, 276, 314
Litigation, 270–276
Long-lead items, 205

M

Management, 24, 26, 33, 117
Marriott Corporation v. Dasta Construction Company, 276
Materials, 56, 58, 71, 111, 114, 146, 167, 169, 172, 255, 279, 294
Mean value, 314
Mobilization, 78, 169, 210, 240
Money, 20, 32, 164–165
    cash flow analysis and, 165, 172
    interest rates and, 165–166
Monitoring and control project status in, 21, 28, 201–207, 214, 233–234, 236, 243, 278, 309
Monte Carlo simulation, 321–322
Morale, 265

N

Network interaction limit (NIL), 183, 184, 186–190
New York Supreme Court, 275
Nodes, 326
Notice to proceed, 100–101

O

Overhead, 177, 294

Overtime, 103, 246
Owners, 20, 88, 168, 209, 230

P

Pathman Construction Co. v. Hi-Way Electric Co., 272
Payments, 112, 169
    final, 113–114, 116, 169–170, 173
    progress, 24, 93, 95
    schedules of, 112, 168
Performance Measurement System (PMS) (U.S. Dept. of Energy), 233
Physical constraints, 26
Planned percent complete (PPC) measure, 236
Planning, 1, 20
Precedence diagrams, 28–31, 35–71, 326
    activity relationships on, 22, 36–45, 47, 58–60, 226, 352
    finish-to-finish, 36, 38–39, 41, 43
    finish-to-start, 36–37, 40, 43
    start-to-finish, 36, 42–45, 60, 354
    start-to-start, 36, 40–42
Present worth analysis, 172–173
Pretask planning, 293–294
Procurement, 24
Productivity, 246
    learning curve and, 179, 236, 256–257, 259, 263
    overtime and, 91, 103, 166, 179, 246–249, 266, 268, 279, 302
Profit, 177
Program Evaluation and Review Technique (PERT), 9–10, 222, 313–325
    Monte Carlo simulation of, 321–322
Progress bar, 220
Progress payments, 24, 96, 112–114
Progress schedule, 273
Project duration, 199–201
Punchlist, 288–289

Q

Quantity take-off, 73–79

R

Rate of improvement, 258
Recurring tasks, 220
Regulatory constraints, 282

Relationships of activities,
  finish-to-finish, 36, 38–39, 42, 45, 60,
    219, 297, 352–353
  finish-to-start, 36–38, 40, 42, 43, 45, 53, 58–59,
    219, 298, 334, 352
  start-to-finish, 36, 42–45, 60, 219, 352, 354
  start-to-start, 36, 40–42, 45, 59, 219, 297,
    352–353
Resource allocation, 118–138
Resource leveling, 138–146, 229
Resources, 23, 32, 117–118, 136, 146–148, 158
Retainage, 169
Rework, 255, 279

S

Schedule variance (SV), 229, 234, 237
Schedules. *See also* Earned Value
  as-built, 211–212, 214–215, 217, 226, 277,
    280, 285, 309
  as-planned, 229, 277
  baseline, 226
  compression, 183–193
  "horse blanket", 12–13
  matrix, 10–13
  of payments, 112
  uncertainty and, 72, 313–315, 318–319, 323
  updating, 95, 201, 208
Scheduling, 20, 33, 82, 83, 201, 216–217, 246, 297,
    300, 313, 347
  as project control, 15, 201, 208, 214
Short-interval schedules, 281–296
  daily project schedule as, 290
  pre-task plan as, 293
  punch list as, 107, 170
Software for scheduling, 7, 23, 27, 35–37, 49, 55,
    70–71, 102, 202, 217, 218, 230–231, 327
Sorting, 220, 228
Southern Fireproofing Company, 273
Specifications, 115
Staggering deliveries, 84
Standard deviation, 86, 314, 319
Standard normal curve, 86
Subcontractors, 91–92, 108, 167
Submittals, 111–112
Substantial completion, 107
Summary tasks, 220
SureTrak Project Manager software, 221–222
Surveys, foreman delay, 253–254
Suspension, 114–115

T

Termination of the contract, 115–116
Time, 31, 82, 90–91, 99–102, 105, 109–110, 117,
    164–165, 176, 180–181, 199, 203, 232,
    252, 294
  arrow diagrams and, 28, 35, 326–327, 330,
    332–334, 336, 352, 354
  buffers of, 307
  in procurement activities, 18, 24, 27, 205, 330
  lead, 36, 43–45, 48, 50, 103, 219, 265, 278, 330
  of payments, 95, 97, 112–114, 116, 166–170,
    173, 240
  on bar charts, 2–5, 7, 10, 21, 204, 272, 283, 296, 307
  production rate diagrams and, 306, 309
  start and finish, 29, 57, 130, 327, 340, 345
Time in contracts
  avoidable delays and, 108–109
  completion time as, 95, 119, 268, 345
  final payments and, 170
  float ownership as, 88, 99–100
  progress payments and, 24, 93–95
  progress schedule and, 91, 93–95, 97–99, 276
  submittals and, 96, 111
  substantial completion and, 102, 107, 169–170,
    173, 288
  suspensions as, 270
  termination and, 115
  unavoidable delays and, 109

U

U.S. Army Corps of Engineers, 287
U.S. Bureau of Reclamation, 287
Unavoidable delays, 109
Uncertainty, 85, 313–314, 318, 320
  in scheduling, 314
Updating the schedule 207–208, 226

V

Variance, 234, 237–238, 319–320

W

Weather, 84–85, 106
Work breakdown structure (WBS), 13–20, 202,
    224, 239, 243
Working days, 102–104
Weather, 84–85, 106
Working days, 102–104